An Introduction to the Theory of Smooth Dynamical Systems

An Introduction to the Theory of Smooth Dynamical Systems

W. Szlenk

Institute of Mathematics
University of Warsaw

A Wiley–Interscience Publication

JOHN WILEY & SONS
Chichester • New York • Brisbane • Toronto • Singapore

PWN–POLISH SCIENTIFIC PUBLISHERS
Warszawa

Translation from the Polish original
Wstęp do teorii gładkich układów dynamicznych
published in 1981 by Państwowe Wydawnictwo Naukowe, Warszawa

Translated by *Marcin E. Kuczma*

Graphic design: *Zygmunt Ziemka*

Library of Congress Cataloging in Publication Data:

Szlenk, W.
An introduction to the theory of smooth dynamical systems.

Translation of: Wstęp do teorii gładkich układów dynamicznych.
1. Differentiable dynamical systems. I. Title.
QA614.8.S9413 1984 516.3′6 82-23771
ISBN 0 471 90117 2

British Library Cataloguing in Publication Data:

Szlenk, W.
An introduction to the theory of smooth dynamical systems.
1. Differentiable dynamical systems
I. Title II. Wstęp do teorii gładkich układów dynamicznych *English*
516.3′6 QA614.8
ISBN 0 471 90117 2

Printed in Poland by W.D.N.

To the memory of
Professor Stanisław Mazur

Contents

Introduction

The theory of dynamical systems has its origin in astronomy. To be more accurate: in celestial mechanics. The general laws of motion of planets about the Sun have been fairly well known since the time of Newton and Kepler, and the same can be said of the shapes of orbits in the simplest cases. According to Newton's second law of dynamics, the force acting on a mobile body is directly proportional to acceleration: in symbols

$$\vec{F} = m\frac{\mathrm{d}^2\vec{x}}{\mathrm{d}t^2},$$

where $\vec{x} = (x^1, x^2, x^3)$ denotes the position of the body in the space, $\vec{F} = (f^1, f^2, f^3)$ is the force and m is a scalar coefficient called the mass of the body. The easiest to handle is the case of a single mobile body subjected, in accordance with the general law of gravitation, to the action of the force

$$\vec{F} = -\mathrm{G}\frac{mM\vec{x}}{\|\vec{x}\|^3},$$

G denoting the gravitation constant, M being the mass of a large "immobile" body which attracts the body considered, and m standing for the mass of the latter ($M \gg m$); $\|\vec{x}\|$ is the length of the vector $\vec{x}$. As a model one can view the case of the sun and the earth. From the mathematical viewpoint, the problem consists in solving a system of three differential equations of the second order, which can easily be reduced to a system of six equations of the first order. An explicit solution is found, and it turns out that the small body runs along a conical curve having the large body at its focus. This conclusion is confirmed by astronomical observation, provided, however, that the time period is not too long. For periods extending to thousands of centuries one must take into consideration the influence of other factors: gravitational effects coming from other planets, the relativity phenomena (change of mass) and so on.

In the case where there are several bodies considered and their mutual gravitation cannot be disregarded, the problem is more complicated. In the case of N bodies the motion of the system is described by N equations

$$\vec{F}_i = m_i \frac{d^2\vec{x}_i}{dt^2}, \qquad i = 1, 2, \ldots, N,$$

in which the forces $\vec{F}_i$ depend on the positions of all these bodies:

$$\vec{F}_i = \vec{F}_i(\vec{x}_1, \vec{x}_2, \ldots, \vec{x}_N) = -\sum_{\substack{j=1 \\ j \neq i}}^{N} G \frac{m_i m_j (\vec{x}_i - \vec{x}_j)}{\|\vec{x}_i - \vec{x}_j\|^3}.$$

Thus we get a system of $3N$ differential equations of the second order, equivalent to a system of $6N$ equations of the first order. Already for $N = 3$ the system is unsolvable in quadratures. All that can be done is to find an approximate solution by numerical methods. Such numerical solutions are of great importance when applied to concrete problems arising in astronomy and technology: forecasting the configuration of celestial bodies, launching space vehicles etc. But they are of little use from the viewpoint of cosmogony of the solar system. They give no information which might permit one to foresee the evolution of the solar system in the distant future. We are facing here a very typical problem of the theory of qualitative properties of ordinary differential equations. The French mathematician Henri Poincaré may be regarded as the initiator of that theory, and certainly the first to apply it in celestial mechanics. In his famous book *Les méthodes nouvelles de la mécanique céleste* (Poincaré, 1892–99) he comprised a large number of ingenious observations and intuitions concerning mechanics and the qualitative theory of differential equations. Even today the book is a source of creative inspiration to many mathematicians.

A system whose motion is described in terms of ordinary differential equations is termed a *classical dynamical system.* A thorough study of such systems shows that they are closely connected with the behaviour of iterates of continuous mappings of metric spaces (and, in particular, diffeomorphisms of manifolds) as well as with the theory of measure preserving transformations (which are the object of research in the so-called *ergodic theory*).

In general, the theory of dynamical systems deals with the action of groups of continuous transformations of topological spaces, a special role being played by groups of diffeomorphisms of manifolds and

1

Basic Concepts and Facts of the Theory of Dynamical Systems

1.1. Introductory Notions and Basic Examples

Let X be a metric space and let $\varphi^\alpha: X \to X$ be a continuous semi-group[1] of continuous mappings of X into X, $\alpha \in \mathfrak{A}$.

Definition 1.1.1 Any such pair (X, φ^α) is called a *topological dynamical system.*

The term: *dynamical system* is also used with regard to another class of objects. Namely:

Definition 1.1.2 By a *metric dynamical system* we mean a quadruple $(X, \mathfrak{M}, \mu, \varphi^\alpha)$, where $(X, \mathfrak{M}, \mu)$ is a normalized measure space (X denoting the space, $\mathfrak{M}$ a σ-algebra and μ a measure with $\mu(X) = 1$) and φ^α denotes a continuous semi-group of measure preserving measurable mappings of X into X, $\alpha \in \mathfrak{A}$.

Of course, in either case, the semi-group can actually be a group (with all φ^α invertible).

Throughout this book we will be concerned only with topological dynamical systems, and specifically with smooth systems, i.e., systems in which X is a differentiable manifold and φ^α are differentiable mappings. Therefore the term "*dynamical system*" will in the sequel always stand for a "topological dynamical system".

We shall distinguish two principal types of dynamical systems, according as the index semi-group $\mathfrak{A}$ is discrete or continuous:

1. $\mathfrak{A} = Z^+ = \{0, 1, 2, \ldots\}$ or $\mathfrak{A} = Z = \{0, \pm 1, \pm 2, \ldots\}$, φ^α being

groups of measure preserving transformations; the ergodic theory may be regarded as a domain within that general theory.

Henri Poincaré, who can justly be called the creator of the theory of dynamical systems, had numerous successors, whose contribution to that theory cannot be overestimated: G. Birkhoff, E. Hopf, A. N. Kolmogorov, V. I. Arnold, D. V. Anosov, J. Moser, Y. G. Sinai, S. Smale and Ch. Pugh, to list the most prominent.

The readers who wish to gain a more profound knowledge of the theory of dynamical systems are invited to consult original papers and survey articles: Abraham and Marsden (1967), Abraham and Robbin (1967), Coddington and Levinson (1955), Hartman (1964), Katok (1967), Smale (1967).

Preface

The theory of dynamical systems has arisen from an attempt at an adequate description of phenomena in the surrounding world. Therefore it is very intimately connected with various domains of knowledge stemming directly from the needs of reality; we mean thereby mechanics, statistical physics (classical and quantum), electrodynamics, hydromechanics, even biology (the problem of evolution of populations).

Owing to its universal character, the theory uses methods that resort to nearly all branches of mathematical science. The interplay between the diverse methods in all their variety goes unexpectedly deep, providing a large number of very specific techniques in the investigation of dynamical systems. But also, conversely, problems posed in the theory of dynamical systems penetrate other mathematical theories, giving them a fresh impulse and opening new problems. The methodology is so widespread in mathematics that, for instance, dynamical systems are not given a separate heading in the *Mathematical Reviews*. The closest relation exists between the theory of dynamical systems and the theory of ordinary differential equations, functional analysis, differential geometry, differential topology, algebraic topology, measure theory, probability theory, and number theory (especially diophantine approximation).

The present book is not a monograph, though certain topics are treated in it with considerable generality. Most of the material presented here was included in lectures held by the author at the Mathematical Institute of the Aarhus University in 1971/72.

The book is addressed to readers who are familiar with standard undergraduate courses in mathematics. Thus the general prerequisites are: linear algebra, calculus in several real variables, abstract measure and integral, basic concepts of differential geometry (differentiable manifolds, tangent spaces, geodesic lines), basic concepts of functional analysis (Banach spaces, linear operators and functionals), elements of topology and, last but not least, ordinary differential equations within the usual elementary scope. At the end of the book the reader will find a list of notes containing the definitions which are omitted in the text, together with brief explanatory remarks. It is not to be expected that anyone would learn those concepts from the notes. They are meant as a reminder; the basic and operational knowledge of the underlying theories should be gathered from other sources. Of a similar type is the first section in Chapter III, containing a brief survey of facts and notions of functional analysis needed in the exposition of the material of that chapter. On the whole they exceed the customary scope of university courses; their presentation in Section 3.1, however, can hardly serve as a unique source of information. The reader should turn to specialized literature.

The book contains a number of exercises. If the reader really wishes to learn something, he ought to devote some energy to solving them; one of the reasons is that they often provide proofs omitted in the main text. Exercises marked by a capital A contain such complementary proofs (or parts of proofs) to be used in the sequel; they are referred to in subsequent considerations. Those marked by a capital B serve just as practice material giving the reader the necessary skill in handling the theory. Almost all the exercises are accompanied by hints. Those marked with an asterisk are relatively difficult, in spite of the hints.

If the reader wants to get some idea as to the kind of mathematics which is cultivated in this book we recommend him to read Section 1.2 through. This may encourage him to further reading; and perhaps it will just save him from unnecessary loss of time and effort.

It is the author's pleasant duty to thank Dr. M. Krych and Dr. M. Misiurewicz for their valuable criticism and comments, which have helped a great deal in the preparation of this book.

W. Szlenk

invertible. Then the semi-group $\{\varphi^\alpha, \alpha \in \mathfrak{A}\}$ is generated by $\varphi^1 = \varphi$, $\varphi^0 =$ identity, and the system can be denoted shortly by (X, φ). Such a system is called a *cascade.*

2. $\mathfrak{A} = \boldsymbol{R}^+$ (the non-negative reals) or $\mathfrak{A} = \boldsymbol{R}$ (the reals), φ^α being invertible. Such a system is called a *flow* and is written down simply in the form (X, φ_t).

Examples of dynamical systems

1.1.1. Let S^1 denote the unit circle and let $\varphi\colon S^1 \to S^1$ be the rotation by an angle $2\pi\alpha$:

$$\varphi(z) = z_0 z, \qquad z = e^{2\pi i x} \in S^1, \qquad z_0 = e^{2\pi i \alpha}$$

(*multiplicative notation*).

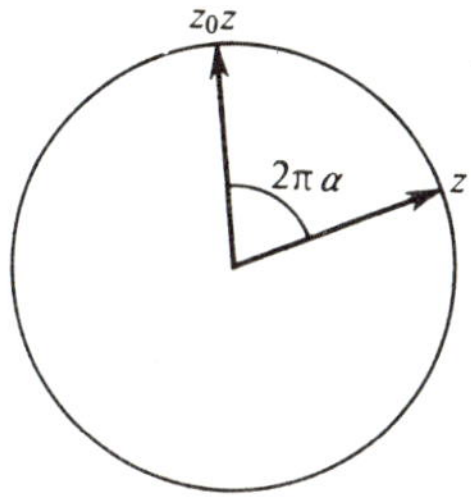

FIG. 1.1.1

The circle is homeomorphic to a segment with endpoints identified. In other words, the circle can be viewed as the real line, numbers differing by an integer identified; i.e., $S^1 = \boldsymbol{R}/\boldsymbol{Z}$. We say that the real line $\boldsymbol{R}$ is a *covering space*[2] for the circle and we call the assignment $x \mapsto p(x) = e^{2\pi i x}$ *projection of* $\boldsymbol{R}$ *onto* S^1.

The mapping φ induces, in a natural way, a mapping $\hat{\varphi}$ of $\boldsymbol{R}$ into $\boldsymbol{R}$ (this is the so-called *operation of lifting*; the map $\hat{\varphi}$ is termed: the *lift* of φ to the real line[3]). Then $\hat{\varphi}$ is of the form:

$$\hat{\varphi}(x) = x + \alpha + m,$$

m denoting an arbitrary integer. Regarding the circle as the interval $\langle 0, 1\rangle$ with endpoints identified, we may write the rotation φ alternatively as:

$$\varphi(x) = x + \alpha \pmod 1.$$

(*additive notation*).

1.1.2. Using the considerations of the preceding example, we can examine the mapping $\varphi\colon S^1 \to S^1$ given by:

$$\varphi(x) = \lambda \sin 2\pi x \pmod 1,$$

i.e.

$$\varphi(z) = e^{2\pi i \lambda \sin 2\pi x}, \qquad z = e^{2\pi i x} \in S^1.$$

Fig. 1.1.2 shows the graph of the function $x \mapsto \lambda \sin 2\pi x$; it also shows how the values of φ are to be read.

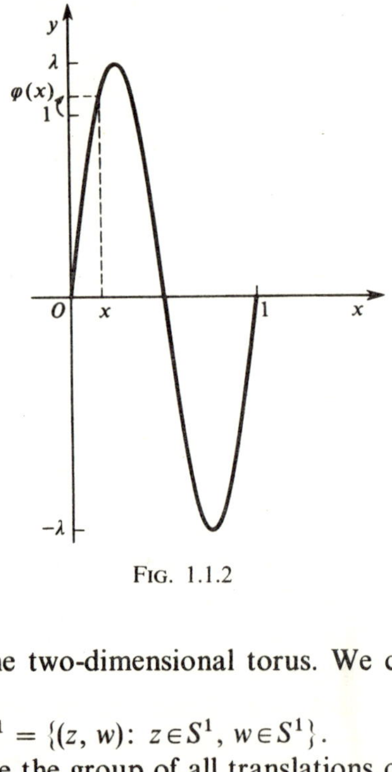

FIG. 1.1.2

1.1.3. Let T^2 be the two-dimensional torus. We can define T^2 in two equivalent ways:

1° $T^2 = S^1 \times S^1 = \{(z, w)\colon z \in S^1, w \in S^1\}$.

2° Let Γ denote the group of all translations of $\boldsymbol{R}^2$ by vectors with integer coordinates:

$$\Gamma = \{\gamma = \gamma_{(n,m)}\colon \gamma\colon \boldsymbol{R}^2 \to \boldsymbol{R}^2,\ \gamma(x^1, x^2) = (x^1 + n, x^2 + m),\ n, m \in Z\}.$$

We define an equivalence relation $\sim$ in $\boldsymbol{R}^2$ as follows:

$$x \sim y \Leftrightarrow \exists \gamma \in \Gamma\colon \gamma(x) = y.$$

Clearly, Γ is isomorphic to $Z \times Z$.

Denoting by $[x]$ the coset (mod $\sim$) containing a point $x = (x^1, x^2)$, we see that

$$[x] = \{(x^1 + n, x^2 + m): n, m \in Z\}.$$

(In Fig. 1.1.3, marked points belong to $[x]$.)

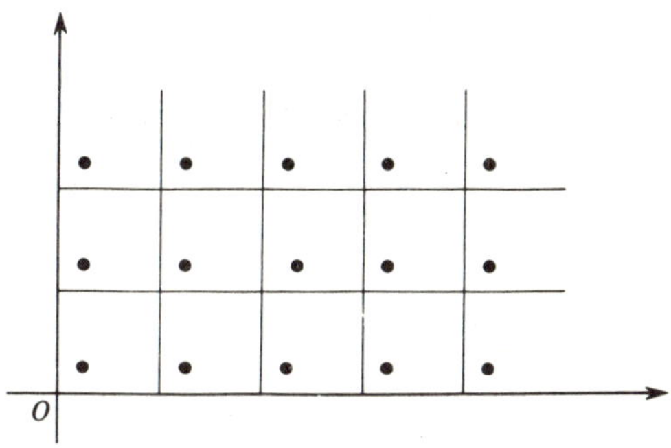

FIG. 1.1.3

Dividing $\boldsymbol{R}^2$ by $\sim$ we obtain the torus

$$\boldsymbol{R}^2/\sim \; = \boldsymbol{R}^2/\Gamma = (\boldsymbol{R} \times \boldsymbol{R})/(Z \times Z) = \boldsymbol{R}/Z \times \boldsymbol{R}/Z = S^1 \times S^1 = T^2.$$

The division by $\sim$ corresponds geometrically to the following procedure (see Fig. 1.1.4): consider the square with vertices (0, 0), (1, 0),

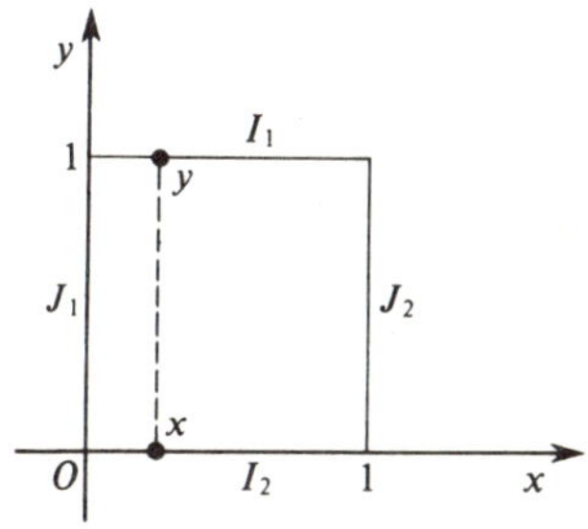

FIG. 1.1.4

(1, 1), (0, 1) and identify pairs of points whose coordinates differ by an integer (e.g. x and y in Fig. 1.1.4; no points in the interior of the square are identified, of course). This means that identification involves opposite sides of the square. This can be visualized sticking first the side I_1 and I_2 (results a cylinder) and then J_1 with J_2, i.e. the sides, or rather the circles, forming the edges of the cylinder. As a result we get a surface resembling that of a tyre-tube.

Just as the line in the case of the circle, the plane $\boldsymbol{R}^2$ is a covering space (universal, in fact [2]) for the torus T^2, with projection p: $\boldsymbol{R}^2 \to T^2$ given by

$$p(x^1, x^2) = (e^{2\pi i x^1}, e^{2\pi i x^2}).$$

Quite analogously one can define the *m-dimensional torus* T^m having $\boldsymbol{R}^m$ as its covering space.

Let φ: $T^2 \to T^2$ be a mapping defined as follows:

$$\varphi(z, w) = (z_0 z, w_0 w) \qquad \text{for } (z, w) \in T^2,$$

where

$$z_0 = e^{2\pi i\alpha}, \qquad w_0 = e^{2\pi i\beta}.$$

Such a mapping is called a *translation* (on the torus).

It is easy to see that a lift of φ to the plane is given by

$$\hat{\varphi}(x, y) = (x+\alpha, y+\beta).$$

Regarding T^2 as $\boldsymbol{R}^2/\Gamma$ we rewrite the mapping φ itself as

$$\varphi(x, y) = (x+\alpha, y+\beta)(\text{mod } 1).$$

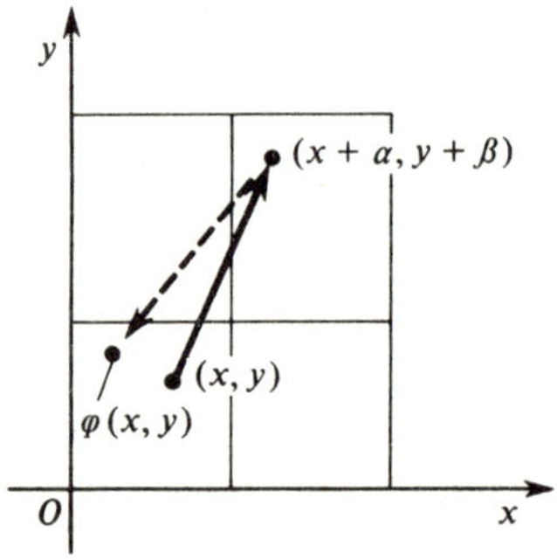

FIG. 1.1.5

1.1.4. Let S^2 be the two-dimensional sphere. Consider the stereographic projection π onto $\boldsymbol{R}^2$ with centre at the "north pole" N. Let A: $\boldsymbol{R}^2 \to \boldsymbol{R}^2$ be the linear transformation defined by the matrix

$$A = \begin{bmatrix} \lambda & 0 \\ 0 & \lambda \end{bmatrix}, \qquad \lambda \neq 0.$$

Then the mapping φ given by:

$$\varphi(x) = \pi^{-1} \circ A \circ \pi(x) \quad \text{for } x \in S^2,\ x \neq N,$$

$$\varphi(N) = N$$

is a diffeomorphism of S^2 onto itself.

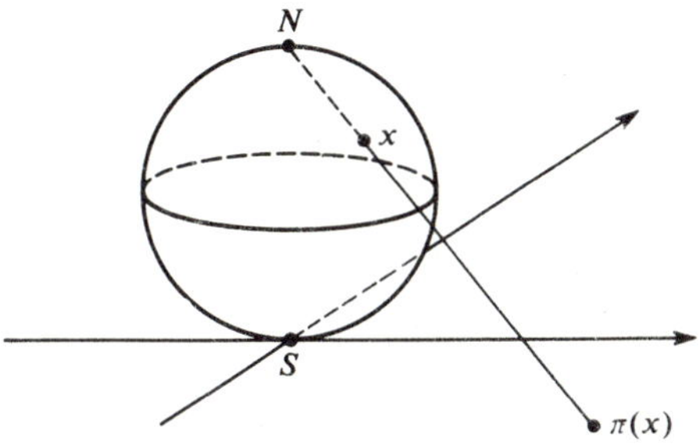

FIG. 1.1.6

1.1.5. Let $X = T^2$ and define

$$\varphi_t(z, w) = (e^{2\pi i\alpha t} z,\ e^{2\pi i\beta t} w).$$

Then (X, φ_t) is a flow. Regarding T^2 as $\boldsymbol{R}^2/\Gamma$ we rewrite φ_t as:

$$\varphi_t(x^1, x^2) = (x^1 + \alpha t,\ x^2 + \beta t)(\text{mod } 1)$$

and denoting by $\hat{\varphi}_t$ the lift of φ_t to the plane, we have

$$\hat{\varphi}_t(x^1, x^2) = (x^1 + \alpha t,\ x^2 + \beta t).$$

Now, let (X, φ^α) be a dynamical system and let $x \in X$ be an arbitrary point.

Definition 1.1.3 The set

$$\gamma(x) = \{y\colon \exists \alpha \in \mathfrak{A}\colon y = \varphi^\alpha(x)\}$$

is called the *trajectory* or the *orbit* of the point x (or: the orbit passing through x).

Let $\mathfrak{A} = Z^+$ or $\mathfrak{A} = Z$.

Definition 1.1.4 A point $x \in X$ is called a *periodic point* of a dynamical system (X, φ) iff there exists an integer $p \geqslant 1$ such that

$$\varphi^p(x) = x.$$

Any such p is called a *period* of x and the least such p is called the *prime period* of x. The set of all points periodic with respect to the transformation φ will be denoted by $\mathrm{Per}(\varphi)$.

Now let $\mathfrak{A} = \boldsymbol{R}^+$ or $\mathfrak{A} = \boldsymbol{R}$.

Definition 1.1.5 (1) A point $x \in X$ is a *fixed point* (or: *critical point*) of the flow (X, φ_t) iff

$$\varphi_t(x) = x \qquad \text{for all } t \in \boldsymbol{R}^+ \ [t \in \boldsymbol{R}],$$

i.e.,

$$\gamma(x) = \{x\}.$$

(2) The trajectory $\gamma(x)$ is called *periodic* iff there exists a number $\tau > 0$ such that

$$\varphi_{t+\tau}(x) = \varphi_t(x) \qquad \text{for all } t \in \boldsymbol{R}^+ \ [t \in \boldsymbol{R}]. \tag{1.1.1}$$

Any such τ is called a *period* of the trajectory and the least such τ is called the *prime period* (provided it exists i.e., x is not a fixed point).

A point x for which (1.1.1) is satisfied with some $\tau > 0$ is called *periodic*.

Examples

1.1.6. Let (X, φ) be as in Example 1.1.1. If α is rational (i.e., z_0 is a root of unity), then every point $z \in S^1$ is periodic. Indeed, let $\alpha = p/q$, $p, q \in Z$. Then

$$\varphi^q(z) = z_0^q z = (\mathrm{e}^{2\pi i p/q})^q z = z,$$

and so φ^q is the identity. The number q is a common period for all z; it is a prime period if the ratio p/q is irreducible.

If α is irrational, there are no periodic points. Moreover, every trajectory is a dense subset of S^1. We shall prove this fact.

Let $I = (a, b)$ be any open arc of the circle. Write $z_n = \varphi^n(z)$. The sequence z_n has a cluster point in S^1, and so there are two terms z_k and z_l such that $|z_k - z_l| < |b-a|$ (the symbol $|x-y|$ denoting the length of the arc (x, y) with endpoints x and y). Denote $J = (z_l, z_k)$ and put $\psi = \varphi^{k-l}$. Since $\psi(z_l) = \varphi^{k-l}(\varphi^l(z)) = \varphi^k(z) = z_k$, the intervals J and $\psi(J)$ are adjacent. So are $\psi(J)$ and $\psi^2(J)$ etc. (see Fig. 1.1.7). All the $\psi^n(J)$ have the same length; thus their closures cover the whole circle. Since their common length is less than $|b-a|$, there is an endpoint of some $\psi^n(J)$ which belongs to (a, b). That means, the trajectory $\gamma(z)$ intersects I, and this ends the proof.

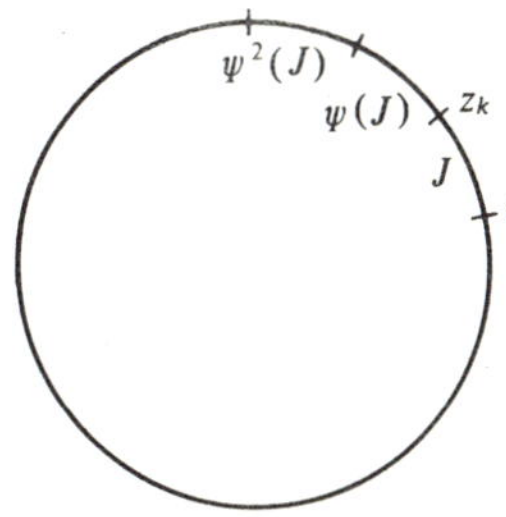

FIG. 1.1.7

1.1.7. Let (X, φ) be as in Example 1.1.3. Let z_0 and w_0 be roots of unity, thus there are integers $p, q > 0$ such that $z_0^p = w_0^q = 1$. Then

$$\varphi^{pq}(z, w) = (z_0^{pq}z, w_0^{pq}w) = (z, w),$$

which means that every point (z, w) is periodic.

Suppose that z_0 and w_0 are *algebraically independent*; we mean thereby that

$$z_0^n w_0^m = 1 \Rightarrow n = m = 0.$$

If $z_0 = e^{2\pi i\alpha}$, $w_0 = e^{2\pi i\beta}$, then z_0 and w_0 are algebraically independent iff α and β are independent over the ring Z:

$$n\alpha + m\beta \in Z \Rightarrow n = m = 0.$$

It can be proved that then every trajectory is a dense subset of T^2 (see Exercise 1.2.2 on p. 22).

1.1.8. Let (X, φ_t) be as in Example 1.1.5. If β/α is rational, then every trajectory is periodic and all the prime periods are equal. If β/α is irrational, then every trajectory is dense in T^2 (see Exercise 1.2.5 on p. 23).

1.1.9. Let K^1 be the unit disc:

$$K^1 = \{(x^1, x^2) \colon (x^1)^2 + (x^2)^2 \leqslant 1\}.$$

Consider the following system of differential equations (in polar coordinates):

$$\frac{dr}{dt} = r(1-r), \qquad \frac{d\omega}{dt} = a, \qquad a > 0.$$

The general solution of this system is:

$$r(t) = \frac{r_0 e^t}{1 - r_0 + r_0 e^t}, \qquad \omega(t) = at + \omega_0,$$

where $0 < r_0 \leqslant 1$, $r(0) = r_0$, $\omega(0) = \omega_0$. Obviously, $(r(t), \omega(t)) \in K^1$ for all t. The group of mappings φ_t defined by

$$\varphi_t(r_0, \omega_0) = \left(\frac{r_0 e^t}{1 - r_0 + r_0 e^t}, at + \omega_0\right), \qquad 0 < r_0 \leqslant 1,$$

$$\varphi_t(0) = 0$$

is a flow on K^1.

The dynamical system (K^1, φ_t) has one fixed point 0 and one closed orbit $\gamma = S^1$. Any other orbit spirals anti-clockwise, approaching S^1 as $t \to +\infty$ and approaching 0 as $t \to -\infty$.

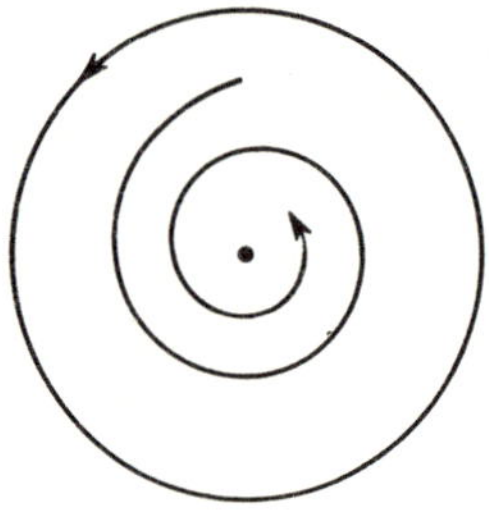

FIG. 1.1.8

From the point of view of applications in mechanics, the most important dynamical systems are flows. Certain problems of dynamics are described in terms of autonomous [22] systems of ordinary differential equations, and any such system defines a flow (under certain additional assumptions).

Indeed, let

$$\frac{dx^i}{dt} = f^i(x), \qquad x \in \mathbf{R}^m, \ i = 1, \ldots, m,$$

be an autonomous system of differential equations admitting, for any initial value $x_0 \in \boldsymbol{R}^m$, a unique solution defined on the whole real line (uniqueness is ensured, for instance, if f^i are of class C^1). For a given initial value x_0 let $x(t)$ denote the solution with $x(0) = x_0$. It follows from the uniqueness assumption that the one-parameter group of transformations φ_t: $\boldsymbol{R}^m \to \boldsymbol{R}^m$ given by

$$\varphi_t(x_0) = x(t), \qquad t \in \boldsymbol{R},$$

is a flow.

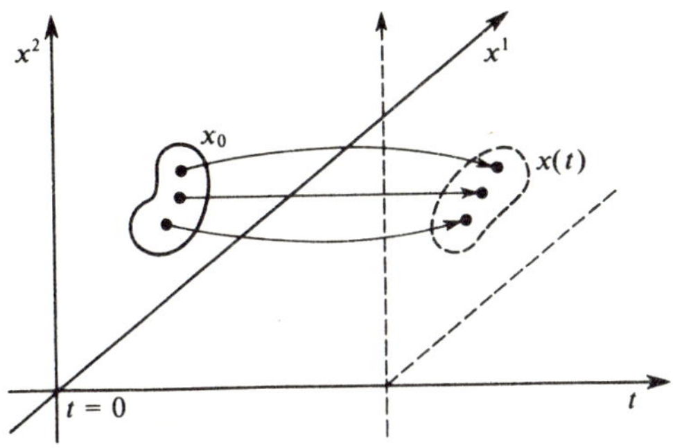

FIG. 1.1.9

Going a step further, we can consider differential equations not only in $\boldsymbol{R}^m$ but also on any finite-dimensional manifolds.

Let M be a finite-dimensional C^∞-manifold[11]. Let $\Phi(x)$, $x \in M$, be a vector field on M. This means, intuitively, that at every point $x \in M$ there is attached a selected tangent vector $\Phi(x) \in T_x(M)$. We say that a differentiable curve γ: $I \to M$, $I = \langle a, b \rangle \subset \boldsymbol{R}$, $\gamma(t) \subset M$, is an integral curve of the field $\Phi(x)$ iff

$$\dot{\gamma}(t) = \frac{d\gamma(t)}{dt} = \Phi(\gamma(t)), \tag{1.1.2}$$

$\dot{\gamma}(t)$ denoting the tangent vector to γ at $\gamma(t)$. If $\Phi(x)$ satisfies some additional conditions, e.g. is of class C^1 (i.e. $\Phi(x)$ depends on x smoothly in local coordinates), then the equation (1.1.2) has, for any given $x_0 \in M$, a unique solution satisfying the initial condition $\gamma(0) = x_0$. (This is a consequence of an analogous theorem for regions in $\boldsymbol{R}^m$.) Denote this solution by $\gamma(x_0, t)$. If, moreover, M is compact, then the solution $\gamma(t)$ is defined for all $t \in \boldsymbol{R}$ (see Theorem 1.4.2). The group of mappings

$$\varphi_t(x_0) = \gamma(x_0, t)$$

is a flow on M which is said to be generated by the vector field $\Phi(x)$. An equation of the form (1.1.2) is termed a *differential equation* on M. The curves $\varphi_t(x_0)$ are *solutions* of that equation.

Now, suppose we are given a system of particles in a stationary field of forces. Assume that the particles are not free, i.e., that there are some constraints imposed upon their positions. E.g.: a) the particles have to be in contact with a surface, or: b) the distance between any two particles has to be invariant (the particles are then said to constitute a *rigid body*).

Any configuration of the particles in the space at a given moment can be specified in terms of a certain number of numerical quantities $q = (q_1, \ldots, q_m)$, which we call the *coordinates* of the system. As a rule, the space of all possible configurations (the so called *configuration space*) forms a differentiable manifold. The parameters $q_1, \ldots, q_m$ may then be regarded as chart coordinates. The number m, equal to the dimension of the manifold, is called the *number of degrees of freedom*.

Examples

1.1.10. Consider the motion of a flat pendulum. Each state is determined by an angle $\omega \in \langle 0, 2\pi \rangle$. The configuration space is a circle.

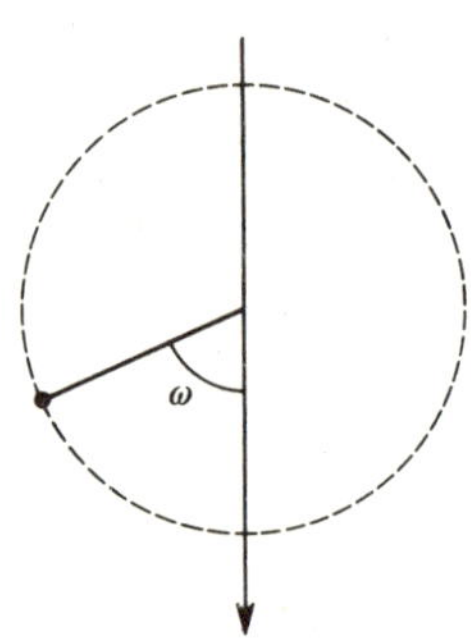

FIG. 1.1.10

1.1.11. For a spherical pendulum the configuration space is a sphere (of radius equal to the length of the pendulum).

1.1.12. Now consider the so-called double flat pendulum. Each state is described by two angles ω, ψ, as shown in Fig. 1.1.11. Thus the configuration space can be identified with the product of two circles, i.e., a two-dimensional torus.

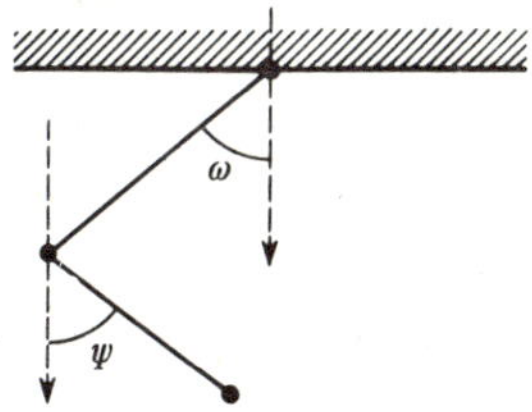

FIG. 1.1.11

Returning to general considerations, let M be a manifold identified with the configuration space and let $q(t) \in M$ denote the state of the system at a moment t (then $t \to q(t)$ represents the "path" of the system in time). The state $q(t)$ can be written in the coordinates involved:

$$q(t) = \big(q^1(t), \ldots, q^m(t)\big).$$

The quantities

$$\frac{\mathrm{d}}{\mathrm{d}t} q^i(t) = v^i(t), \qquad i = 1, \ldots, m,$$

denote the component velocities (in coordinates q^i); the vector

$$v(t) = \frac{\mathrm{d}q(t)}{\mathrm{d}t}$$

is a tangent vector to the path $q(t)$. The motion of our system is fully determined by prescribing the initial values of position $q_0 = q(0)$ and velocity $v_0 = v(0)$. The pair (q_0, v_0) is an element of the tangent bundle $T(M)$. The equations of motion are

$$\frac{\mathrm{d}q(t)}{\mathrm{d}t} = v(t), \qquad \frac{\mathrm{d}v(t)}{\mathrm{d}t} = F(q, v),$$

$F(\cdot, \cdot)$ denoting the vector of force. For any $x = (q, v) \in T(M)$ let us write $\Phi(x) = \big(v, F(q, v)\big) \in T_x\big(T(M)\big)$. Then the motion is described by the single equation

$$\frac{\mathrm{d}x(t)}{\mathrm{d}t} = \Phi\big(x(t)\big),$$

which determines a flow on $T(M)$; motion occurs on the integral curves of the above equation. In the most commonly encountered dynamical systems of mechanics motion is energy preserving (e.g. in Hamiltonian systems, see formula (4.10.2) on p. 250).

Subsets of $T(M)$ which are invariant with respect to the flow and such that the energy of the system is constant at each of their points are, as a rule, compact submanifolds of $T(M)$. Therefore special attention will be paid to flows on compact manifolds.

In the case of a single flat pendulum the tangent bundle $T(M)$ can be identified (in fact is diffeomorphic) with a cylinder, one of its cross-sections standing for the configuration space; see Fig. 1.1.12, where all possible types of trajectories are shown. There are two fixed points

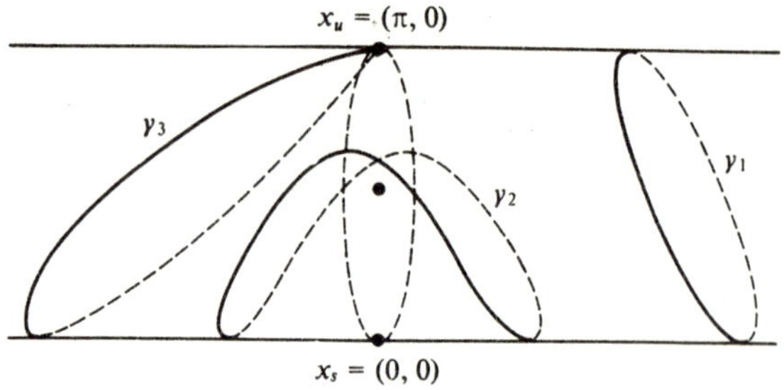

FIG. 1.1.12

(*equilibrium* points): $x_s = (0, 0)$ (*stable* equilibrium) and $x_u = (\pi, 0)$ (*unstable* equilibrium – the pendulum "sticks up"). If the speed is great enough, the pendulum swings round without changing direction; its motion is rotation (clockwise or anti-clockwise). The trajectory is of type γ_1 (Fig. 1.1.12). If the pendulum oscillates about the stable equilibrium point x_s, the trajectory is of type γ_2. There is one more type of trajectory: running from the point x_u (starting at the moment $-\infty$), via x_s, again to x_u (attaining it at $+\infty$) – this is a curve of type γ_3.

In mechanics it is often more convenient to consider the manifold (q, p), p denoting the impulse vector, instead of (q, v). The manifold (q, p) is called the *phase space*.

1.2. Survey of the Topics Studied in This Book

We now list the main topics which will be dealt with in this study. They belong to the most important problems considered in the theory of dynamical systems.

1. Classification of dynamical systems from the viewpoint of the topological properties of their trajectories (the concept of topological conjugacy).
2. Generic properties.
3. Stability of systems.
4. Theory of entropy.
5. Specific properties of certain classes of dynamical systems.
6. Invariant measures for smooth dynamical systems.

Let us now briefly comment on the particular topics.

1.2.1. Classification of dynamical systems

Definition 1.2.1 Two dynamical systems (X, φ^α) and (Y, ψ^α), $\alpha \in \mathfrak{A}$, are said to be *topologically conjugate* iff there exists a homeomorphism $h: X \to Y$ such that the following diagram commutes (for all $\alpha \in \mathfrak{A}$):

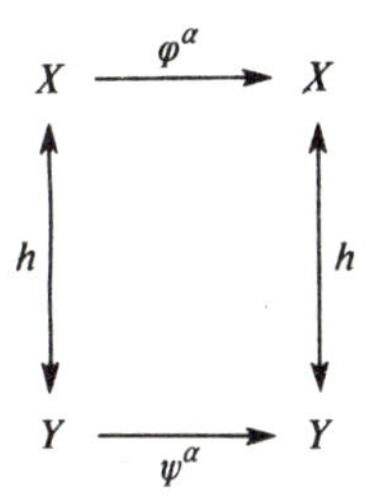

FIG. 1.1.13

This means that $\varphi^\alpha = h^{-1} \circ \psi^\alpha \circ h$ or $h \circ \varphi^\alpha = \psi^\alpha \circ h$ for all α. We then say that h *conjugates* the two systems, or that h conjugates φ^α with ψ^α.

Obviously, if $\mathfrak{A}$ is Z^+ or Z, the above condition is equivalent to

$$\varphi = h^{-1} \circ \psi \circ h \qquad \text{or} \qquad h \circ \varphi = \psi \circ h$$

(φ and ψ denote the generators).

Examples

1.2.1. Let X and Y be segments on the real line: $X = \langle a, b \rangle$, $Y = \langle c, d \rangle$. Let φ and ψ be orientation-preserving homeomorphisms, $\varphi: \langle a, b \rangle \to \langle a, b \rangle$, $\psi: \langle c, d \rangle \to \langle c, d \rangle$, having no fixed points in the interior of these intervals. Then the systems (X, φ) and (Y, ψ) are topologically conjugate.

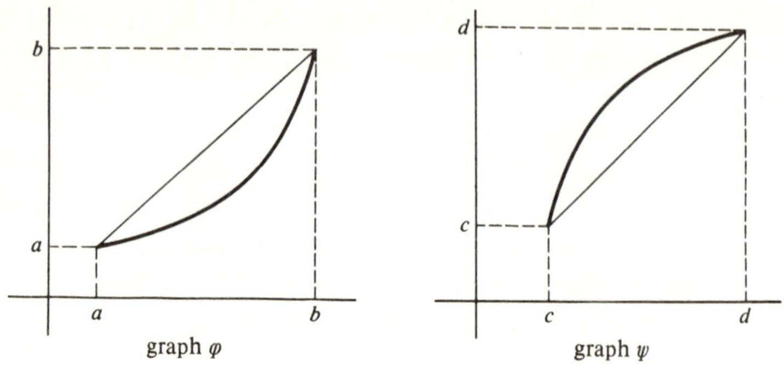

FIG. 1.2.1

Proof Let φ and ψ be as in Fig. 1.2.1. Choose $x_0 \in (a, b)$, $y_0 \in (c, d)$ and write $x_n = \varphi^n(x_0)$, $y_n = \psi^n(y_0)$. We see without difficulty that

$$\langle a, b \rangle = \bigcup_{n=-\infty}^{+\infty} (x_{n+1}, x_n \rangle \cup \{a, b\},$$

$$\langle c, d \rangle = \bigcup_{n=-\infty}^{+\infty} \langle y_n, y_{n+1}) \cup \{c, d\},$$

$$\varphi\big((x_{n+1}, x_n\rangle\big) = (x_{n+2}, x_{n+1}\rangle, \qquad \psi\big(\langle y_n, y_{n+1})\big) = \langle y_{n+1}, y_{n+2}).$$

Let $y = l(x)$ be the linear function taking $(x_1, x_0\rangle$ onto $\langle y_0, y_1)$, with $l(x_0) = y_0$ and $l(x_1) = y_1$. We define $h(x) = y$ as follows:

$$h(x) = \begin{cases} d & \text{if } x = a, \\ c & \text{if } x = b, \\ l(x) & \text{if } x \in (x_1, x_0\rangle, \\ \psi^n \circ l \circ \varphi^{-n}(x) & \text{if } x \in (x_{n+1}, x_n\rangle. \end{cases}$$

We leave to the reader the verification of the following details: that the definition is correct, that it indeed defines a homeomorphism, and that $h \circ \varphi = \psi \circ h$. It is shown in Fig. 1.2.2 how h actually acts.

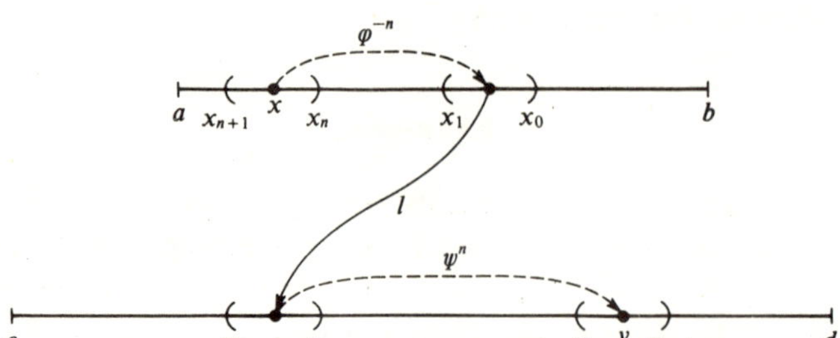

FIG. 1.2.2

Remark 1.2.1 Even if φ and ψ are of class C^∞, a *differentiable* h conjugating φ with ψ does not necessarily exist (see Exercise 1.2.7 on p. 24).

Remark 1.2.2 Let X and Y be as in the above example and suppose that φ and ψ are now as in Fig. 1.2.3.

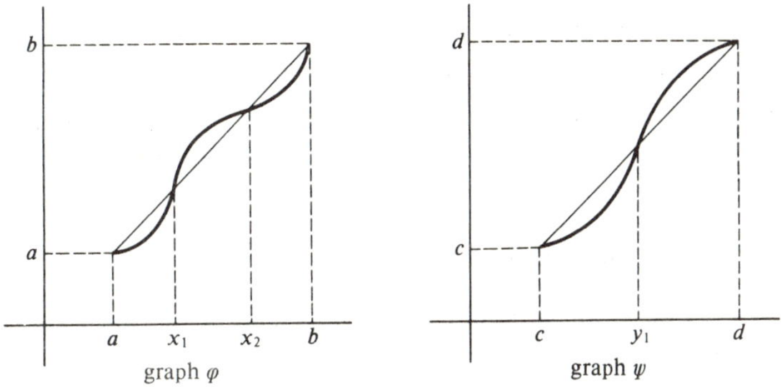

FIG. 1.2.3

The systems (X, φ) and (Y, ψ) are not topologically conjugate. Indeed, note that if x_0 is a fixed point of φ, then $y_0 = h(x_0)$ is a fixed point of ψ:

$$\varphi(y_0) = \psi \circ h(x_0) = h \circ \varphi(x_0) = h(x_0) = y_0.$$

Hence the number of fixed points is an invariant of topological conjugacy. In the situation now considered, φ has 4 fixed points, a, x_1, x_2, b, whereas ψ has 3 fixed points, c, y_1, d, and so the two systems are not topologically conjugate.

1.2.2. Generic properties

Which properties deserve being called "generic"? There exist quite regular manifolds admitting diffeomorphisms with trajectories behaving very wildly. The following question arises: how "frequently" does such a situation occur. To what extent are such diffeomorphisms "typical"?

In certain classes of mappings from X into X we can introduce natural topologies. For instance, let M be a compact manifold without boundary and let $\mathrm{Diff}^r(M)$ denote the class of all C^r-diffeomorphisms $\varphi: M \to M$, $r \geqslant 1$. The natural topology for the class $\mathrm{Diff}^r(M)$ is that of uniform C^r-convergence of mappings together with the inverse mappings [10]. In that topology $\mathrm{Diff}^r(M)$ is a complete metric space.

Let D be a complete metric space consisting of certain continuous mappings of X into X.

Definition 1.2.2 A property $\mathscr{P}$, which can be shared by mappings in D, is called *generic* iff the set

$$A = \{\varphi \in D\colon\ \varphi \text{ does not have property } \mathscr{P}\}$$

is meagre, i.e. is of first Baire category.

E.g., let $X = \langle 0, 1 \rangle = I$, $D = \mathrm{Diff}^1(I)$ with topology of uniform C^1-convergence. The property "$\varphi \in D$ has finitely many fixed points" is generic. In fact, the set A in this case is closed and nowhere dense (see Exercise 1.2.6 on p. 24).

1.2.3. Stability of dynamical systems

Let (X, φ) be a given dynamical system and D a family of mappings of X into X; assume that D is equipped with a topology. A mapping $\varphi \in D$ is called structurally stable in the class D if for every $\psi \in D$ sufficiently close to φ the systems (X, φ) and (X, ψ) are topologically conjugate. This means, speaking more intuitively, that a small perturbation of φ does not influence the topological character of trajectories in their totality.

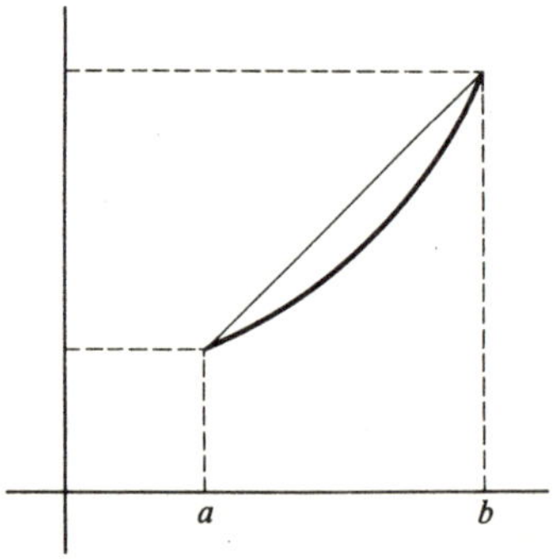

FIG. 1.2.4

For instance, let $X = I = \langle a, b \rangle$, $D = \mathrm{Diff}^1(I)$ and let φ be as in Fig. 1.2.4. Since φ has no fixed points inside (a, b), $\varphi'(a) \neq 1$, $\varphi'(b) \neq 1$, every ψ close to φ in C^1-topology has no fixed points in (a, b), either. It follows from the considerations of Example 1.2.1 that φ and ψ are then topologically conjugate, and thus φ is structurally stable.

1.2.4. Theory of entropy

Entropy is a certain numerical quantity which characterizes the intensiveness with which the points in X are "mixed" by a transformation φ. In fact, there are two notions of entropy: 1) topological entropy, 2) metric entropy.

Metric entropy is considered in the measure-theoretic case and is defined for measure preserving transformations. For more information on this subject see Section 5.7. We now restrict our attention to the concept of topological entropy.

Let X be, as usual, a compact metric space. Given a family $\xi = \{U_i\}$ of open sets covering X, (i.e., a *cover* of X), let $N(\xi)$ denote the least number equal to the cardinality of a subcover of ξ. $N(\xi)$ is finite, by the compactness of X. If $\xi = \{U_i\}$ and $\eta = \{V_j\}$ are two covers of X, there is automatically defined a third cover which we denote by $\xi \vee \eta$:

$$\xi \vee \eta = \{U_i \cap V_j\}.$$

Let φ: $X \to X$ be a continuous mapping; then $\varphi^{-1}(\xi)$, defined as the family $\{\varphi^{-1}(U_i)\}$, also is a cover of X. Every cover $\xi = \{U_i\}$ then induces a sequence of covers of X:

$$\xi \vee \varphi^{-1}(\xi) = \{U_{i_1} \cap \varphi^{-1}(U_{i_2})\}$$

and by induction

$$\begin{aligned}\xi_n &= \xi \vee \varphi^{-1}(\xi) \vee \dots \vee \varphi^{-n+1}(\xi)\\ &= \{U_{i_1} \cap \varphi^{-1}(U_{i_2}) \cap \dots \cap \varphi^{-n+1}(U_{i_n})\},\end{aligned}$$

$i_1, \dots, i_n$ ranging over all possible n-tuples of indices.

It can be proved that the following limit always exists:

$$h(\varphi, \xi) = \lim_{n \to \infty} \frac{\log N(\xi_n)}{n}.$$

The topological entropy of φ is defined by:

$$h(\varphi) = \sup_{\xi} h(\varphi, \xi)$$

(cf. Section 5.7).

Exercises

1.2.1.A. Show that the following conditions are equivalent:

(a) $e^{2\pi i\alpha}$ is not a root of unity.

(b) For any Riemann-integrable function $f\colon S^1 \to C$ and any point $z \in S^1$ we have:

$$\lim_{n\to\infty} \frac{1}{n}\sum_{k=0}^{n-1} f(e^{2\pi i k\alpha}z) = \frac{1}{2\pi}\int_0^{2\pi} f(e^{it})\mathrm{d}t.$$

(c) Let I be an open arc on S^1, let $z \in S^1$ and let $N_n(I, z)$ be the number of all integers k, $0 \leqslant k \leqslant n$, such that $e^{2\pi i k\alpha}z \in I$; then

$$\lim_{n\to\infty} \frac{1}{n} N_n(I, z) = \frac{1}{2\pi}|I|$$

($|\cdot|$ is the Lebesgue measure on S^1).

(d) The set $\{e^{2\pi i k\alpha}z\}$ is dense in S^1, for any $z \in S^1$.

Hint. First prove (a)$\Rightarrow$(b) for polynomials, then apply the fact that polynomials are dense in the family of continuous functions.

1.2.2.B. Let T^2 be the two-dimensional torus and let $z_0, w_0 \in S^1$. Show that the following conditions are equivalent:

(a) $z_0^r w_0^s = 1 \Rightarrow r = s = 0$ $(r, s \in Z)$.

(b) For any Riemann-integrable function $f\colon T^2 \to C$ and any $z, w \in S^1$ we have:

$$\lim_{n\to\infty} \frac{1}{n}\sum_{k=0}^{n-1} f(z_0^k z, w_0^k w) = \frac{1}{4\pi^2}\int_0^{2\pi}\int_0^{2\pi} f(e^{ix}, e^{iy})\,dx\,dy.$$

(c) For any measurable set $A \subset T^2$ and any $z, w \in S^1$ we have

$$\lim_{n\to\infty} \frac{1}{n} N_n\big(A, (z, w)\big) = \frac{1}{4\pi^2}|A|,$$

where $N_n\big(A, (z, w)\big)$ is the number of all integers k, $0 \leqslant k \leqslant n$, for which $(z_0^k z, w_0^k w) \in A$; $|\cdot|$ is the Lebesgue measure on T^2.

(d) The set $\{(z_0^k z, w_0^k w)\colon k \in Z\}$ is dense in T^2, for any $z, w \in S^1$.

1.2.3.B. Imagine a square billiard table Q with sides of length 1. A billiard ball is hit at a point $x \in \operatorname{int} Q$. Suppose that it moves with a constant scalar velocity and rebounds from the sides of Q according to the usual rule of reflexion (angle of incidence equals angle of reflexion). Assume that it never gets to a corner. Prove that:

(a) The trajectory $\gamma(x)$ is a closed curve iff $\tan\alpha$ is rational; α denotes the angle of incidence onto a side of Q.

(b) If $\tan\alpha$ is irrational, $\gamma(x)$ is dense in Q.

Hint. (a) Reflect Q symmetrically with respect to the side which $\gamma(x)$ meets first; the symmetric image of the successive piece of $\gamma(x)$ will be colinear with the first piece. Continuing, we get a straight line on the plane.

(b) If $\tan\alpha$ is irrational, this straight line (taken mod 1) fills Q densely; apply Exercise 1.2.1.

1.2.4.A. Let φ: $\boldsymbol{R}^2 \to \boldsymbol{R}^2$ be a linear mapping given by a matrix $\varphi = \begin{bmatrix} a & b \\ c & d \end{bmatrix}$. Show that:

(a) φ induces a mapping of $T^2 = \boldsymbol{R}^2/\Gamma$ (see Example 1.1.3) into itself iff $a, b, c, d \in Z$.

(b) φ induces a bijection of T^2 iff $\det\varphi = \pm 1$. In this case φ is called an *algebraic automorphism* of T^2.

Hint. Let $Q = \langle 0, 1 \rangle \times \langle 0, 1 \rangle$; the area of $\varphi(Q)$ has to be equal to 1 to make φ a bijection of T^2.

1.2.5.A. Let $\varphi = \begin{bmatrix} a & b \\ c & d \end{bmatrix}$ be an algebraic automorphism of the torus T^2 (i.e. $a, b, c, d \in Z$, $\det\varphi = \pm 1$; see the preceding exercise). Suppose that φ has no eigenvalue of modulus 1.

(a) Show that there are two real eigenvalues λ and μ with $|\lambda| > 1$, $|\mu| < 1$.

Let w and v be the corresponding eigenvectors:

$$\varphi(w) = \lambda w, \qquad \varphi(v) = \mu v$$

and consider the straight lines L^+ and L^- on the plane $\boldsymbol{R}^2$:

$$L^+ = \{tw\colon t \in \boldsymbol{R}\}, \qquad L^- = \{tv\colon t \in \boldsymbol{R}\}.$$

Let p: $\boldsymbol{R}^2 \to T^2$ be the natural projection:

$$p(x, y) = (e^{2\pi i x}, e^{2\pi i y}), \qquad (x, y) \in \boldsymbol{R}^2.$$

(b) Show that $p(L^+)$ and $p(L^-)$ are dense subsets of T^2. We may visualize $p(L^+)$ and $p(L^-)$ as immersed curves winding densely about the torus like thread on a screw.

Hint. (b) See Hint (b) to Exercise 1.2.3; apply Exercise 1.2.1.

1.2.6.A. Let $X = I = \langle 0, 1\rangle$. Show that the set

$$A' = \{\varphi \in \mathrm{Diff}^1(I)\colon \varphi \text{ has finitely many fixed points}\}$$

is open and dense in $\mathrm{Diff}^1(I)$ (see Section 1.2.2).

1.2.7.A. Let $\varphi, \psi\colon I = \langle 0, 1\rangle \to I$ be two mappings having no fixed points in the interior of I and suppose that

$$\left.\frac{\mathrm{d}^n\varphi(x)}{\mathrm{d}x^n}\right|_{\substack{x=0\\x=1}} = \left.\frac{\mathrm{d}^n\psi(x)}{\mathrm{d}x^n}\right|_{\substack{x=0\\x=1}} \qquad \text{for all } n = 0, 1, 2, \ldots$$

Prove that then φ and ψ are topologically conjugate (cf. Example 1.2.1 on p. 17).

Give an example of two mappings φ and ψ with the above properties, being of class C^∞ and admitting no conjugating function of class C^1.

1.3. Survey of Basic Facts of the Theory of Ordinary Differential Equations

In this section we recall some important facts of the theory of ordinary differential equations, which we will constantly use in the sequel. Let us consider a system of differential equations

$$\dot{x} = \frac{\mathrm{d}x}{\mathrm{d}t} = f(x, t) \tag{1.3.1}$$

where $x = (x^1, \ldots, x^m) \in \boldsymbol{R}^m$ and $f(x, t) = (f^1(x, t), \ldots, f^m(x, t)) \in \boldsymbol{R}^m$.

Theorem 1.3.1 Let $f(x, t)$ be a continuous function on a parallelepiped

$$Q = \{(x, t)\colon t_0 - \eta \leqslant t \leqslant t_0 + \eta,\ |x^i - x_0^i| \leqslant a,\ i = 1, \ldots, m\}$$

and suppose that $f(x, t)$ satisfies the Lipschitz condition with respect to x. Then there exists a $\delta > 0$ such that equation (1.3.1) *has a unique solution $x(t)$ in the interval $t \in (t_0 - \delta, t_0 + \delta)$ satisfying $x(t_0) = x_0$; δ is given by:*

$$\delta = \min\left(\eta, \frac{a}{M}, \frac{1}{L}\right), \qquad M = \sup_{(x,t)\in Q} \|f(x, t)\|,$$

L = the Lipschitz constant

(see Coddington and Levinson, 1955; Hartman, 1964).

Theorem 1.3.2 *Let $f(x, t)$ be of class C^1 on $V = D \times I$ where D is an open subset of $\boldsymbol{R}^m$ and $I = (-a, a)$. Denote by $x(t, x_0)$ the unique solution of (1.3.1) with the initial condition $x(0, x_0) = x_0$. Then $x(t, x_0)$ is of class C^1 in V. Moreover, the second order mixed derivatives $\partial^2 x(t, x_0)/\partial t \partial x_0$ and $\partial^2 x(t, x_0)/\partial x_0 \partial t$ exist and are continuous* (see Coddington and Levinson, 1955; Hartman, 1964).

Theorem 1.3.3 *Let $A(t)$ be a continuous $m \times m$-matrix on an interval $\langle a, b \rangle$; i.e., $A(t)$ is an $m \times m$-matrix-valued function, $A(t) = (a_{ij}(t))_{i,j=1}^m$, where $a_{ij}(t)$ are continuous functions. Then:*

(1) *The system of linear differential equations*

$$\dot{x} = A(t)x \tag{1.3.2}$$

has a unique solution for any initial value $x(t_0) = x_0$.

(2) *The space E of all solutions of* (1.3.2) *is of dimension m.*

(3) *Let $x_1, \ldots, x_m$ be a basis for E, $x_j = (x_j^1, \ldots, x_j^m)$, $j = 1, \ldots, m$; the matrix $Y = (x_j^i(t))$, called a fundamental matrix, satisfies the equation*

$$\dot{Y} = A \cdot Y.$$

It can be chosen so that $Y(t_0) = I$ (the unit matrix).

(4) *Any solution of* (1.3.2) *is of the form*

$$y(t) = Y(t) \cdot c, \tag{1.3.3}$$

where $c = (c^1, \ldots, c^m)$ is a constant vector; conversely, any $y(t)$ of this form is a solution of (1.3.2); *moreover, we have* $\det Y(t) \neq 0$ *for all t.*

Remark 1.3.1 If $Y_1(t)$ and $Y_2(t)$ are two fundamental matrices, then there is a constant matrix C such that

$$Y_2 = Y_1 \cdot C.$$

Let us now consider a linear system

$$\dot{x} = Ax \tag{1.3.4}$$

where A is a constant real 2×2-matrix and $x = (x^1, x^2) \in \boldsymbol{R}^2$. Clearly, the function $x(t) \equiv 0$ is a solution of (1.3.4), i.e., the origin is a fixed point of the flow $\varphi_t(x_0) = x(t, x_0)$. We are going to examine the behaviour of solutions in a neighbourhood of 0.

We may assume that the matrix A is written in the Jordan real canonical form (cf. Gantmacher, 1960; Nemytskii and Stepanov, 1960);

the matrix A can thus be of 6 possible types (see Coddington and Levinson, 1955).

I. $A = \begin{bmatrix} \lambda & 0 \\ 0 & \lambda \end{bmatrix}$, where $\lambda \neq 0$. Then

$$x^1(t) = C^1 e^{\lambda t}, \qquad x^2(t) = C^2 e^{\lambda t}.$$

Hence

$$\frac{x^2}{x^1} = \frac{C^2}{C^1} = \text{const}.$$

The graph of the integral curves is shown in Fig. 1.3.1. In this case the point (0, 0) is called a *proper node.*

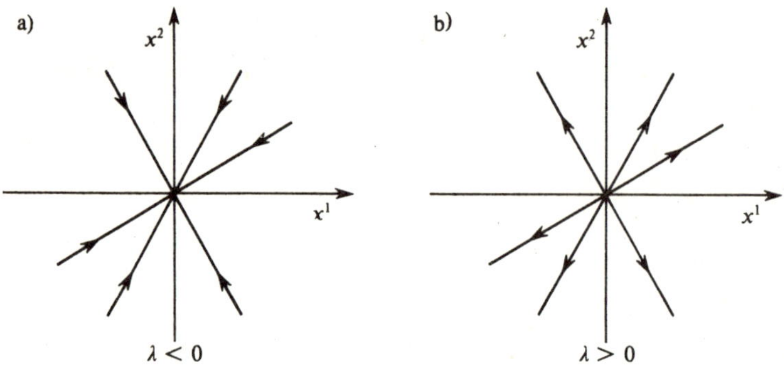

FIG. 1.3.1

II. $A = \begin{bmatrix} \lambda & 0 \\ 0 & \mu \end{bmatrix}$, where $\lambda\mu > 0$. Then

$$x^1(t) = C^1 e^{\lambda t}, \qquad x^2(t) = C^2 e^{\mu t}.$$

The graph of integral curves is shown in Fig. 1.3.2 for negative λ, μ (for positive λ, μ, reverse the arrows: in Fig. 1.3.2a for $0 < \mu < \lambda$ and in Fig. 1.3.2b for $0 < \lambda < \mu$). In this case the point (0, 0) is called an *improper node.*

III. $A = \begin{bmatrix} \lambda & 0 \\ \gamma & \lambda \end{bmatrix}$, where $\lambda \neq 0$, $\gamma > 0$. Then

$$x^1(t) = C^1 e^{\lambda t}, \qquad x^2(t) = (C^2 + C^1 \gamma t)\, e^{\lambda t}.$$

The graph of the integral curves is shown in Fig. 1.3.3. Also in this case the point (0, 0) is called an *improper node.*

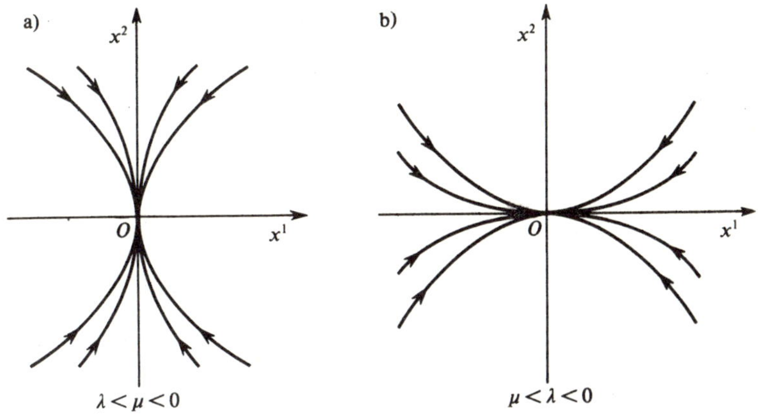

FIG. 1.3.2

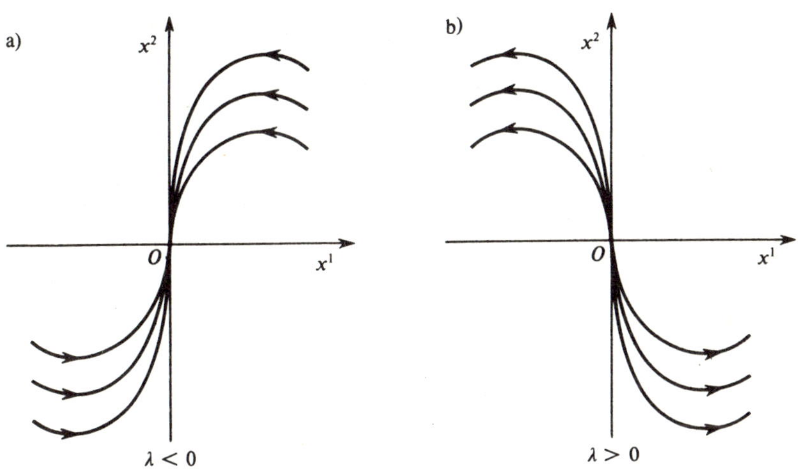

FIG. 1.3.3

IV. $A = \begin{bmatrix} \lambda & 0 \\ 0 & \mu \end{bmatrix}$, where $\lambda\mu < 0$. Then

$$x^1(t) = C^1 e^{\lambda t}, \qquad x^2(t) = C^2 e^{\mu t}.$$

The graph of integral curves is shown in Fig. 1.3.4 for $\lambda < 0 < \mu$ (for $\mu < 0 < \lambda$, reverse the arrows). The point (0, 0) is called a *saddle*.

The remaining two possibilities concern the case where the eigenvalues λ, μ are not real. The Jordan real form of A is then as follows: $A = \begin{bmatrix} \alpha & -\beta \\ \beta & \alpha \end{bmatrix}$ where $\lambda = \alpha + i\beta$, $\mu = \alpha - i\beta$, $\beta \neq 0$.

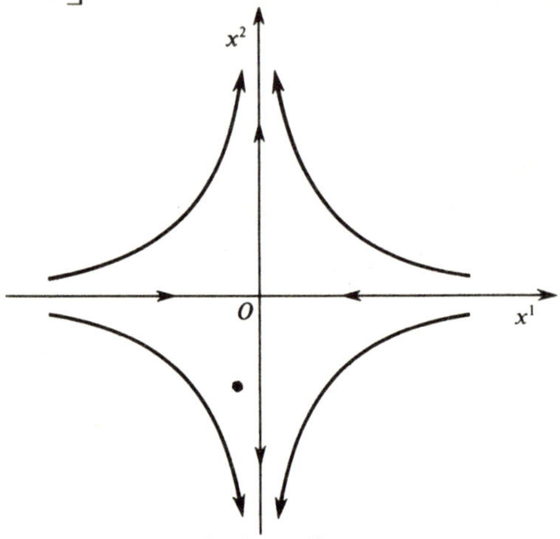

$(\lambda < 0, \mu > 0)$

FIG. 1.3.4

V. λ and μ are neither real nor purely imaginary: $\alpha \neq 0$, $\beta \neq 0$. Then

$$x^1(t) = e^{\alpha t}(C^1 \cos\beta t + C^2 \sin\beta t),$$
$$x^2(t) = e^{\alpha t}(-C^2 \cos\beta t + C^1 \sin\beta t).$$

In the polar coordinates (ϱ, ω) the solution has the form:

$$\varrho(t) = \varrho_0 e^{\alpha t}, \qquad \omega(t) = \beta t + \omega_0.$$

The graph of the integral curves is shown in Fig. 1.3.5 for $\alpha < 0$ (for $\alpha > 0$ reverse the arrows). In this case the point (0, 0) is called a *focus*.

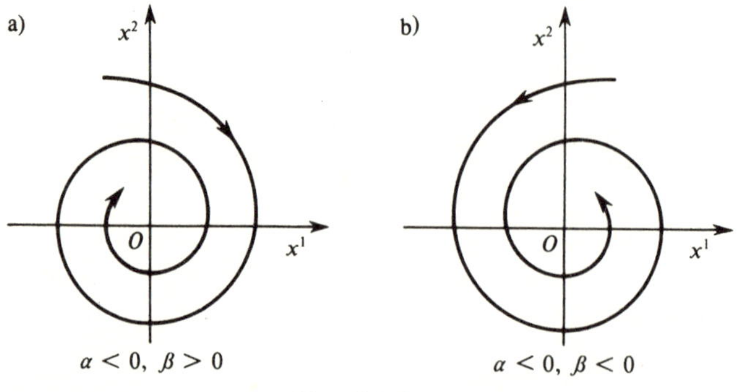

FIG. 1.3.5

VI. λ and μ are purely imaginary: $\lambda = \beta \mathrm{i}$, $\mu = -\beta \mathrm{i}$, $\beta \neq 0$. Then $A = \begin{bmatrix} 0 & -\beta \\ \beta & 0 \end{bmatrix}$ and

$$x^1(t) = C^1 \cos\beta t + C^2 \sin\beta t, \qquad x^2(t) = -C^2 \cos\beta t + C^1 \sin\beta t.$$

Hence

$$(x^1)^2 + (x^2)^2 = \text{const}.$$

The graph of the integral curves is shown in Fig. 1.3.6. The point (0, 0) is called a *centre*.

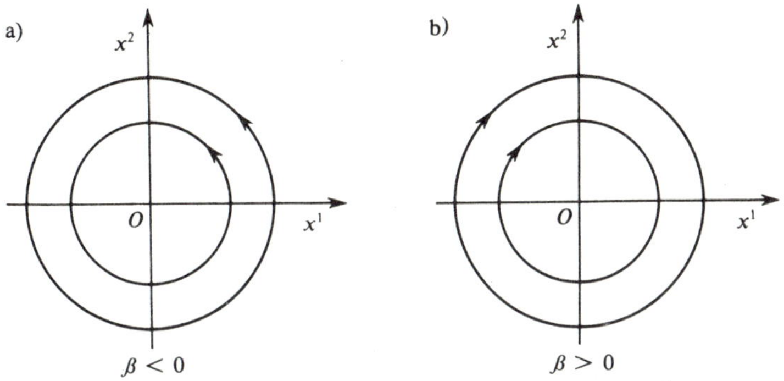

FIG. 1.3.6

Exercises

1.3.1.B. Let A, B be two linear operators from $\boldsymbol{R}^m$ to $\boldsymbol{R}^m$. Show that if $AB = BA$, then $\mathrm{e}^{A+B} = \mathrm{e}^A \mathrm{e}^B$; the operator e^A is defined by $\mathrm{e}^A = \sum_{n=0}^{\infty} \frac{A^n}{n!}$. Is this necessarily true if $AB \neq BA$?

Hint. In the noncommuting case try finding A and B of the form $\begin{bmatrix} \alpha & \gamma \\ 0 & \beta \end{bmatrix}$.

1.3.2.A. Let $f(\lambda) = \sum_{n=0}^{\infty} a_n \lambda^n$ be an entire function and let A: $\boldsymbol{R}^m \to \boldsymbol{R}^m$ be a linear operator. We define:

$$f(A) = \sum_{n=0}^{\infty} a_n A^n,$$

the series converging absolutely: $\sum \|a_n A^n\| \leqslant \sum |a_n| \, \|A\|^n < \infty$, since f is entire. Thus $f(A)$ is a well defined linear operator from $\boldsymbol{R}^m$ to $\boldsymbol{R}^m$.

Show that μ is an eigenvalue of $f(A)$ if any only if there is an eigenvalue λ of A such that $\mu = f(\lambda)$.

Hint. Write A in the Jordan form. Decompose A into the sum $A = D + E$ with D diagonal and E having zeros on the diagonal. Then $E^p = 0$ for some p.

1.3.3.B. Let A be an $m \times m$-matrix. Show that if $\det A \neq 0$, there is a matrix B such that $A = e^B$.

Hint. Apply the identity

$$\sum_{n=0}^{\infty} \frac{1}{n!} \left(\sum_{k=1}^{\infty} (-1)^{k+1} \frac{t^k}{k} \right)^n = 1 + t \qquad \text{for } |t| < 1,$$

i.e. $\exp \log(1+t) = 1 + t$; insert E in place of t.

1.3.4.A. Let A: $\boldsymbol{R}^m \to \boldsymbol{R}^m$ be a linear operator. Show that the following two conditions are equivalent:

(a) $\lim_{n \to \infty} A^n x = 0$ for any $x \in \boldsymbol{R}^m$;

(b) if λ is an eigenvalue of A, then $|\lambda| < 1$.

Hint. Write A in the Jordan form.

1.3.5.A. Let f: $\boldsymbol{R}^m \to \boldsymbol{R}^m$ be a C^1-mapping satisfying $f(0) = 0$. Prove that if all eigenvalues of the derivative $\mathrm{d}f(0)$ are less than 1 in absolute value, then there is an ellipsoid B with centre at 0 such that for any $x \in B$ we have:

$$f^n(x) \in B \text{ for } n \in Z^+ \qquad \text{and} \qquad \lim_{n \to \infty} f^n(x) = 0.$$

Hint. For any sufficiently small $\varepsilon > 0$ there is $\delta > 0$ such that $f(x) = \mathrm{d}f(0)x + \varrho(x)$ where $\|\varrho(x)\| \leqslant \varepsilon \|x\|$ for $\|x\| < \delta$. Apply Exercise 1.3.4.

1.3.6*.B. Let π denote the stereographic projection of the m-sphere S^m onto $\boldsymbol{R}^m$ having N as the north pole. Let A be a linear operator of $\boldsymbol{R}^m$ into $\boldsymbol{R}^m$. Define a mapping φ: $S^m \to S^m$ by:

$$\varphi = \pi^{-1} \circ A \circ \pi.$$

(a) Prove that φ is a diffeomorphism of S^m iff A is a similarity transformation of $\boldsymbol{R}^m$.

(b) Give examples of A for which: 1° φ is discontinuous, 2° φ is continuous but is not a diffeomorphism.

Hint. Calculate the derivative $d\varphi$ at the point N.

1.4. Vector Fields and Their Integral Curves on Manifolds

Let M be a C^∞-manifold (without boundary).

Definition 1.4.1 By a *vector field of class* C^r on M ($r \geqslant 1$) we mean any C^r-section [8] Φ of the tangent bundle $T(M)$:

$$\Phi: M \to T(M).$$

The set of all C^r-vector fields on M will be denoted by $\mathscr{F}^r(M)$. To denote vector fields we shall use Greek capitals Φ, Ψ. The symbol $\Phi(x)$ will denote the value of the field at the point x.

Roughly speaking, a vector field on M is the set of pairs $(x, \Phi(x))$, where $x \in M$ and each $\Phi(x)$ is a vector tangent to M at x. The field is of class C^r iff the assignment is of class C^r in every chart.

Definition 1.4.2 Let Φ be a vector field of class C^r on M, $r \geqslant 1$. By an *integral curve* of Φ passing through a point $x \in M$ we mean a curve $\gamma(t)$ on M, $t \in (a, b)$, such that $\gamma(t_0) = x$ for a certain $t_0 \in (a, b)$ and

$$\dot{\gamma}(t) = \Phi(\gamma(t)) \qquad \text{for all } t \in (a, b); \tag{1.4.1}$$

$\dot{\gamma}(t)$ is the tangent vector to $\gamma(t)$ at t.

We may always assume that the time interval is of type $(-a, a)$ with $t_0 = 0$. The existence and uniqueness of integral curves is a direct consequence of suitable theorems of differential equations: at any point $x \in M$ we choose a chart (U, α), $U \ni x$, and transfer equation (1.4.1) to an open set of $\boldsymbol{R}^m$. Let $c(t) = \alpha \circ \gamma(t)$ and $f(y) = (d\alpha)(x)(\Phi(x))$ for $y = \alpha(x)$. Then (1.4.1) induces the equation:

$$\dot{c}(t) = f(c(t)),$$

in which $c(t) \in \boldsymbol{R}^m, f(y) \in \boldsymbol{R}^m$, and one can apply theorems formulated in the preceding section.

Definition 1.4.3 Let $\Phi \in \mathscr{F}^r(M)$ and let $I(x) = (a_x, b_x)$ denote the maximal interval in which the integral curve passing through x is defined. The vector field Φ is called *complete* iff $I(x) = \boldsymbol{R}$ for any $x \in M$, i.e., $a_x = -\infty$, $b_x = +\infty$.

The above definition and considerations together with the theorems of the preceding section imply:

Theorem 1.4.1 *If $\Phi \in \mathscr{F}^r(M)$ is complete, $r \geq 1$, then:*

(1) *for every point $x \in M$ there exists a unique integral curve $\gamma(t) = \gamma_x(t)$ such that $\gamma_x(0) = x$;*

(2) *writing $\varphi_t(x) = \varphi(x, t) := \gamma_x(t)$, we have $\varphi_0(x) = x$; the mapping $\varphi: M \times \mathbf{R} \to M$ is of class C^r;*

(3) *for every $t \in \mathbf{R}$ the mapping $\varphi_t: M \to M$ is a diffeomorphism;*

(4) *the family of mappings φ_t is a group:*

$$\varphi_{t+s} = \varphi_t \circ \varphi_s, \qquad t, s \in \mathbf{R}.$$

If the completeness assumption is omitted, assertion (2) has to be reformulated as follows:

(2′) *for every $x_0 \in M$ there are a neighbourhood $U \ni x_0$ and an interval $I = (-a, a)$ such that $\varphi(x, t) = \varphi_t(x) := \gamma_x(t)$ is a C^r-mapping from $U \times I$ into M.*

Assertions (3) and (4) are not necessarily true in that case.

Theorem 1.4.2 *Let $\Phi \in \mathscr{F}^r(M)$ be a vector field on a compact manifold. Then Φ is complete.*

Proof Suppose that for some $x_0 \in M$ the trajectory $\gamma(t) = \varphi_t(x_0)$ with $\gamma(0) = x_0$ is defined in the maximal existence interval $t \in (\alpha, \beta)$ where, say, $\beta < +\infty$. Let $t_n \in (\alpha, \beta)$, $t_n \nearrow \beta$ as $n \to \infty$. M being compact, the sequence $(\gamma(t_n))$ contains a convergent subsequence; we may assume for simplicity that $\gamma(t_n) = x_n$ itself converges to a limit $y \in M$. Let U be a neighbourhood of y and $I = (-a, a)$ an interval with properties stated in assertion (2′). Then $x_n \in U$ for n sufficiently large, and so the trajectory passing through x_n exists for $t \in (-a, a)$. Since $x_n = \gamma(t_n)$, the curve $\gamma(t)$ is defined, in particular, for all t with $t_n < t < t_n + a$; and since $t_n + a \to \beta + a$, we see that $\gamma(t)$ can be extended on the right beyond β, contrary to the definition of β. The proof is the same if $\alpha > -\infty$.

Let $\Phi \in \mathscr{F}^r(M)$ be a vector field such that $\Phi(x_0) \neq 0$ at some point $x_0 \in M$. Then there exists a chart (U, α), $U \ni x_0$, such that Φ induces via α a field of parallel vectors in $\mathbf{R}^m$.

To see this, take a submanifold S of codimension 1, passing through x_0 and such that $\Phi(x_0) \notin T_{x_0}(S)$. Then $\Phi(x) \notin T_x(S)$ for all x in a neighbourhood B of x_0 in S. Now select a chart (U_1, α_1) in which

$B = \{x = (x^1, \ldots, x^m)\colon x^m = 0\}$, $\Phi(x_0) = (0, \ldots, 0, r)$, $\Phi^m(x) > 0$ for $x \in B$ ($\Phi^m(x)$ denotes the m-th coordinate of $\Phi(x)$ in this chart). Let $(-a, a)$ be a time interval such that $\Phi^m(\varphi_t(x)) > 0$ for $x \in B$, $t \in (-a, a)$. To define the desired chart (U, α), it suffices to take $B \times (-a, a)$ as U and (x, t) as local coordinates. It is then easy to see that in these coordinates the field trajectories are straight line segments parallel to the time axis $0t$; transition from coordinates (x, t) to the original coordinates is diffeomorphic; we leave the details to the reader.

Definition 1.4.4 Any chart with the property that the field trajectories are straight line segments in that chart is called a *rectifying* chart or a flow box.

Remark 1.4.1 If M is compact, then every vector field $\Phi \in \mathscr{F}^1(M)$ generates a flow of diffeomorphisms φ_t on M twice differentiable in t; and conversely, every flow of diffeomorphisms φ_t of class C^2 in t induces a vector field $\Phi \in \mathscr{F}^1(M)$ according to the formula:

$$\Phi(x) = \frac{\mathrm{d}}{\mathrm{d}t} \varphi_t(x)\bigg|_{t=0}.$$

Examples

1.4.1. Let $M = \boldsymbol{R}^n$ and let $\Phi(x) = [\alpha, \beta]$ be a constant vector field (see Fig. 1.4.1). Integral curves are straight lines in $\boldsymbol{R}^2$.

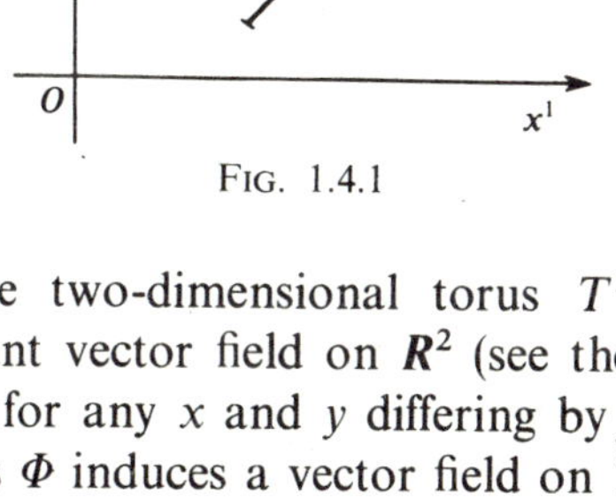

FIG. 1.4.1

1.4.2. Let M be the two-dimensional torus $T^2 = \boldsymbol{R}^2/\Gamma$ and let $\Phi = [\alpha, \beta]$ be a constant vector field on $\boldsymbol{R}^2$ (see the preceding example). Clearly, $\Phi(x) = \Phi(y)$ for any x and y differing by a vector with integer components and thus Φ induces a vector field on T^2. Integral curves of the induced field coincide with projections of straight lines in $\boldsymbol{R}^2$ onto T^2 (cf. Exercise 1.2.5 on p. 23). If α and β are linearly independent over the rationals, the integral lines wind densely round the torus; if α and β are linearly dependent, the integral curves are closed.

1.4.3. Let $M = S^2 = \{x = (x^1, x^2, x^3) \in \boldsymbol{R}^3 \colon (x^1)^2 + (x^2)^2 + (x^3)^2 = 1\}$. The metric in $\boldsymbol{R}^3$ induces a Riemannian metric [15] on S^2. Let $h(x) = -mgx^3$ (the *gravitational potential*). Then [21] $\Phi(x) = \operatorname{grad} h(x)$ is a C^∞-vector field on S^2. Suppose that a particle has to be in contact with the surface S^2, but is able to move all over it; its motion is determined by the vector field of forces $\Phi(x)$ (no other forces are assumed to act upon it). Then the integral curves are meridians, and, besides, two curves reduce to fixed points: the north and the south poles.

The above example is a particular case of a more general situation.

Let M be a Riemannian manifold. [15] A rather simple type of vector fields are the so-called gradient fields:

Definition 1.4.5 A vector field $\Phi \in \mathscr{F}^r(M)$ is termed a *gradient field* iff there exists a real-valued function $h(x)$ such that $\Phi(x) = \operatorname{grad} h(x)$; the flow φ_t generated by Φ is called a *gradient flow*.

One can distinguish 3 possible types of integral curves of a complete vector field on a manifold M.

1° Let $x \in M$ be a point at which $\Phi(x) = 0$. Then the constant curve: $\gamma(t) = x$ for all $t \in \boldsymbol{R}$, is the only integral curve passing through x. The point x itself is a fixed point of the flow φ_t generated by Φ. Any such point is called a *critical point* for the field Φ (or for the flow φ_t).

Now suppose $x \in M$, $\Phi(x) \neq 0$, and let $\gamma(t)$ be the integral curve passing through x, $\gamma(0) = x$. Since $\dot{\gamma}(0) = \Phi(x) \neq 0$, there is an interval $I \subset \boldsymbol{R}$ containing 0 and such that γ (restricted to I) is a homeomorphism (in fact a diffeomorphism) of I onto the arc $\gamma(I)$. In other words, γ is a local homeomorphism from $\boldsymbol{R}$ onto $\gamma(\boldsymbol{R})$. We have two possibilites:

2° There exist numbers $t_1, t_2, t_1 \neq t_2$, such that $\gamma(t_1) = \gamma(t_2)$. Then $\gamma(t)$ is periodic in t and the set $\gamma(\boldsymbol{R})$ is homeomorphic to the circle S^1. In this case the trajectory $\gamma(t)$ is said to be *closed* or *periodic*.

3° γ maps $\boldsymbol{R}$ onto $\gamma(\boldsymbol{R})$ in a one-to-one way. Then γ is an injective immersion (see Definition 1.6.1) of $\boldsymbol{R}$ into M; however, the set $\gamma(\boldsymbol{R})$ with topology induced from M is not necessarily homeomorphic to $\boldsymbol{R}$ (cf. Example 1.4.2 on p. 33).

Definition 1.4.6 Let $\Phi \in \mathscr{F}^r(M)$. By a *critical element* of Φ we mean either a critical point or a closed orbit. The set of all critical or periodic points will be denoted by $\mathrm{Per}(\Phi)$.

In our subsequent considerations we shall distinguish between a curve $\gamma(t)$ (which, formally, is a function from an interval $I \subset \boldsymbol{R}$ into a manifold) and the set $\hat{\gamma} = \gamma(I)$ of all values assumed by γ. $\hat{\gamma}$ is simply a certain geometrical figure; speaking of $\hat{\gamma}$, we disregard parametrization. The field Φ induces, in a natural way, the orientation of each orbit.

Definition 1.4.7 Let $\Phi \in \mathscr{F}^r(M)$. The set $P(\Phi, M)$ of all oriented orbits $\hat{\gamma}$ of Φ is called the *phase portrait* of Φ.

Definition 1.4.8 Let M_1, M_2 be manifolds and let $\Phi_i \in \mathscr{F}^r(M_i)$, for $i = 1, 2$. The fields Φ_1 and Φ_2 are said to have *equivalent phase portraits* iff there exists a homeomorphism h of M_1 onto M_2 such that

$$\hat{\gamma} \in P(\Phi_1, M_1) \quad \text{iff} \quad h(\hat{\gamma}) \in P(\Phi_2, M_2).$$

Examples

1.4.4. Let $\Phi \in \mathscr{F}^r(M)$ and let $\alpha(x)$ be a C^1-function on M, $\alpha(x) > 0$ for $x \in M$. Then the fields Φ and $\alpha\Phi$ have equivalent phase portraits; it suffices to take the identity for h.

1.4.5. Let $M = T^2 = \boldsymbol{R}^2/\Gamma$ and let $\Phi_1 = [\alpha_1, \beta_1]$, $\Phi_2 = [\alpha_2, \beta_2]$ be two constant vector fields on T^2. If there exists a linear mapping $A = \begin{bmatrix} a & b \\ c & d \end{bmatrix}$, $a, b, c, d \in Z$, with $\det A = \pm 1$ and such that $A([\alpha_1, \beta_1]) = k[\alpha_2, \beta_2]$, k being a constant, $k \neq 0$, then Φ_1 and Φ_2 have equivalent phase portraits.

In comparing the phase portraits of two vector fields, we look upon their oriented orbits as sets of points and disregard their particular parametrization. We are only interested in the sense of orientation. It is of no significance with what speed this or that segment of the trajectory is passed by a mobile point. It is the topological structure of the orbit that is important. This is well illustrated by Example 1.4.4.

The notion of phase portrait equivalence is weaker than that of topological conjugacy. If two flows are topologically conjugate, then, clearly, they have equivalent phase portraits. The converse does not hold. For instance, any two flows on S^1 without critical points and such that at each point of S^1 they define tangent vectors of coincident direction have equivalent phase portraits (there is only one orbit, the entire circle); but the flows $\varphi_t(z) = e^{2\pi i t\alpha}z$ and $\psi_t(z) = e^{2\pi i t\beta}z$ are not in general topologically conjugate (see Corollary 2.4.1 on p. 105).

Exercises

1.4.1.B. Let Φ_1, Φ_2 be two constant vector fields on the torus T^2. Denote by ϑ_i, $i = 1, 2$, the angle between the direction of Φ_i and the axis $0x^1$ (in the covering plane). Show that if the numbers $\tan\vartheta_1$, $\tan\vartheta_2$ are rational, then Φ_1 and Φ_2 have equivalent phase portraits.

Hint. There exists an algebraic automorphism of the torus establishing the desired equivalence.

1.4.2.A. Let $\Phi \in \mathscr{F}^r(M)$, M being compact. Show that, if γ is a trajectory such that $\hat{\gamma}$ is a closed subset of M, then $\hat{\gamma}$ is a submanifold of dimension 0 or 1.

Hint. Apply the Baire theorem.

1.4.3.B. Show that if $\Phi \in \mathscr{F}^r(M)$ is a gradient vector field, then Φ has no periodic orbits.

Hint. If $\Phi = \operatorname{grad} h$, then h increases along trajectories.

1.4.4.A. Consider the torus T^2 embedded in the three-dimensional Euclidean space; T^2 arises by rotation of a circle about an axis. Denote by P the *main plane of symmetry* for T^2, i.e., the only plane of symmetry perpendicular to the rotation axis. T^2 can be situated in $\boldsymbol{R}^3$ in three possible positions (relative to a fixed coordinate system in $\boldsymbol{R}^3$):

(a) vertical position: the coordinate axis $0x^3$ is contained in P (we may assume $P = \{x\colon x^2 = 0\}$; Fig. 1.4.2);

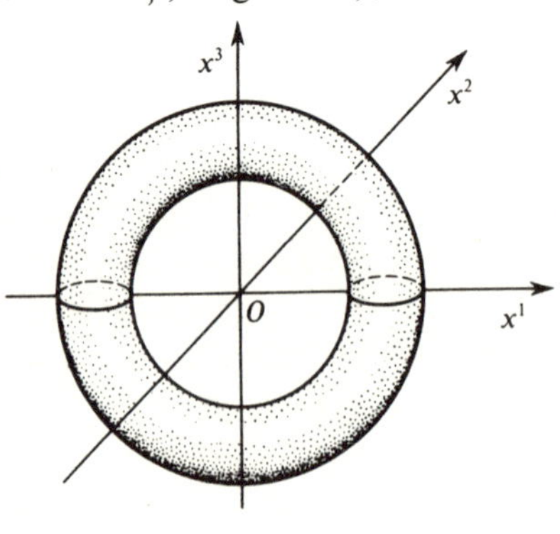

FIG. 1.4.2

(b) horizontal position: $0x^3 \perp P$ $(P = \{x\colon x^3 = 0\}$; Fig. 1.4.3);

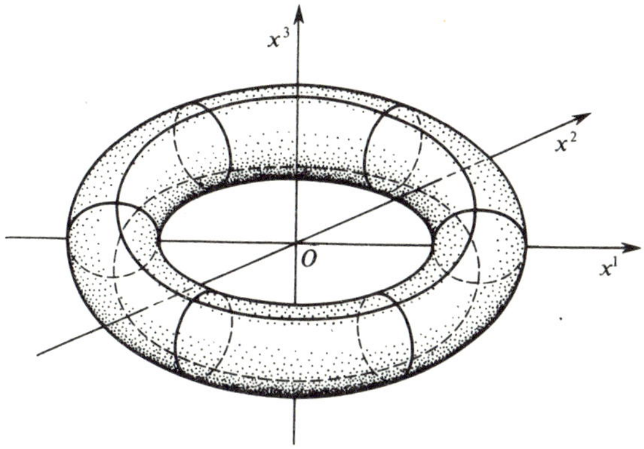

FIG. 1.4.3

(c) tilted position: the angle α between $0x^3$ and P belongs to the interval $(0, \pi/2)$ (Fig. 1.4.4).

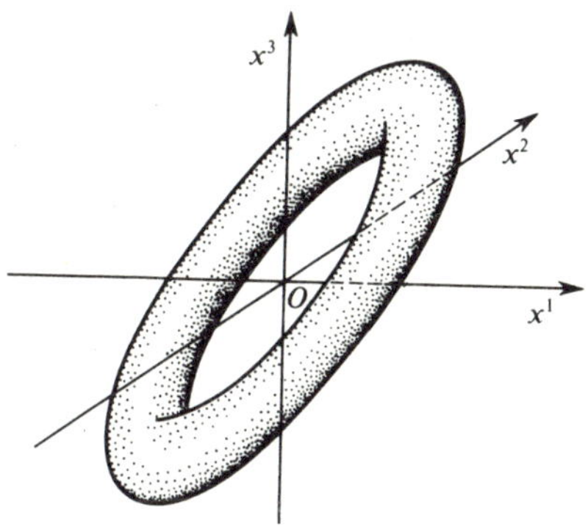

FIG. 1.4.4

Denote by M_a, M_b, M_c the torus T^2 in positions (a), (b), (c), respectively. Let $h(x) = -mgx^3$ and let Φ_a, Φ_b, Φ_c denote the gradient fields $\operatorname{grad} h(x)$ on M_a, M_b, M_c. Show that the phase portraits of no two of the fields Φ_a, Φ_b, Φ_c are equivalent.

Hint. For Φ_a and Φ_b the sets of critical points are not homeomorphic; the same holds for Φ_b and Φ_c. In the case of Φ_a and Φ_c the systems of trajectories joining critical points are not equivalent.

1.4.5.*B. Let $\Phi_i = [\alpha_i, \beta_i]$, $i = 1, 2$, be two constant vector fields on T^2 and suppose that Φ_1, Φ_2 have equivalent phase portraits. Show that there exists a matrix $A = \begin{bmatrix} a & b \\ c & d \end{bmatrix}$ with $a, b, c, d \in Z$, $\det A = \pm 1$, and such that

$$A([\alpha_1, \beta_1]) = k[\alpha_2, \beta_2],$$

where k is a constant.

1.5. The Hadamard–Perron Theorem

In this section we discuss the behaviour of trajectories of a dynamical system in a neighbourhood of a critical point.

Let E be a finite-dimensional vector space over the complex field C and let A: $E \to E$ be a linear operator.

Definition 1.5.1 We say that $0 \in E$ is a *hyperbolic point* (for A) iff all eigenvalues $\lambda_1, \ldots, \lambda_k$ of A have moduli different from unity: $|\lambda_j| \neq 1$ for $j = 1, \ldots, k$.

Suppose 0 is hyperbolic for A. It follows from classical theorems of linear algebra (e.g., the Jordan form for matrices) that E admits decomposition into the direct sum of two invariant subspaces E^+ and E^-,

$$E = E^+ \oplus E^-$$

such that all eigenvalues of $A|E^+$ are less than 1 in absolute value, while the eigenvalues of $A|E^-$ are greater than 1 in absolute value. The subspace E^+ is said to be *contracted* and the subspace E^- *expanded* by A.

Let φ: $M \to M$ be a diffeomorphism of a manifold M, $\dim M = m$, and let x_0 be a fixed point:

$$\varphi(x_0) = x_0.$$

Definition 1.5.2 The fixed point x_0 is called *hyperbolic* (for φ) iff 0 is hyperbolic for the derivative $d\varphi(x_0)$ in the tangent space $T_{x_0}(M)$, i.e., iff every eigenvalue λ of the operator $d\varphi(x_0)$: $T_{x_0}(M) \to T_{x_0}(M)$ fulfils: $|\lambda| \neq 1$.

The diffeomorphism φ is said to be *contracting* at x_0 iff all eigenvalues of $d\varphi(x_0)$ satisfy: $|\lambda| < 1$; it is said to be *expanding* at x_0 iff all eigenvalues of $d\varphi(x_0)$ satisfy: $|\lambda| > 1$.

We are going to show that, given any manifold M, any diffeomorphism $\varphi: M \to M$ and any hyperbolic fixed point x_0, $\varphi(x_0) = x_0$, there exist two smooth submanifolds in a neighbourhood of x_0, both containing x_0, such that φ restricted to one of them is contracting and restricted to the other is expanding. This is undoubtedly one of the most important theorems of the theory of smooth dynamical systems. It was proved, in various versions, by several authors, but finally gained renown in mathematics as the Hadamard–Perron theorem. We shall retain this name although the version which we present is due to Sternberg (1955).

Theorem 1.5.1 (Hadamard–Perron; abbreviated to H–P theorem) Let $\varphi: M \to M$ be a diffeomorphism of a manifold M and let x_0 be a hyperbolic fixed point of φ: $\varphi(x_0) = x_0$, $x_0 \in M$. Then there exists an arbitrary small neighbourhood U of x_0 such that the set

$$W^+_{\text{loc}} := \{x \in U:\ \varphi^n(x) \in U \ \text{for}\ n \in Z^+\}$$

is a C^1-submanifold of M with the property:

$$T_{x_0}(W^+_{\text{loc}}) = E^+,$$

where E^+ denotes the subspace of $T_{x_0}(M)$ contracted by $d\varphi(x_0)$; moreover, we have:

$$W^+_{\text{loc}} = \{x \in U:\ \varphi^n(x) \to x_0 \ \text{as}\ n \to +\infty\}.$$

Hadamard and Perron proved that W^+_{loc} is of class C^0; Sternberg (1955) showed that W^+_{loc} is of class C^1; later it was proved (Abraham and Robbin, 1967) that W^+_{loc} is of class C^r, $r \geqslant 1$, provided so is φ.

We first prepare a lemma, which we will need in further reasoning.

Lemma 1.5.1 Let A: $\mathbf{R}^m \to \mathbf{R}^m$ be a linear operator with eigenvalues $\lambda_1, \ldots, \lambda_s$. Suppose that

$$|\lambda_i| < 1 \qquad \text{for} \quad i = 1, \ldots, s.$$

Then there exists an inner product $\langle \cdot, \cdot \rangle$ in $\mathbf{R}^m$ such that

$$|A| < 1,$$

where $|\cdot|$ denotes the operator norm induced by $\langle \cdot, \cdot \rangle$.

Proof To begin with, we show that the sequence of norms $(\|A^n\|)$ tends to 0 exponentially, i.e. that there are numbers $\alpha > 0$, $0 < b < 1$ such that $\|A^n\| \leqslant \alpha b^n$, $n = 0, 1, 2, \ldots$ (here $\|\cdot\|$ is the usual operator norm). Indeed, write

$$a = \max\{|\lambda_i|: i = 1, \ldots, s\} = \max\{|\lambda|: \lambda \in \mathrm{sp}A\}$$

($\mathrm{sp}A$ is the spectrum of A). Then, by assumption, $a < 1$. Let us consider the function

$$f(\lambda) = (A - \lambda I)^{-1}$$

defined for all $\lambda \notin \mathrm{sp}A$. This function is holomorphic in the set $C - \mathrm{sp}A$. To see this, take $h \in C$ with $|h|$ sufficiently small, and examine the difference quotient:

$$\frac{1}{h}(f(\lambda+h) - f(\lambda)) = \frac{1}{h}[(A - (\lambda+h)I)^{-1} - (A - \lambda I)^{-1}]$$
$$= (A - (\lambda+h)I)^{-1}(A - \lambda I)^{-1}.$$

The last expression tends to a limit as $h \to 0$, and thus the complex derivative $f'(\lambda)$ exists.

For $|\lambda| > \|A\|$ we have

$$f(\lambda) = (A - \lambda I)^{-1} = -\frac{1}{\lambda}\sum_{n=0}^{\infty}\frac{A^n}{\lambda^n}, \tag{1.5.1}$$

the series converging absolutely: $\left\|\frac{A^n}{\lambda^n}\right\| \leqslant \left(\frac{\|A\|}{|\lambda|}\right)^n$. The values of $f(\lambda)$ are $m \times m$-matrices. The analyticity of $f(\lambda)$ means the analyticity of all functions which are entries of the matrix $f(\lambda)$. Since $a \leqslant \|A\|$, $f(\lambda)$ is analytic in the whole annulus $|\lambda| > a$. Hence, by the well known theorem on analytic functions, equality (1.5.1) holds for all λ with $|\lambda| > a$, and the series on the right of (1.5.1) converges for all such λ.

Take a number b such that $a < b < 1$. Then the series (1.5.1) with $\lambda = b$ converges, and so $\|A^n/b^n\| \to 0$ as $n \to \infty$. Consequently, $\|A^n\| < b^n$ for n large enough. Thus there exists a number $\alpha > 0$ such that $\|A^n\| < \alpha b^n$ for all $n = 0, 1, 2, \ldots$

We define the desired inner product $\langle\cdot,\cdot\rangle$ as follows:

$$\langle x, y\rangle = (x, y) + (Ax, Ay) + \ldots + (A^{k-1}x, A^{k-1}y), \tag{1.5.2}$$

where k is chosen so that $\|A^k\| < 1$; $(\cdot,\cdot)$ denotes the original

(Euclidean) inner product. It remains to show that the operator norm $|\cdot|$ induced by $\langle \cdot, \cdot \rangle$ has the property asserted. Indeed,

$$\begin{aligned}
|Ax|^2 &= (Ax, Ax) + \ldots + (A^k x, A^k x) \\
&= \big((x, x) + \ldots + (A^{k-1}x, A^{k-1}x)\big) - (x, x) + (A^k x, A^k x) \\
&\leqslant |x|^2 - \|x\|^2 + \|A^k\|^2 \|x\|^2 = |x|^2\left(1 - (1 - \|A^k\|^2)\frac{\|x\|^2}{|x|^2}\right) \\
&= |x|^2\left(1 - (1 - \|A^k\|^2)\frac{\|x\|^2}{\|x\|^2 + \|Ax\|^2 + \ldots + \|A^{k-1}x\|^2}\right) \\
&\leqslant |x|^2\left(1 - (1 - \|A^k\|^2)\frac{\|x\|^2}{\|x\|^2 + \|A\|^2\|x\|^2 + \ldots + \|A^{k-1}\|^2\|x\|^2}\right) \\
&= |x|^2\left(1 - \frac{1 - \|A^k\|^2}{1 + \|A\|^2 + \ldots + \|A^{k-1}\|^2}\right).
\end{aligned}$$

Hence

$$|A| \leqslant 1 - \frac{1 - \|A^k\|^2}{1 + \|A\|^2 + \ldots + \|A^{k-1}\|^2} < 1,$$

which ends the proof of the lemma.

Proof of the H–P theorem We first outline the idea of the proof. Since M is locally diffeomorphic to $\boldsymbol{R}^m$ and the theorem is of local character, we may assume that a certain neighbourhood of x_0 in M is just a piece of $\boldsymbol{R}^m$ and that $x_0 = 0$, $\varphi(0) = 0$, and φ maps U_1, a neighbourhood of 0 in $\boldsymbol{R}^m$, onto U_2, another neighbourhood of 0. Let E^+ and E^- be the subspaces of $\boldsymbol{R}^m$ contracted and expanded (respectively) by the derivative $\mathrm{d}\varphi(0)$. We will prove the existence of a function $g \in C^1(E^+, E^-)$, $g(0) = 0$, $\mathrm{d}g(0) = 0$, such that W^+_{loc} coincides with the set

$$\{(u, g(u)) \colon u \in E^+\} \cap U,$$

where U is a certain (arbitrarily small) neighbourhood of 0. Hence it easily follows (see Exercise 1.4.4a) that W^+_{loc} is a C^1-submanifold of $\boldsymbol{R}^m$ and that $T_0(W^+_{\mathrm{loc}}) = E^+$. The function g will be defined as the fixed point of a certain contraction [4] $F \colon \bar{\Omega} \to \bar{\Omega}$, where $\bar{\Omega}$ denotes a suitable closed subset of the space $C(E^+, E^-)$. The smoothness of g will be deduced from the fact that if $k \in \Omega$ is of class C^1, then the derivatives

$d(F^n(k))(u)$, $u \in E^+$, form an equicontinuous family. The proof will be split into several steps.

Step I. Let us write φ in the form:

$$\varphi(x) = d\varphi(0)x + f(x), \tag{1.5.3}$$

where $f(0) = 0$, $df(0) = 0$. By assumption, the eigenvalues of $d\varphi(0)$ are different from 1 in absolute value; thus there exists a basis in $\boldsymbol{R}^m$ in which the operator $d\varphi(0)$ has the form of a matrix:

$$d\varphi(0) = \begin{bmatrix} A & 0 \\ 0 & B \end{bmatrix}$$

composed of two matrices (operators) A and B such that the eigenvalues of A are less than 1 and those of B are greater than 1 (in absolute value of course). Then we have

$$d\varphi(0)|_{E^+} = A, \qquad d\varphi(0)|_{E^-} = B$$

and

$$\boldsymbol{R}^m = E^+ \oplus E^-.$$

According to Lemma 1.5.1, we may assume that

$$a = \|A\|_{E^+} < 1, \qquad \frac{1}{b} = \|B^{-1}\|_{E^-} < 1.$$

For any vector $x \in \boldsymbol{R}^m$, let u and v denote its components belonging to E^+ and E^-, respectively:

$$x = (u, v), \qquad u \in E^+, \qquad v \in E^-.$$

Then (1.5.3) can be rewritten as:

$$\varphi(u, v) = \big(Au + f_1(u, v),\ Bv + f_2(u, v)\big), \tag{1.5.4}$$

where

$$f = (f_1, f_2), \quad f_i(0, 0) = 0, \quad \partial_u f_i(0, 0) = \partial_v f_i(0, 0) = 0$$

for $i = 1, 2$.

The symbols ∂_u, ∂_v denote the derivatives of f_i regarded as functions of u and v, respectively.

As has been assumed φ maps U_1 onto U_2. Since U_2 is not necessarily contained in U_1, iteration of φ is not possible. However, we are only interested in the behaviour of φ near 0. Therefore we can avoid the trouble of iterating by means of the following trick: we construct a mapping $\tilde{\varphi}$ defined on the whole of $\boldsymbol{R}^m$ and coinciding with φ in a certain neighbourhood of 0 (contained in U_1 of course). The procedure is standard: we take a function α: $\boldsymbol{R}^m \to \boldsymbol{R}$ of class C^1 and such that $\alpha(x) = 1$ in a ball $K(0, r_1)$ (the ball with centre at 0 and radius r_1) and $\alpha(x) = 0$ outside some larger ball $K(0, r_2)$, $0 < r_1 < r_2$. Put

$$\tilde{f}(x) = \alpha(tx) f(x),$$

where $t > 0$ is so large that $K(0, r_2/t) \subset U_1$. The factor $\alpha(tx)$ is zero for $x \notin K(0, r_2/t)$ and thus the function $\tilde{f}$ may be regarded as defined on the whole of $\boldsymbol{R}^m$. For $x \in K(0, r_1/t)$ we have $\tilde{f}(x) = f(x)$.

We define $\tilde{\varphi}$ as follows:

$$\tilde{\varphi}(x) = \mathrm{d}\varphi(0)x + \tilde{f}(x).$$

The mappings φ and $\tilde{\varphi}$ are identical inside $K(0, r_1/t)$. Moreover,

$$\mathrm{d}\tilde{f}(x) = t\,\mathrm{d}\alpha(tx) f(x) + \alpha(tx)\mathrm{d}f(x).$$

Since $\mathrm{d}f(0) = 0$, for any $\vartheta > 0$ we can find $t > 0$ such that, by applying the above procedure and writing the resulting $\tilde{f}$ as $\tilde{f} = (\tilde{f}_1, \tilde{f}_2)$, we have

$$\max(\|\partial_u \tilde{f}_1(u, v)\|, \|\partial_v \tilde{f}_2(u, v)\|) < \vartheta$$

for all $x = (u, v) \in \boldsymbol{R}^m$.

This reasoning shows that without loss of generality we may assume, to simplify notation, that the function f itself has the properties just derived with regard to $\tilde{f}$. And so we assume that

$$f_1(u, v) = 0, \qquad f_2(u, v) = 0 \qquad \text{for } (u, v) \notin K, \tag{1.5.5}$$

K denoting a certain ball with centre at 0 and

$$\max(\|\partial_u f_1(u, v)\|, \|\partial_v f_2(u, v)\|) < \vartheta, \qquad i = 1, 2, \tag{1.5.6}$$

for all $x = (u, v) \in \boldsymbol{R}^m$. Here the number ϑ will be subject to the condition

$$\frac{\vartheta}{b - a - 3\vartheta} < 1. \tag{1.5.7}$$

The definition of W_{loc}^+ imposes the inclusion $\varphi(W_{\text{loc}}^+) \subset W_{\text{loc}}^+$. Therefore the function g should be chosen so that its graph be invariant under φ, i.e., the inclusion

$$\varphi(\{u, g(u): u \in E^+\}) \subset \{u, g(u): u \in E^+\} \tag{1.5.8}$$

has to hold. This condition will be satisfied provided that

$$Bg(u) + f_2(u, g(u)) - g(Au + f_1(u, g(u))) = 0$$

for all u, which means that

$$g(u) = B^{-1}\big(g(Au + f_1(u, g(u))) - f_2(u, g(u))\big).$$

Step II. Let $C_0^1(E^+, E^-)$ denote the class of C^1-mappings from E^+ to E^- vanishing at infinity. For any $g \in C_0^1(E^+, E^-)$ we define another mapping $F(g)$, also from E^+ to E^-, by the formula

$$F(g)(u) = B^{-1}\big(g(Au + f_1(u, g(u))) - f_2(u, g(u))\big) \qquad \text{for } u \in E^+.$$

Lemma 1.5.2 The assignment $g \mapsto F(g)$ *has the following properties:*

(1) *There exists a number* σ, $0 < \sigma < 1$, *such that the condition*

$$\|dg(u)\| \leqslant \sigma \qquad \textit{for } u \in E^+$$

implies

$$\|d(F(g))(u)\| \leqslant \sigma \qquad \textit{for } u \in E^+.$$

(2) *Let* $h_g(\delta)$ *and* $k(\delta)$ *be defined by:*

$$h_g(\delta) = \sup_{\|u_1 - u_2\| \leqslant \delta} \|dg(u_1) - dg(u_2)\|,$$

$$k(\delta) = \sup_{\substack{\|y_1 - y_2\| \leqslant \delta \\ \|z_1 - z_2\| \leqslant \delta}} \max_{\substack{i = 1,2 \\ j = 1,2}} \|\partial_j f_i(y_1, z_1) - \partial_j f_i(y_2, z_2)\|;$$

here ∂_1 *stands for* ∂_y *and* ∂_2 *for* ∂_z. *Then there exists a number* $c > 0$ *such that the inequality*

$$h_g(\delta) \leqslant ck(\delta) \qquad \textit{for all } \delta > 0$$

implies

$$h_{F(g)}(\delta) \leqslant ck(\delta) \qquad \textit{for all } \delta > 0.$$

(3) F *is a contraction on the space*

$$\Omega = C_0^1(E^+, E^-) \cap \{g: \|dg(u)\| \leqslant \sigma\}$$

with the norm of uniform convergence (Ω *is not complete!*).

The distance $\|p-q\|$ is estimated as follows:

$$\begin{aligned}\|p-q\| &= \|(Au+f_1(u, g(u)))-(Aw+f_1(w, g(w)))\| \\ &\leqslant \|A\|\cdot\|u-w\|+\|f_1(u, g(u))-f_1(u, g(w))+f_1(u, g(w))- \\ &\quad -f_1(w, g(w))\| \\ &\leqslant \|A\|\cdot\|u-w\|+\|\partial_v f_1\|\cdot\|g(u)-g(w)\|+\|\partial_u f_1\|\cdot\|u-w\| \\ &\leqslant \|A\|\cdot\|u-w\|+\vartheta\|dg\|\cdot\|u-w\|+\vartheta\|u-w\| \\ &\leqslant (\|A\|+\vartheta\sigma+\vartheta)\,\|u-w\|.\end{aligned}$$

Now assume that ϑ is so small that

$$\|A\|+\vartheta(\sigma+1) = \|A\|+\vartheta\left(\frac{\vartheta}{b-a-3\vartheta}+1\right) < 1. \tag{1.5.10}$$

Then $\|p-q\| < \|u-w\|$ and

$$\begin{aligned}&\|\mathrm{d}(F(g))(u)-\mathrm{d}(F(g))(w)\| \\ &\qquad\leqslant \frac{1}{b}\big(h(\|u-w\|)(a+2\vartheta+2\vartheta\sigma)+k(\|u-w\|)(\sigma+1)^2\big) \\ &\qquad\leqslant \frac{1}{b}\big(h(\|u-w\|)(a+4\vartheta)+4k(\|u-w\|)\big).\end{aligned}$$

Suppose that g fulfils the condition

$$h_g(\delta) \leqslant ck(\delta)$$

(where c is still to be defined). Then

$$\begin{aligned}\|\mathrm{d}(F(g))(u)-\mathrm{d}(F(g))(w)\| &\leqslant \frac{1}{b}\big(ck(\|u-w\|)(a+4\vartheta)+4k(\|u-w\|)\big) \\ &= k(\|u-w\|)\frac{1}{b}\big(c(a+4\vartheta)+4\big).\end{aligned}$$

Thus let $c > 0$ be such that

$$\frac{1}{b}\big(c(a+4\vartheta)+4\big) \leqslant c,$$

i.e.,

$$c \geqslant \frac{4}{b-a-4\vartheta};$$

we postulate that the denominator be positive:

$$b-a-4\vartheta > 0. \tag{1.5.11}$$

Then

$$\|\mathrm{d}(F(g))(u)-\mathrm{d}(F(g))(w)\| \leqslant ck(\|u-w\|),$$

and so

$$h_{F(g)}(\delta) = \sup_{\|u-w\|\leqslant\delta} \|\mathrm{d}(F(g))(u)-\mathrm{d}(F(g))(w)\| \leqslant ck(\delta).$$

(3) Let $g_1, g_2 \in \Omega$. This means:

$$g_1, g_2 \in C_0^1(E^+, E^-) \qquad \text{and} \qquad \|\mathrm{d}g_i(u)\| \leqslant \sigma < 1, \qquad i = 1, 2.$$

By the definition of norm in Ω,

$$\|g_1-g_2\| = \sup_{w\in E^+} \|g_1(w)-g_2(w)\|.$$

Let us estimate $\|F(g_1)(u)-F(g_2)(u)\|$:

$$\begin{aligned}
&\|F(g_1)(u)-F(g_2)(u)\| \\
&= \Big\|B^{-1}\big(g_1(Au+f_1(u, g_1(u)))-g_2(Au+f_1(u, g_2(u)))- \\
&\quad -(f_2(u, g_1(u))-f_2(u, g_2(u)))\big)\Big\| \\
&\leqslant \|B^{-1}\|\Big(\big\|g_1(Au+f_1(u, g(u)))-g_2(Au+f_1(u, g_2(u)))\big\|+ \\
&\qquad +\|\partial_v f_2\|\cdot\|g_1(u)-g_2(u)\|\Big) \\
&\leqslant \frac{1}{b}\Big(\big\|g_1(Au+f_1(u, g_1(u)))-g_1(Au+f_1(u, g_2(u)))\big\|+ \\
&\quad +\big\|g_1(Au+f_1(u, g_2(u)))-g_2(Au+f_1(u, g_2(u)))\big\|+ \\
&\quad +\|\partial_v f_2\|\cdot\|g_1(u)-g_2(u)\|\Big) \\
&\leqslant \frac{1}{b}\Big(\|\mathrm{d}g_1\|\cdot\big\|f_1(u, g_1(u))-f_1(u, g_2(u))\big\|+ \\
&\quad +\sup_w\|g_1(w)-g_2(w)\|+\vartheta\sup_u\|g_1(u)-g_2(u)\|\Big) \\
&\leqslant \frac{1}{b}(\sigma\|\partial_v f_1\|\cdot\|g_1(u)-g_2(u)\|+(1+\vartheta)\|g_1-g_2\|) \\
&\leqslant \frac{1}{b}(\sigma\vartheta+\vartheta+1)\|g_1-g_2\| \leqslant \frac{1}{b}(2\vartheta+1)\|g_1-g_2\|.
\end{aligned}$$

Let, finally, ϑ be so small that

$$d = \frac{1}{b}(2\vartheta + 1) < 1. \tag{1.5.12}$$

Then

$$\|F(g_1) - F(g_2)\| = \sup_u \|F(g_1)(u) - F(g_2)(u)\| \leqslant d\|g_1 - g_2\|$$

and $d < 1$. Thus F is a contraction on Ω.

Step III. Let ϑ satisfy (1.5.7), (1.5.10), (1.5.11), (1.5.12). Let $g_0(u) = 0$ for all $u \in E^+$. Obviously, g_0 fulfils all conditions occurring in Lemma 1.5.2. We define inductively a sequence (g_n):

$$g_{n+1} = F(g_n).$$

Since the space $C_0(E^+, E^-)$ is complete, Lemma 1.5.2 yields the following conclusions:

1° The sequence $(g_n(u))$ converges uniformly (assertion (3) of the lemma).

2° The mappings $\mathrm{d}g_n(u)$ are equicontinuous (assertion (2) of the lemma).

From now on the symbol $g(u)$ will denote the limit

$$g(u) = \lim_{n \to \infty} g_n(u).$$

By 1°, $g(u)$ is continuous. By 2° and the Arzela–Ascoli theorem, there is a subsequence (g_{n_k}) such that $(\mathrm{d}g_{n_k})$ converges uniformly. Thus the limit

$$\mathrm{d}g(u) = \lim_{k \to \infty} \mathrm{d}g_{n_k}(u)$$

exists and is a continuous function. Since $g_n(0) = 0$ and $\mathrm{d}g_n(0) = 0$ for $n = 1, 2, \ldots$, we have $g(0) = 0$ and $\mathrm{d}g(0) = 0$.

Step IV. We claim that

$$\{(u, g(u)) \colon u \in E^+\} \cap U = \{x \colon \varphi^n(x) \in U, n = 0, 1, \ldots\}$$

for any sufficiently small ball U with centre at 0.

Let $(u, g(u))$ be any point of the graph of g lying in U. Write $(u_n, v_n) = \varphi^n(u, g(u))$, $n = 0, 1, \ldots$ Using (1.5.4) we get

$$u_1 = Au + f_1(u, g(u)), \qquad v_1 = Bg(u) + f_2(u, g(u)).$$

Applying to f_1 the Mean Value Theorem together with estimate (1.5.6), we obtain

$$\|u_1\| \leqslant \|A\|\,\|u\| + \vartheta\big(\|u\| + \|g(u)\|\big).$$

Since $\mathrm{d}g(0) = 0$ and $g(0) = 0$, there exists a ball $V \subset E^+$ centred at 0 such that $\|g(u)\| < \|u\|$ for $u \in V$. Hence

$$\|u_1\| \leqslant (\|A\| + 2\vartheta)\,\|u\| = (a + 2\vartheta)\,\|u\|.$$

Taking ϑ sufficiently small, we have $a_1 = a + 2\vartheta < 1$ and $\|u_1\| \leqslant a_1\,\|u\|$. Consequently, $\|u_n\| \leqslant a_1^n\,\|u\|$, and so $u_n \to 0$ as $n \to \infty$, whence

$$\varphi^n(u, g(u)) = (u_n, g(u_n)) \to (0, 0) \qquad \text{as } n \to \infty. \tag{1.5.13}$$

This gives inclusion $\subset$ in our claim, provided that U is small enough.

To obtain the opposite inclusion, assume that (u, v) is a point such that $\varphi^n(u, v) = (u_n, v_n) \in U$ for $n = 0, 1, \ldots$ Just as before, $u_n \to 0$. Now we estimate $\|v_{n+1} - g(u_{n+1})\|$ from below, $n \in Z^+$. This time we apply to f_2 the Mean Value Theorem together with estimate (1.5.6). We get

$$\begin{aligned}
&\|v_{n+1} - g(u_{n+1})\| \\
&= \big\|Bv_n + f_2(u_n, v_n) - Bg(u_n) - f_2(u_n, g(u_n))\big\| \\
&\geqslant \big\|B(v_n - g(u_n))\big\| - \vartheta\,\|v_n - g(u_n)\|.
\end{aligned}$$

Since $\|B^{-1}\| = 1/b < 1$ and $\|B^{-1}\|\,\|Bv\| \geqslant \|B^{-1}Bv\| = \|v\|$, it follows that $\|Bv\| \geqslant b\,\|v\|$ for any v. Taking ϑ sufficiently small, we have $b_1 = b - \vartheta > 1$, and so

$$\begin{aligned}
\|v_{n+1} - g(u_{n+1})\| &\geqslant b\,\|v_n - g(u_n)\| - \vartheta\,\|v_n - g(u_n)\| \\
&= b_1\,\|v_n - g(u_n)\| \geqslant b_1^{n+1}\,\|v - g(u)\|.
\end{aligned} \tag{1.5.14}$$

By assumption, $(u_n, v_n) \in U$ for all n; the sequence $(g(u_n))$ is bounded (in fact convergent to zero). Thus there is a constant C such that

$$\|v_n - g(u_n)\| \leqslant C \qquad \text{for } n = 0, 1, \ldots$$

This yields, in view of (1.5.14),

$$C \geqslant b_1^n\,\|v - g(u)\| \qquad \text{for } n = 0, 1, \ldots$$

Since $b_1 > 1$, we obtain $v = g(u)$. This settles the desired inclusion $\supset$ and ends the proof of our claim.

This claim, together with relation (1.5.13), results in the following set equality:

$$W_{\text{loc}}^{+} = \{x\colon\ \varphi^n(x) \in U,\ n \in Z^+,\ \lim_{n\to\infty} \varphi^n(x) = 0\}.$$

Thus the proof of the Hadamard–Perron theorem is complete.

On replacing φ by φ^{-1} in the H–P theorem we obtain

Corollary 1.5.1 *Let x_0 be a hyperbolic fixed point of a diffeomorphism $\varphi \in \mathrm{Diff}^1(M)$. Then there exists an arbitrarily small neighbourhood U of x_0 such that the set*

$$W_{\text{loc}}^{-} := \{x \in U\colon\ \varphi^n(x) \in U \quad \textit{for } n \in Z^-\}$$

is a C^1-submanifold of M with the property

$$T_{x_0}(W_{\text{loc}}^{-}) = E^-,$$

where E^- denotes the subspace of $T_{x_0}(M)$ expanded by $\mathrm{d}\varphi(x_0)$; moreover, we have

$$W_{\text{loc}}^{-} = \{x \in U\colon\ \varphi^n(x) \to x_0 \textit{ as } n \to -\infty\}.$$

Exercise

1.5.1.A. Let E and F be linear subspaces of $\boldsymbol{R}^m$ such that $\boldsymbol{R}^m = E \oplus F$. Let $g \in C^r(E, F)$. Show that the graph of g, i.e., the set $\{(u, g(u))\colon\ u \in E\}$ is a C^r-submanifold of $\boldsymbol{R}^m$ $(r \geqslant 1)$.

1.6. The Stable Manifold and the Unstable Manifold of a Hyperbolic Critical Point. The Case of a Cascade

Let M and N be two manifolds.

Definition 1.6.1 A C^1-mapping $f\colon\ M \to N$ is called an *immersion* iff the rank of the linear map $\mathrm{d}f(x)\colon\ T_x(M) \to T_{f(x)}(N)$ is equal to the dimension of M for every $x \in M$:

$$\operatorname{rank}\ \mathrm{d}f(x) = \dim M \quad \text{for } x \in M.$$

For instance, if $M = \boldsymbol{R}$ and N is any manifold, $\dim N > 0$, then every curve $\gamma(t)$ in N, $t \in \boldsymbol{R}$, such that $\dot{\gamma}(t) \neq 0$ for all t is an immersion of the line into N.

Theorem 1.6.1 (Smale) Let M be a manifold and let $\varphi \in \mathrm{Diff}^1(M)$. For every hyperbolic fixed point $x_0 \in M$, $\varphi(x_0) = x_0$, there exists an injective immersion ψ: $\boldsymbol{R}^k \to M$ such that:

(1) $\psi(0) = x_0$;

(2) $\psi(\boldsymbol{R}^k) = W^+ := \{x \in M\colon \lim_{n\to\infty} \varphi^n(x) = x_0\}$;

(3) $T_{x_0}(\psi(\boldsymbol{R}^k)) = E^+$,

where $k = \dim E^+$ and E^+ is the subspace of $T_{x_0}(M)$ contracted by $d\varphi(x_0)$.

Proof It is easy to see that

$$W^+ = \bigcup_{n=-\infty}^{+\infty} \varphi^{-n}(W^+_{\mathrm{loc}}),$$

where W^+_{loc} is defined in the H–P theorem. Let V be a neighbourhood of 0 in $\boldsymbol{R}^k$ and let ψ_V: $V \to W^+_{\mathrm{loc}}$ be a diffeomorphism which maps V onto a neighbourhood of x_0 in W^+_{loc}. We will not restrict generality by assuming that the image $\psi_V(V)$ is the entire W^+_{loc}. Such a ψ_V certainly exists (Exercise 1.5.1). We may assume $\psi_V(0) = x_0$. Define

$$g = \psi_V^{-1} \circ \varphi \circ \psi_V.$$

Since

$$dg(0) = (d\psi_V)^{-1}(x_0) \circ d\varphi(x_0) \circ d\psi_V(0),$$

the operators $dg(0)$ and $d\varphi(x_0)|_{E^+}$ have identical eigenvalues. By the definition of E^+, the operator $d\varphi(x_0)|_{E^+}$ has eigenvalues with moduli less than 1; consequently, so are all the eigenvalues of $dg(0)$. In virtue of Lemma 1.5.1 one can introduce an inner product in $\boldsymbol{R}^k$ in which $dg(0)$ is less than 1 in norm:

$$\|dg(0)\| < 1.$$

Let $V_0 \subset V$ be a ball with centre at 0 such that

$$\|dg(x)\| \leqslant \alpha < 1 \qquad \text{for } x \in V_0;$$

here α is any number in the interval $\|dg(0)\| < \alpha < 1$. Let $\hat{g}$ be an extension of g to the entire $\boldsymbol{R}^k$ satisfying the condition:

$$\|d\hat{g}(x)\| \leqslant \alpha \qquad \text{for all } x \in \boldsymbol{R}^k$$

(see Exercise 1.8.1 on p. 69). By the Mean Value Theorem $\hat{g}$ is a contraction on $\mathbf{R}^k$. We define $\psi(x)$ by

$$\psi(x) = \varphi^{-s} \circ \psi_V \circ \hat{g}^s(x),$$

where $s > 0$ is an integer such that $\hat{g}^s(x) \in V_0$; it is not difficult to notice that this definition does not depend on the choice of s. The resulting mapping ψ: $\mathbf{R}^k \to M$ has all the properties asserted; the verification is left to the reader.

On replacing φ by φ^{-1} in the H–P theorem and in the last theorem, we obtain:

Corollary 1.6.1 *For every hyperbolic fixed point* $x_0 \in M$ *of a diffeomorphism* $\varphi \in \mathrm{Diff}^1(M)$ *there exists an injective immersion* ϑ: $\mathbf{R}^{m-k} \to M$, $m = \dim M$, *such that*:

(1) $\vartheta(0) = x_0$;

(2) $\vartheta(\mathbf{R}^{m-k}) = W^- := \{x \in M: \lim_{n \to -\infty} \varphi^n(x) = x_0\}$;

(3) $T_{x_0}(\vartheta(\mathbf{R}^{m-k})) = E^-$,

where $\dim E^- = m - k$ *and* E^- *is the subspace of* $T_{x_0}(M)$ *expanded by* $\mathrm{d}\varphi(x_0)$ (*i.e., contracted by* $\mathrm{d}\varphi^{-1}(x_0)$).

Definition 1.6.2 Let x_0 be a hyperbolic fixed point for a mapping $\varphi \in \mathrm{Diff}^1(M)$. The sets W^+ and W^- defined in Theorem 1.6.1 and Corollary 1.6.1 are termed, respectively, the *stable manifold* and the *unstable manifold* of x_0 (although, in general, they are not necessarily submanifolds of M; see Example 1.6.3).

Examples

1.6.1. Let $M = \mathbf{R}^2$, $\varphi(x^1, x^2) = (\lambda x^1, \mu x^2)$, $\lambda > 1$, $0 < \mu < 1$. The origin is a hyperbolic fixed point. The stable and the unstable manifolds at (0, 0) coincide with the coordinate axes $0x^2$ and $0x^1$, respectively (see Fig. 1.6.1):

$$W^+ = \{(x^1, x^2): x^1 = 0\}, \qquad W^- = \{(x^1, x^2): x^2 = 0\}.$$

1.6.2. Let $M = S^2$ and let φ be as in Example 1.1.4 on p. 8:

$$\varphi(x) = \pi^{-1} \circ A \circ \pi(x), \qquad A = \begin{bmatrix} \lambda & 0 \\ 0 & \lambda \end{bmatrix}, \qquad \lambda > 1,$$

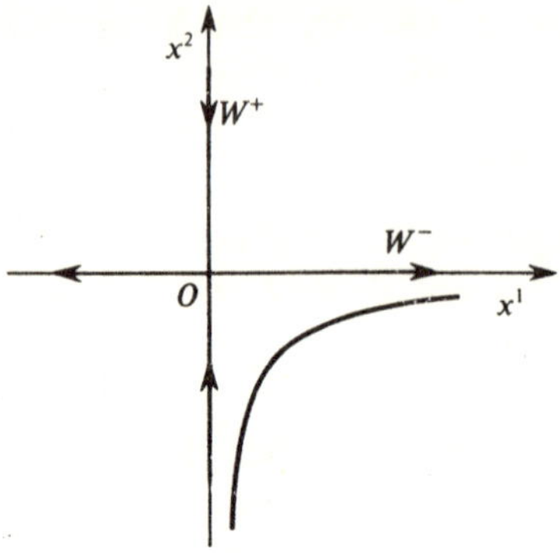

FIG. 1.6.1

where π denotes the stereographic projection. The north and south poles (denoted by x_N and x_S) are hyperbolic fixed points for φ. Since $\lambda > 0$, the submanifolds W^+ and W^- for x_N and x_S are as follows (see Fig. 1.6.2):

$$W^+(x_S) = \{x_S\}, \qquad W^-(x_S) = S^2 - \{x_N\},$$

$$W^+(x_N) = S^2 - \{x_S\}, \qquad W^-(x_N) = \{x_N\}.$$

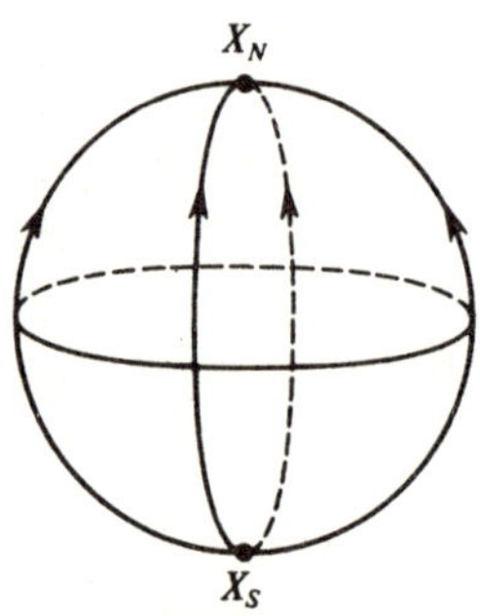

FIG. 1.6.2

1.6.3. Let $M = T^2$ and let $\varphi = \begin{bmatrix} a & b \\ c & d \end{bmatrix}$ be an algebraic automorphism of the torus; the eigenvalues of φ are assumed to be different from 1 in absolute value. The point (0, 0) is a hyperbolic fixed point for φ (in $\boldsymbol{R}^2$ as well as in T^2). In $\boldsymbol{R}^2$ the stable and the unstable manifolds at (0, 0) are:

$$W^+ = \{tu\colon t \in \boldsymbol{R}\}, \qquad W^- = \{tv\colon t \in \boldsymbol{R}\},$$

u and v denoting the eigenvectors

$$\varphi(u) = \lambda u, \qquad \varphi(v) = \mu v, \qquad |\lambda| > 1, \qquad |\mu| < 1.$$

(The last two inequalities may be assumed because of $|\lambda\mu| = |\det\varphi| = 1$.) Consequently, the stable and the unstable manifolds for φ on T^2 result by projecting these lines onto T^2. According to Exercise 1.2.5 on p. 23 the resulting sets are dense in T^2.

Now, let x_0 be a periodic point for a diffeomorphism $\varphi \in \mathrm{Diff}^1(M)$; i.e., suppose there is a $p \in Z^+$ such that

$$\varphi^p(x_0) = x_0.$$

Definition 1.6.3 The periodic point x_0 is called *hyperbolic* iff it is a hyperbolic fixed point for φ^p. (Then also every point $x_k = \varphi^k(x_0)$, $k = 0, 1, \ldots, p-1$, is so.) Let γ be the periodic trajectory of φ passing through x_0, i.e., let $\gamma = \{x_0, \varphi(x_0), \ldots, \varphi^{p-1}(x_0)\}$. The *stable* and the *unstable manifolds* of γ are defined as follows

$$W^+(\gamma) = \bigcup_{k=0}^{p-1} W^+(x_k, \varphi^p), \qquad W^-(\gamma) = \bigcup_{k=0}^{p-1} W^-(x_k, \varphi^p),$$

where the symbols $W^\pm(x_k, \varphi^p)$ denote the corresponding manifolds of x_k for the diffeomorphism φ^p.

Clearly, the stable and unstable manifolds of a given hyperbolic periodic point depend on the diffeomorphism φ. For the development of the theory it is important to know the nature of that dependence. This is made clear by the following theorem.

Theorem 1.6.2 Let (φ_k) be a sequence of diffeomorphisms on a compact manifold M, convergent in C^1-topology to a diffeomorphism φ_0 (i.e., let $\mathrm{d}\varphi_n(x) \to \mathrm{d}\varphi_0(x)$ uniformly in $x \in M$). Assume that $x_0 \in M$ is a hyperbolic fixed point for φ_0. Then:

(1) *There exists a neighbourhood U of x_0 such that, for n large enough, each of the φ_n has exactly one hyperbolic fixed point x_n in U* (see Exercise 3.4.1 on p. 160), *and we have* $\dim W^\pm(x_n, \varphi_n) = \dim W^\pm(x_0, \varphi_0)$; *the symbols $W^\pm(x_n, \varphi_n)$ denote the stable and unstable manifolds of x_n in the system (M, φ_n), $n = 0, 1, 2, \ldots$*

(2) *There exists a sequence of injective immersions ψ_n: $\boldsymbol{R}^s \to M$, where $s = \dim W^+(x_0, \varphi_0)$, such that $\psi_n(\boldsymbol{R}^s) = W^+(x_n, \varphi_n)$, $n = 0, 1, 2, \ldots$; the sequence (ψ_n) converges to (ψ_0) in C^1-topology, uniformly on compact subsets of $\boldsymbol{R}^s$.*

The proof of this theorem is difficult and therefore omitted here. We refer the reader to Hirsch, Pugh and Shub (1977).

1.7. The Stable Manifold and the Unstable Manifold of a Hyperbolic Critical Point. The Case of a Flow

In this section we prove certain theorems on flows, which can be regarded as counterparts to the theorems of the preceding section, concerning a cascade.

Let $\Phi \in \mathscr{F}^r(M)$ be a given vector field on a manifold M and let φ_t be the flow generated by Φ. Suppose that $x_0 \in M$ is a critical point for Φ, i.e., $\Phi(x_0) = 0$. Then $\varphi_t(x_0) = x_0$ for all $t \in \boldsymbol{R}$. Since

$$\varphi_t \circ \varphi_s(x) = \varphi_{t+s}(x) \qquad \text{for } x \in M,\ t,\ s \in \boldsymbol{R},$$

we have at the point x_0

$$\mathrm{d}\varphi_t(x_0) \circ \mathrm{d}\varphi_s(x_0) = \mathrm{d}\varphi_{t+s}(x_0).$$

Thus the family of derivatives $\mathrm{d}\varphi_t(x_0)$ constitutes a continuous group of linear operators on the space $T_{x_0}(M)$. It follows (see Exercise 1.8.4) that there exists a unique linear operator $B\colon T_{x_0}(M) \to T_{x_0}(M)$ such that

$$\mathrm{d}\varphi_t(x_0) = \mathrm{e}^{tB} \qquad \text{for } t \in \boldsymbol{R}. \tag{1.7.1}$$

Definition 1.7.1 The operator B in formula (1.7.1) is called the *Hessian* of the vector field Φ at x_0; the Hessian will be denoted by the symbol $\dot{\Phi}(x_0)$.

Our present objective is to represent the Hessian in a given coordinate system.

Let (U, α) be a chart, $x_0 \in U$, $\alpha(x_0) = y_0 \in \boldsymbol{R}^m$. The differential equation on M,

$$\frac{\mathrm{d}\varphi_t(x)}{\mathrm{d}t} = \Phi(\varphi_t(x)),$$

is transferred locally, via α, to $\alpha(U)$, a subset of $\boldsymbol{R}^m$. Write $y = \alpha(x)$ and let $\psi_t(y)$ be the flow generated by the vector field $\Psi(y)$ on $\alpha(U)$ into which the field $\Phi(x)$ is carried by α. We have

$$\frac{\mathrm{d}\psi_t(y)}{\mathrm{d}t} = \Psi(\psi_t(y))$$

and $\Psi(y_0) = 0$. Differentiating this equation with respect to y and changing the order of differentiation (we may do this on account of

Theorem 1.3.2 on p. 25), we obtain

$$\frac{\partial}{\partial y}\frac{\mathrm{d}\psi_t(y)}{\mathrm{d}t} = \frac{\partial}{\partial y}\Psi(\psi_t(y)),$$

$$\frac{\mathrm{d}}{\mathrm{d}t}\frac{\partial \psi_t}{\partial y}(y) = \mathrm{d}\Psi(\psi_t(y))\frac{\partial \psi_t}{\partial y}(y).$$

We may view $\mathrm{d}\Psi$ as a matrix; this is just the Jacobian matrix of the vector Ψ. For $y = y_0$ we get

$$\frac{\mathrm{d}}{\mathrm{d}t}\frac{\partial \psi_t}{\partial y}(y_0) = \mathrm{d}\Psi(y_0)\frac{\partial \psi_t}{\partial y}(y_0).$$

The matrix $\dfrac{\partial \psi_t}{\partial y}(y_0)$ then satisfies the linear differential equation (just obtained) with constant coefficients, together with the "initial condition":

$$(\partial \psi_0/\partial y)(y_0) = \text{identity}.$$

Thus

$$\frac{\partial \psi_t}{\partial y}(y_0) = \mathrm{e}^{\mathrm{d}\Psi(y_0)t}.$$

If now $\Psi(y)$ is written in local coordinates as

$$\Psi(y) = [\Psi^1(y), \ldots, \Psi^n(y)],$$

then

$$\mathrm{d}\Psi(y_0) = \left(\frac{\partial \Psi^i(y_0)}{\partial y^j}\right)_{i,j}.$$

Hence the Jacobian matrix $\mathrm{d}\Psi(y_0)$ is the required representation of the Hessian $\dot{\Phi}(x_0)$ in local coordinates.

Definition 1.7.2 A critical point $x_0 \in M$ of a vector field $\Phi \in \mathscr{F}^r(M)$ is called *non-degenerate* iff the Hessian $\dot{\Phi}(x_0)$: $T_{x_0}(M) \to T_{x_0}(M)$ is a surjection.

Theorem 1.7.1 If x_0 is a non-degenerate critical point of a vector field $\Phi \in \mathscr{F}^r(M)$, then x_0 is an isolated point of the set of all critical points of Φ.

Proof The theorem is of local character (it involves the behaviour of Φ in a neighbourhood of a given point). Therefore we may assume that M is simply a subset of $\boldsymbol{R}^m$, $x_0 = 0$, and that Φ is a C^1-vector field defined in a neighbourhood U of 0. Then Φ can be regarded as a mapping $\Phi: U \to \boldsymbol{R}^m$, $\Phi(0) = 0$. Since the point 0 is non-degenerate, $\det \dot{\Phi}(0) \neq 0$. Thus, by the Inverse Mapping Theorem, there is a neighbourhood V of 0 such that $\Phi(x) \neq 0$ for all $x \in V$, $x \neq 0$.

Definition 1.7.3 A critical point $x_0 \in M$ of a vector field $\Phi \in \mathscr{F}^r(M)$ is called *hyperbolic* iff the Hessian $\dot{\Phi}(x_0)$ does not have purely imaginary complex numbers as eigenvalues.

Remark 1.7.1 If x_0 is a hyperbolic critical point of Φ (in the sense of Definition 1.7.3), then x_0 is a hyperbolic fixed point of each φ_t, for $t \neq 0$ (in the sense of Definition 1.5.2).

Proof Since

$$d\varphi_t(x_0) = e^{t\dot{\Phi}(x_0)},$$

every eigenvalue λ of the operator $d\varphi_t(x_0)$ is of the form

$$\lambda = e^{t\mu},$$

μ being an eigenvalue of $\dot{\Phi}(x_0)$. Since $\operatorname{Re}\mu \neq 0$,

$$|\lambda| = e^{t\operatorname{Re}\mu} \neq 1.$$

Definition 1.7.4 The eigenvalues $\mu_1, \ldots, \mu_m$ of the operator $\dot{\Phi}(x_0)$ are called the *characteristic exponents* at the critical point x_0; the numbers $e^{\mu_1}, \ldots, e^{\mu_m}$ are called the *characteristic multipliers* at x_0.

Theorem 1.7.2 Let $\Phi \in \mathscr{F}^r(M)$ be a vector field and φ_t the flow generated by Φ. Suppose that x_0 is a hyperbolic critical point of Φ. Then there exists an injective immersion $\Psi: \boldsymbol{R}^k \to M$ such that:

(1) $\Psi(0) = x_0$,

(2) $\Psi(\boldsymbol{R}^k) = W^+ := \{x: \lim_{t \to +\infty} \varphi_t(x) = x_0\}$,

(3) $T_{x_0}(\Psi(\boldsymbol{R}^k)) = E^+$, *where $E^+ \subset T_{x_0}(M)$ is the eigenspace of the operator $\dot{\Phi}(x_0)$ corresponding to the eigenvalues μ such that* $\operatorname{Re}\mu < 0$.

Proof In virtue of the H–P theorem, it suffices to show that $W^+ = W_1^+ := \{x: \lim_{n \to \infty} \varphi_n(x) = x_0\}$. Obviously,

$$W^+ \subset W_1^+.$$

To obtain the opposite inclusion, take an $x \in W_1^+$ and let (U, α) be a chart for W_{loc}^+ (at x_0): $x_0 \in U$, $\alpha(x_0) = 0$, $\alpha(U) \subset \boldsymbol{R}^k$. Let V be an open convex bounded subset of $\boldsymbol{R}^k$ such that $0 \in V \subset \bar{V} \subset \alpha(U)$ and let $\varphi_n(x) \in \alpha^{-1}(V)$ for $n > n_0$. The flow $\varphi_t(x)$ is transferred by α to $\alpha(U)$; denote by $\psi_t(y)$ the resulting flow in $\alpha(U)$. Since the assignment $(t, y) \mapsto \psi_t(y)$ is of class C^1 in $\boldsymbol{R} \times \alpha(U)$, there is a constant L such that

$$\|\mathrm{d}\psi_t(\xi)\| \leqslant L \qquad \text{for } (t, \xi) \in \langle 0, 1\rangle \times \bar{V}. \tag{1.7.2}$$

The sequence $\psi_n(y)$ tends to 0 ($y = \alpha(x)$, $x \in W_1^+$ fixed) and we have $\psi_n(y) \in V$ for $n > n_0$. Write $\psi_{n_0}(y) = \bar{y}$. By the Mean Value Theorem and in view of (1.7.2) we get:

$$\begin{aligned}\|\psi_t(\bar{y})\| &= \|\psi_t(\bar{y}) - \psi_t(0)\| \\ &= \left\|\psi_{t-E(t)}\left(\psi_{E(t)}(\bar{y})\right) - \psi_{t-E(t)}\left(\psi_{E(t)}(0)\right)\right\| \\ &\leqslant \sup_{\xi \in \bar{V}} \|\mathrm{d}\psi_{t-E(t)}(\xi)\| \cdot \|\psi_{E(t)}(\bar{y}) - \psi_{E(t)}(0)\| \\ &\leqslant L\|\psi_{E(t)}(\bar{y})\|\end{aligned}$$

($E(t)$ denotes the greatest integer which does not exceed t). The last expression tends to 0 as $t \to +\infty$. Hence $W^+ = W_1^+$.

Remark 1.7.2 We define, analogously,

$$W^- := \{x\colon \lim_{t \to -\infty} \varphi_t(x) = x_0\}.$$

This set is obtained by an injective immersion of the space $\boldsymbol{R}^{m-k}$ into M ($m = \dim M$) and we have $T_{x_0}(W^-) = E^-$, where $E^- \subset T_{x_0}(M)$ is the eigenspace of $\dot{\Phi}(x_0)$ corresponding to eigenvalues μ with $\mathrm{Re}\,\mu > 0$.

Example

1.7.1. Let $M = \boldsymbol{R}^2$, $\Phi(x) = [2x^2, 4x^1 - 4(x^1)^3]$. The flow φ_t is then given by the following system of differential equations:

$$\frac{\mathrm{d}}{\mathrm{d}t} x^1 = 2x^2, \qquad \frac{\mathrm{d}}{\mathrm{d}t} x^2 = 4x^1 - 4(x^1)^3.$$

The first integral is of the form:

$$(x^2)^2 - 2(x^1)^2 + (x^1)^4 = c = \text{const.}$$

Integral curves are as shown in Fig. 1.7.1.

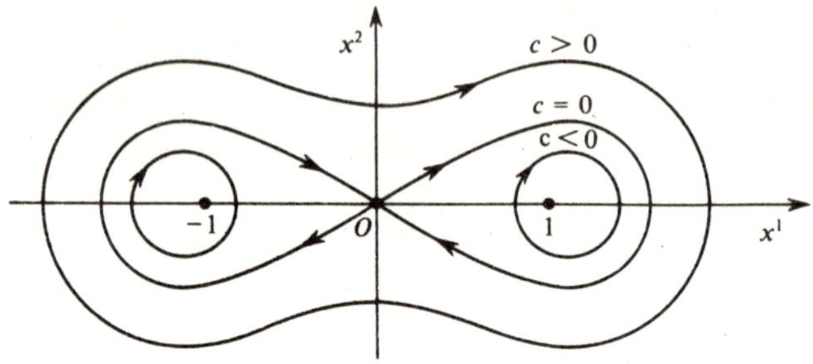

FIG. 1.7.1

There are 3 critical points for Φ: $(-1, 0)$, $(1, 0)$ and $(0, 0)$. The Hessian is given by the matrix

$$\dot{\Phi}(x) = \begin{bmatrix} 0 & 2 \\ 4-12(x^1)^2 & 0 \end{bmatrix}.$$

The characteristic exponents are:

$$\lambda = \pm 2\sqrt{2} \qquad \text{for } x = (0, 0),$$

$$\lambda = \pm 4\mathrm{i} \qquad \text{for } x = (1, 0) \text{ and } x = (-1, 0).$$

The point $x_0 = (0, 0)$ is hyperbolic and $W^+ = W^-$ at x_0; the points $(-1, 0)$ and $(1, 0)$ are not hyperbolic.

The manifold W^+ can be equipped, in a natural way, with either of the following two topologies: the topology inherited from $\boldsymbol{R}^2$ and the topology induced from $\boldsymbol{R}$ by immersion. Observe that these topologies are not equivalent.

1.8. Linearization of a Dynamical System in a Neighbourhood of a Hyperbolic Critical Point

In this section we present another theorem of capital importance for the theory of smooth dynamical systems. This theorem was proved independently by two authors, Hartman and Grobman, in the early sixties.

Theorem 1.8.1 (Grobman–Hartman) Let M be a manifold and let $\varphi \in \mathrm{Diff}^r(M)$. *Suppose that* $x_0 \in M$ *is a hyperbolic fixed point of* φ: $\varphi(x_0) = x_0$. *Then there exist a neighbourhood U of* x_0 *in M and a neighbourhood V of* 0 *in* $T_{x_0}(M)$ *such that the dynamical systems* (U, φ) *and* $(V, \mathrm{d}\varphi(x_0))$ *are topologically conjugate.*

An analogous assertion is valid for flows: if x_0 is a hyperbolic critical point for a flow φ_t, then there exist a neighbourhood U of x_0 in M and a neighbourhood V of 0 in $T_{x_0}(M)$ such that the flows (U, φ_t) and $(V, e^{t\Phi(x_0)})$ are topologically conjugate.

This theorem is often called the *linearization theorem.* It asserts, in fact, that in a neighbourhood of a hyperbolic fixed point the trajectories of the system considered behave like the trajectories of a certain linear map in a neighbourhood of 0 in $\boldsymbol{R}^m$. In the case of flows, the trajectories of a vector field near a hyperbolic critical point x_0 are topologically equivalent to the trajectories of a certain system of linear differential equations with constant coefficients. In other words, given a system of differential equations on M, one can reduce it in a neighbourhood of x_0 (a hyperbolic critical point) to a system of linear differential equations with constant coefficients, using a suitable homeomorphism (i.e., change of variables). Hence comes the term linearization.

The idea of the proof which we present below comes from Pugh (1969). The original proofs of Grobman and Hartman are technically much more complicated (see Hartman, 1964). However, the principal idea is the same in the three proofs.

Proof of Theorem 1.8.1

I. The case of a cascade The theorem is of local character; therefore we may freely assume that M is locally a part of $\boldsymbol{R}^m$, $x_0 = 0$, $\varphi(0) = 0$. Just as in the proof of the H–P theorem, we decompose $\boldsymbol{R}^m$ into the direct sum $E^+ \oplus E^-$, i.e., we write any $x \in \boldsymbol{R}^m$ as $x = (u, v)$, $u \in E^+$, $v \in E^-$; E^+ is the subspace of $\boldsymbol{R}^m = T_0(\boldsymbol{R}^m)$ contracted by the operator $d\varphi(0)$, E^- is the expanded subspace.

In a sufficiently small neighbourhood U_1 of 0 we can write φ in the form

$$x_1 = \varphi(x) = Lx + f(x) = \big(Au + f_1(u, v)\big) + \big(Bv + f_2(u, v)\big),$$

where

$$d\varphi(0) = L = \begin{bmatrix} A & 0 \\ 0 & B \end{bmatrix}, \qquad f = (f_1, f_2)$$

and f_1, f_2 are such that

$$f_i(0, 0) = 0, \qquad \partial_u f_i(0, 0) = 0, \qquad \partial_v f_i(0, 0) = 0, \qquad i = 1, 2.$$

The eigenvalues of A are, in modulus, less than 1 (though different from 0); those of B are greater than 1. In virtue of Lemma 1.5.1 we may assume that

$$\|A\| < 1, \qquad \|B^{-1}\| < 1.$$

Putting

$$\delta = \min\{|\lambda|:\ \lambda \in \operatorname{sp} L\},$$

we have

$$\|Lx\| \geqslant \frac{1}{2}\delta \|x\| \qquad \text{for all } x \in \boldsymbol{R}^m.$$

Let $\varepsilon > 0$ be such that the quantity c (which we now define) is less than 1,

$$c = \max(\|A\|, \|B^{-1}\|) + \varepsilon + \varepsilon\|B^{-1}\| < 1 \tag{1.8.1}$$

and besides let $2\varepsilon < \delta$. Take a neighbourhood U of 0 such that $\bar{U} \subset U_1$ and

$$\max(\|\partial_u f_i(x)\|, \|\partial_v f_i(x)\|) \leqslant \varepsilon \qquad \text{for } x \in \bar{U},\ i = 1, 2.$$

We extend $f = (f_1, f_2)$ to a map $\tilde{f} = (\tilde{f}_1, \tilde{f}_2)$ defined on the whole of $\boldsymbol{R}^m$ and satisfying the following conditions:

$$\max(\|\partial_u \tilde{f}_i(x)\|, \|\partial_v \tilde{f}_i(x)\|) \leqslant \varepsilon \qquad \text{for } x \in \boldsymbol{R}^m,\ i = 1, 2, \tag{1.8.2}$$

and

$$\tilde{f}_i(x) = 0 \qquad \text{for } x \notin U_2,\ i = 1, 2,$$

where U_2 is a sufficiently large open set in $\boldsymbol{R}^m$, $U_2 \supset \bar{U}_1$. This can be done; see Exercise 1.8.1 on p. 69.

The mapping

$$\tilde{\varphi}(x) = Lx + \tilde{f}(x)$$

is one-to-one: to see this, suppose $\tilde{\varphi}(x_1) = \tilde{\varphi}(x_2)$; then

$$Lx_1 - Lx_2 = \tilde{f}(x_2) - \tilde{f}(x_1)$$

and by the Mean Value Theorem we get

$$\|L(x_1 - x_2)\| = \|\tilde{f}(x_2) - \tilde{f}(x_1)\| \leqslant \varepsilon\|x_1 - x_2\|;$$

on the other hand,

$$\|L(x_1 - x_2)\| \geqslant \frac{1}{2}\delta \|x_1 - x_2\|,$$

and since $2\varepsilon < \delta$, we obtain $x_1 - x_2 = 0$.

The mappings φ and $\tilde{\varphi}$ coincide in U; thus it suffices to prove the theorem for $\tilde{\varphi}$ in place of φ. Since $\tilde{\varphi}$ is defined on the whole of $\boldsymbol{R}^m$, there is no problem with iteration of $\tilde{\varphi}$. To simplify notation from now on we will write φ instead of $\tilde{\varphi}$. In other words, we assume that φ itself has the properties stated for $\tilde{\varphi}$:

$$\max\left(\|\partial_u f_i(x)\|, \|\partial_v f_i(x)\|\right) \leqslant \varepsilon \qquad \text{for } x \in \boldsymbol{R}^m,\ i = 1, 2, \tag{1.8.3}$$

$$f(x) = 0 \qquad \text{for } x \notin U_2. \tag{1.8.4}$$

Our objective is to find a homeomorphism h which establishes topological conjugacy between (U, φ) and $\left(V, \mathrm{d}\varphi(0)\right)$. We will try to find h among mappings of the form

$$h = I + g, \tag{1.8.5}$$

where I is the identity and g satisfies:

$$\lim_{\|x\| \to +\infty} g(x) = 0.$$

Let us consider the following function class H:

$$H = \{g\colon \boldsymbol{R}^m \to \boldsymbol{R}^m,\ g \text{ continuous}, \lim_{\|x\| \to +\infty} g(x) = 0,\ g(0) = 0\}.$$

H becomes a complete metric space if the metric is defined by:

$$\varrho(g, k) = \sup_{x \in \boldsymbol{R}^m}\left(\|g_1(x) - k_1(x)\| + \|g_2(x) - k_2(x)\|\right)$$

for $g = (g_1, g_2)$, $k = (k_1, k_2)$, where $g_1(x), k_1(x) \in E^+$, $g_2(x), k_2(x) \in E^-$

If a homeomorphism h of the form (1.8.5) conjugates φ with $L = \mathrm{d}\varphi(0)$, then

$$\begin{aligned} \varphi \circ (I+g) &= (I+g) \circ L, \\ (L+f) \circ (I+g) &= (I+g) \circ L, \\ L + L \circ g + f \circ (I+g) &= L + g \circ L, \\ g &= L^{-1} \circ \left(g \circ L - f \circ (I+g)\right). \end{aligned} \tag{1.8.6}$$

The last equality written "coordinatewise" assumes the form

$$g_1 = A^{-1} \circ (g_1 \circ L - f_1 \circ (I+g))$$

and

$$g_2 = B^{-1} \circ (g_2 \circ L - f_2 \circ (I+g)).$$

The first of the resulting equalities can be rewritten as

$$g_1 = A \circ g_1 \circ L^{-1} + f_1 \circ (I+g) \circ L^{-1}.$$

We now define a map $F\colon H \to H$ by putting

$$F(g) = (F_1(g), F_2(g)),$$

where

$$F_1(g) = A \circ g_1 \circ L^{-1} + f_1 \circ (I+g) \circ L^{-1},$$
$$F_2(g) = B^{-1} \circ (g_2 \circ L - f_2 \circ (I+g)).$$

According to the preceding considerations, all that we need is to find a $g_0 \in H$ satisfying: $F(g_0) = g_0$.

We first show that F indeed maps H into H: it is readily seen that $F(g)(0) = 0$. To verify that

$$\lim_{\|x\| \to +\infty} \|F(g)(x)\| = 0,$$

suppose that $\|x\| \to +\infty$; then

$$\|L^{-1}x\| \to +\infty \Rightarrow g_1(L^{-1}x) \to 0 \Rightarrow A \circ g_1 \circ L^{-1}x \to 0;$$
$$\|L^{-1}x\| \to +\infty \Rightarrow \|(I+g)(L^{-1}x)\| = \|L^{-1}x + g(L^{-1}x)\| \to +\infty$$

and this, in view of (1.8.4), yields

$$f_1 \circ (I+g) \circ L^{-1}(x) = 0 \text{ for } \|x\| \text{ large enough.}$$

Thus $F_1(g)(x) \to 0$. Further, we have

$$\|Lx\| \to +\infty \Rightarrow g_2 \circ L(x) \to 0 \Rightarrow B^{-1} \circ g_2 \circ L(x) \to 0;$$

$$g(x) \to 0 \Rightarrow \|(I+g)(x)\| \to +\infty,$$

whence (again by (1.8.4))

$$f_2 \circ (I+g)(x) = 0 \text{ for } \|x\| \text{ large enough,}$$

and so $F_2(g)(x) \to 0$. It follows that $F(g) \in H$.

Now we prove that F is a contraction on H. Let $g, k \in H$, $g = (g_1, g_2)$, $k = (k_1, k_2)$. Then

$$\varrho\big(F(g), F(k)\big) = \sup_{x \in R^m}\big(\|F_1(g)(x) - F_1(k)(x)\| + \|F_2(g)(x) - F_2(k)(x)\|\big).$$

On account of (1.8.3) we have

$$\begin{aligned}
&\sup \|F_1(g)(x) - F_1(k)(x)\| \\
&\quad = \sup \|A \circ g_1 \circ L^{-1}(x) - A \circ k_1 \circ L^{-1}(x) + \\
&\qquad\qquad + f_1 \circ (I+g) \circ L^{-1}(x) - f_1 \circ (I+k) \circ L^{-1}(x)\| \\
&\quad \leqslant \|A\| \sup \|g_1(x) - k_1(x)\| + \sup \|\partial f_1\| \cdot \|g \circ L^{-1}(x) - k \circ L^{-1}(x)\| \\
&\quad \leqslant \|A\| \cdot \|g_1 - k_1\| + \varepsilon \|g - k\|\,; \\
&\sup \|F_2(g)(x) - F_2(k)(x)\| \\
&\quad = \sup \|B^{-1} \circ g_2 \circ L(x) - B^{-1} \circ k_2 \circ L(x) + \\
&\qquad\qquad + B^{-1} \circ f_2 \circ (I+g)(x) - B^{-1} \circ f_2 \circ (I+k)(x)\| \\
&\quad \leqslant \|B^{-1}\| \sup \|g_2(x) - k_2(x)\| + \|B^{-1}\| \sup \|\partial f_2\| \cdot \|g(x) - k(x)\| \\
&\quad \leqslant \|B^{-1}\| \cdot \|g_2 - k_2\| + \varepsilon \|B^{-1}\| \cdot \|g - k\|.
\end{aligned}$$

Clearly, the symbols $\|g-k\|$ and $\|g_i - k_i\|$ denote here the sup norm. Since $\|g - k\| \leqslant \varrho(g, k)$, we get

$$\begin{aligned}
\varrho\big(F(g), F(k)\big) &\leqslant \max(\|A\|, \|B^{-1}\|)\varrho(g, k) + (\varepsilon + \varepsilon\|B^{-1}\|)\,\|g - k\| \\
&\leqslant c\varrho(g, k),
\end{aligned}$$

where $c < 1$ is the constant defined by (1.8.1). Thus, by the fixed point theorem, there exists a unique $g_0 \in U$ such that $F(g_0) = g_0$, and this is just what we needed. The mapping $h_0 = I + g_0$ has the desired property:

$$\varphi \circ h_0 = h_0 \circ L.$$

It remains to verify that h_0 is a homeomorphism. Analogously to the construction of h_0 we can construct a mapping

$$h_1 = I + g_1, \qquad g_1 \in H,$$

unique among mappings of this form, which conjugates φ with L in the opposite order:

$$h_1 \circ \varphi = L \circ h_1.$$

Hence

$$(L\circ h_1)\circ h_0 = (h_1\circ\varphi)\circ h_0 = h_1\circ(\varphi\circ h_0) = h_1\circ(h_0\circ L),$$

i.e.,

$$L\circ(h_1\circ h_0) = (h_1\circ h_0)\circ L.$$

The composite mapping $h_1\circ h_0$ is also of the form: identity plus a member of H:

$$h_1\circ h_0 = (I+g_1)\circ(I+g_0) = I+g_1+g_0+g_1\circ g_0 = I+\tilde{g}\,;$$

it is easy to see that $\tilde{g}\in H$.

Now, we have two mappings of the form $h = I+g$, $g\in H$, fulfilling the condition

$$L\circ h = h\circ L.$$

One of them is $h = h_1\circ h_0$, the other is $h = I$. In quite the same way we show that also $h = h_0\circ h_1$ fulfils this condition. The foregoing argument shows that such an h is unique within the class under consideration. Consequently,

$$h_1\circ h_0 = h_0\circ h_1 = I,$$

and so h_0 is a homeomorphism.

II. The case of a flow. We have to prove that in a certain neighbourhood U of x_0 the flow φ_t generated by the vector field Φ is topologically conjugate to the flow ψ_t generated by the vector field $\Psi(x) = \dot{\Phi}(x_0)x$ in a neighbourhood of 0 in $T_{x_0}(M)$. As before, we may assume that $U\subset \boldsymbol{R}^m$, $x_0 = 0$, $T_{x_0}(M) = \boldsymbol{R}^m$.

The flow φ_t satisfies the equation:

$$\frac{\mathrm{d}}{\mathrm{d}t}\varphi_t(x) = \Phi(\varphi_t(x)) = \dot{\Phi}(0)\varphi_t(x)+F(\varphi_t(x)),$$

where $\dot{\Phi}(0)$ is the linear summand in the Taylor expansion of $\Phi(x)$; $F(0) = 0$ and $\mathrm{d}F(0) = 0$. Integration yields:

$$\varphi_t(x) = \mathrm{e}^{t\dot{\Phi}(0)}x+\int_0^t \mathrm{e}^{(t-s)\dot{\Phi}(0)}F(\varphi_s(x))\mathrm{d}s. \tag{1.8.8}$$

Denote:

$$f_t(x) = \int_0^t \mathrm{e}^{(t-s)\dot{\Phi}(0)}F(\varphi_s(x))\mathrm{d}s.$$

Since $d\varphi_t(0) = e^{t\dot{\Phi}(0)}$, we may rewrite (1.8.8) as

$$\varphi_t(x) = d\varphi_t(0)x + f_t(x).$$

For the moment, let us restrict attention to $t \in \langle 0, 1 \rangle$. Let U be neighbourhood of 0 such that

$$\|dF(x)\| < \varepsilon \qquad \text{for } x \in \bigcup_{t \in \langle 0,1 \rangle} \varphi_t(U);$$

ε will be specified later on. By assumption, 0 is a hyperbolic critical point. Thus the operator $\dot{\Phi}(0)$ admits representation

$$\dot{\Phi}(0) = \begin{vmatrix} G & 0 \\ 0 & D \end{vmatrix}$$

in a suitable coordinate system; the real parts of the eigenvalues of G are negative and those of D are positive. There exist constants $\alpha > 0$, $\delta > 0$ and a norm $\|\cdot\|$ such that

$$\|e^{tG}\| \leqslant e^{-\alpha t}, \qquad \|e^{-tD}\| \leqslant e^{-\alpha t},$$
$$\|e^{t\dot{\Phi}(0)}x\| \geqslant \delta \|x\| \tag{1.8.9}$$

(see Exercise 1.8.2 on p. 69).

Differentiating $f_t(x)$ with respect to x, we get

$$df_t(x) = \int_0^t e^{(t-s)\dot{\Phi}(0)} dF(\varphi_s(x)) d\varphi_s(x) ds.$$

Hence

$$\|df_t(x)\| \leqslant \int_0^t \|e^{(t-s)\dot{\Phi}(0)}\| \cdot \|dF(\varphi_s(x))\| \cdot \|d\varphi_s(x)\| ds \leqslant \int_0^t e^{t\|\dot{\Phi}(0)\|} \varepsilon K$$
$$\leqslant tKe^{\|\dot{\Phi}(0)\|}\varepsilon = \varepsilon t K_1,$$

where $K = \sup_x \sup_{0 \leqslant t \leqslant 1} \|d\varphi_t(x)\|$, $K_1 = Ke^{\|\dot{\Phi}(0)\|}$. We introduce a quantity c_t defined by a formula analogous to (1.8.1) with A, B^{-1}, ε replaced by e^{Gt}, e^{-Dt}, $\varepsilon t K_1$:

$$c_t = \max(\|e^{Gt}\|, \|e^{-Dt}\|) + \varepsilon t K_1 + \varepsilon t K_1 \|e^{-Dt}\|.$$

Using (1.8.9), we see that

$$c_t \leqslant e^{-\alpha t} + \varepsilon t K_1 (1 + e^{-\alpha t}),$$

and so

$$c_t < 1 \qquad \text{for all } t \in \langle 0, 1 \rangle,$$

provided that ε is small enough: $\varepsilon < \alpha e^{-\alpha}/2K_1$.

Now we resort to the Grobman–Hartman theorem, namely to the already proved case of cascade: there exists a neighbourhood U_1 of 0, $U_1 \subset U$, which does not depend on t and is such that:

1° inequalities corresponding to (1.8.1), (1.8.2) are satisfied in U_1 for every $t \in \langle 0, 1 \rangle$;

2° for any $t \in \langle 0, 1 \rangle$ there is a homeomorphism h_t of U_1 onto a neighbourhood of 0, unique among mappings of the form (1.8.5) with $g \in H$, such that

$$\varphi_t \circ h_t = h_t \circ L_t, \tag{1.8.10}$$

where $L_t = e^{t\Phi(0)}$.

Identity (1.8.10) yields

$$\varphi_{nt} \circ h_t = h_t \circ L_{nt}$$

for any integer n. On the other hand, if t and n are such that $nt \in \langle 0, 1 \rangle$, then applying (1.8.10) with t replaced by nt we get

$$\varphi_{nt} \circ h_{nt} = h_{nt} \circ L_{nt}.$$

Hence, by uniqueness, $h_{nt} = h_t$ for any such t and n. Consequently, $h_w = h_1 = h$ for every rational $w \in \langle 0, 1 \rangle$:

$$\varphi_w \circ h(x) = h \circ L_w(x) \qquad \text{for } x \in U_1.$$

Fixing $t \in \langle 0, 1 \rangle$ and approaching t by rationals, $w_n \to t$, we finally obtain:

$$\varphi_t \circ h(x) = h \circ L_t(x) \qquad \text{for } x \in U_1.$$

This holds for any $t \in \langle 0, 1 \rangle$, and thus, automatically, for an arbitrary $t \in \boldsymbol{R}$.

Remark 1.8.1 Even for an analytic diffeomorphism φ a homeomorphism h of class C^1 conjugating φ with $d\varphi(x_0)$ does not necessarily exist. This can be checked by the following example: $M = \boldsymbol{R}^3$, $x_0 = 0$,

$$\varphi(x, y, z) = (ax, ac(y + \varepsilon xz), cz),$$

where $0 < c < 1$, $ac > 1$, $\varepsilon > 0$.

Exercises

1.8.1.A. Let U be a closed ball in $\boldsymbol{R}^m$ and let $g\colon U \to \boldsymbol{R}^m$ be of class C^1. Assume that

$$\|dg(x)\| \leqslant \alpha \qquad \text{for } x \in U.$$

Show that there exist balls U_1 and U_2, $U_1 \subset U \subset U_2$, and a C^1-mapping $\hat{g}\colon \boldsymbol{R}^m \to \boldsymbol{R}^m$ such that:

$$\|d\hat{g}(x)\| \leqslant \alpha \qquad \text{for all } x \in \boldsymbol{R}^m,$$

$$\hat{g}(x) = g(x) \qquad \text{for } x \in U_1,$$

$$\hat{g}(x) = 0 \qquad \text{for } x \notin U_2.$$

Hint. Postulate $\hat{g}(x) = g(\alpha(x))\beta(x)$, where $\alpha\colon \boldsymbol{R}^m \to U$, $\beta\colon \boldsymbol{R}^m \to \boldsymbol{R}$.

Remark The procedure described in the above exercise is a standard one. Relevant to it but much more subtle is the Whitney extension theorem.[23]

1.8.2.A. Let $A\colon \boldsymbol{C}^m \to \boldsymbol{C}^m$ be a linear operator and suppose that $\operatorname{sp} A \subset \{\lambda\colon \operatorname{Re}\lambda < 0\}$. Show that one can introduce an inner product in $\boldsymbol{C}^m$ such that, denoting by $\|\cdot\|$ the norm induced by the new product, we have

$$\|e^{tA}\| \leqslant e^{-\alpha t} \qquad \text{for } t \geqslant 0,$$

where α is a positive constant.

Hint. Use the following theorem (see Nemytskii and Stepanov, 1960) on the Jordan form of matrices: for every $\varepsilon > 0$ there is a basis in $\boldsymbol{C}^m$ in which

$$A = \begin{bmatrix} B_1 & & 0 \\ & B_2 & \\ & & \ddots \\ 0 & & & B_k \end{bmatrix}, \qquad \text{where } B_j = \begin{bmatrix} \lambda_j & \varepsilon & & 0 \\ & \lambda_j & \ddots & \\ & & \ddots & \varepsilon \\ 0 & & & \lambda_j \end{bmatrix}.$$

Having chosen a suitable ε, introduce the inner product in which the basis corresponding to ε is orthonormal.

1.8.3.B. Prove that any two contractions of the same space E are topologically conjugate. Assume that $\dim E < +\infty$; the theorem is also true if E is any Banach space (Strelcyn, 1970b).

Hint. Let S be the unit sphere in $E = \boldsymbol{R}^m$ and let g_1, g_2 be the contractions in question. Let V_i denote the "annulus" limited by the surfaces S and $g_i(S)$, $i = 1, 2$. Map V_1 homeomorphically onto V_2; then extend the mapping to a homeomorphism of the entire $\boldsymbol{R}^m$; compare Example 1.2.1.

1.8.4.A. Let $A(t)$ be a continuous group of linear transformations of $\boldsymbol{R}^m$ (i.e. $A(t)\colon \boldsymbol{R}^m \to \boldsymbol{R}^m$ for any $t \in \boldsymbol{R}$, $\det A(t) \neq 0$, $A(t+s) = A(t)A(s)$, $\lim_{t_n \to t} A(t_n) = A(t)$). Show that there exists a linear transformation $B\colon \boldsymbol{R}^m \to \boldsymbol{R}^m$ such that

$$A(t) = e^{tB} \qquad \text{for all } t \in \boldsymbol{R}.$$

Hint. This is a special case of a general theorem on the representation of a continuous semi-group of operators in a Banach space; see (Dunford and Schwartz, 1958; p. 614).

1.9. Behaviour of a Flow in a Neighbourhood of a Closed Orbit

Let M be a manifold and let $\Phi \in \mathscr{F}^r(M)$. The vector field Φ generates the flow $\varphi_t(x) = \varphi(t, x)$. Suppose that $\gamma(t)$ is a closed orbit of Φ, with prime period τ. Let $\hat{\gamma}$ denote the set of points lying on this curve (i.e. the set of values assumed by γ; cf. Section 1.4, p. 35). Each point $x_0 \in \hat{\gamma}$ is a fixed point of the diffeomorphism $\varphi_\tau(x)$. Thus

$$d\varphi_\tau(x_0)\colon T_{x_0}(M) \to T_{x_0}(M).$$

Assume $\gamma(0) = x_0$. Obviously, $\lambda = 1$ is an eigenvalue of the operator $d\varphi_\tau(x_0)$; the tangent vector $\dot{\gamma}(0)$ is the corresponding eigenvector:

$$d\varphi_\tau(x_0)\dot{\gamma}(0) = \dot{\gamma}(\tau) = \dot{\gamma}(0).$$

Definition 1.9.1 (1) By a *local transversal section* (or a *cross-section*) of γ at a point $x_0 \in \hat{\gamma}$ we mean any submanifold $S \subset M$ of codimension 1 such that $x_0 \in S$ and $\Phi(x) \notin T_x(S)$ for all $x \in S$.

Then the tangent space $T_x(M)$, $x \in S$, admits the decomposition

$$T_x(M) = [\Phi(x)] \oplus T_x(S),$$

where $[\Phi(x)]$ denotes the one-dimensional space spanned by $\Phi(x)$.

(2) Let S be a cross-section of γ at x_0 and let U_0 be a sufficiently

small neighbourhood of x_0 in S (see the theorem that follows). We define a mapping $\Theta\colon U_0 \to S$ by:

$$\Theta(x) = \varphi(t(x), x) \qquad \text{for } x \in U_0,$$

where $t(x)$ denotes the least positive number t such that $\varphi(t, x) \in S$. The mapping Θ is called the *Poincaré map* at x_0.

Theorem 1.9.1 (1) *For every cross-section S of a closed orbit γ at a point $x_0 \in \hat{\gamma}$, the Poincaré map $\Theta\colon U_0 \to S$ is correctly defined, provided that the neighbourhood U_0 of x_0 in S is small enough.*

(2) *If S_1 and S_2 are two cross-sections of γ at $x_1, x_2 \in \hat{\gamma}$ (respectively) and if Θ_1, Θ_2 are the corresponding Poincaré maps, then Θ_1 and Θ_2 are diffeomorphically conjugate, i.e., there is a diffeomorphism $h\colon V_1 \to V_2$ (V_i denotes a suitable neighbourhood of x_i in S_i, $i = 1, 2$) satisfying*

$$\Theta_1 \circ h = h \circ \Theta_2.$$

Proof (1) The assertion is of local character. Thus we may assume that S is contained within a single chart. By an appropriate choice of the coordinate system, S can be written (in that chart) as follows:

$$S = \{x \in \boldsymbol{R}^m\colon x^m = 0\},$$

$$x_0 = (0, \ldots, 0), \qquad \dot{\gamma}(0) = [0, \ldots, 0, 1].$$

Let

$$\varphi(t, x) = (\varphi^1(t, x), \ldots, \varphi^m(t, x)).$$

Then $\varphi(t, x) \in S$ iff $\varphi^m(t, x) = 0$.

Now, we have $\varphi^m(\tau, 0) = 0$ and

$$\frac{d\varphi^m}{dt}(\tau, 0) = 1.$$

In virtue of the Implicit Function Theorem there exist a neighbourhood U_0 of 0, an interval $(\tau - \varepsilon, \tau + \varepsilon)$ and a C^1-function $t(x)$ such that

$$\varphi^m(t(x), x) = 0 \qquad \text{and} \qquad t(x) \in (\tau - \varepsilon, \tau + \varepsilon) \text{ for } x \in U_0.$$

For any x, the number $t(x)$ is indeed the least time it takes x to return to S because τ is the prime period. Hence

$$\Theta(x) = \varphi(t(x), x) \qquad \text{for } x \in U_0$$

is a well defined C^1-map.

It is not difficult to see that Θ is in fact a diffeomorphism.

(2) If $x_1 \neq x_2$, then shifting S_1 along trajectories of the flow φ_t we reduce the situation to the one where $x_1 = x_2$. Again the setting is local, and so we may assume that $S_1, S_2 \subset \boldsymbol{R}^m$. Locally, each integral curve intersects S_1 at exactly one point; the same is the case with S_2.

We define h as follows: we choose a neighbourhood V_1 of x_1 in S_1; for any $x \in V_1$ we draw a trajectory through x and define $y = h(x)$ as the point where this trajectory meets S_2; we put $V_2 = h(V_1)$. The neighbourhood V_1 has to be chosen small enough to ensure uniqueness of y and correctness of the definition.

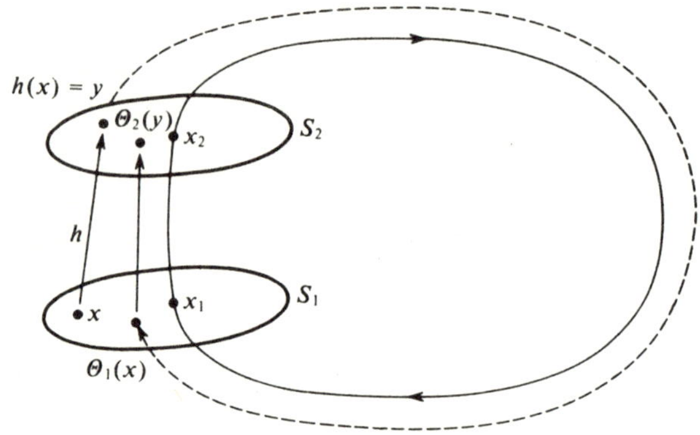

FIG. 1.9.1

Clearly, h conjugates Θ_1 and Θ_2 (see Fig. 1.9.1). By construction, h is a diffeomorphism.

Remark 1.9.1 (1) Let γ, x_i, S_i, Θ_i be as above, $i = 1, 2$. Since Θ_1 and Θ_2 are diffeomorphically conjugate, we have

$$\operatorname{sp}(\mathrm{d}\Theta_1(x_1)) = \operatorname{sp}(\mathrm{d}\Theta_2(x_2)).$$

(2) If S is a cross-section of a closed orbit γ at $x_0 \in \hat{\gamma}$ and if Θ denotes the corresponding Poincaré map, then

$$\operatorname{sp}(\mathrm{d}\Theta(x_0)) \cup \{1\} = \operatorname{sp}(\mathrm{d}\varphi_\tau(x_0)),$$

τ denoting the prime period of γ.

Proof (1) This is an immediate consequence of Theorem 1.9.1 and of the invariance of the spectrum under non-singular linear transformations.

(2) As pointed out in Definition 1.9.1, the tangent space at x_0 admits the decomposition

$$T_{x_0}(M) = [\dot{\gamma}(0)] \oplus T_{x_0}(S),$$

where $[\dot{\gamma}(0)]$ is the span of $\dot{\gamma}(0)$. Thus every vector $\dot{y} \in T_{x_0}(M)$ can be written as:

$$\dot{y} = (\dot{y}_1, \dot{y}_2), \qquad \text{where } \dot{y}_1 \in [\dot{\gamma}(0)],\ \dot{y}_2 \in T_{x_0}(S).$$

The vector $\dot{\gamma}(0)$ being invariant under the operator $d\varphi_\tau(x_0)$, we may represent this operator in the matrix form

$$d\varphi_\tau(x_0) = \begin{bmatrix} 1 & \alpha \\ 0 & A \end{bmatrix}$$

where A is an $(m-1)\times(m-1)$-matrix. Differentiating the identity

$$\Theta(x) = \varphi\big(t(x), x\big)$$

with respect to x and putting $x = x_0$, we obtain for any $\dot{x} \in T_{x_0}(S)$ (calculation carried out in "coordinates" $\dot{y}_1, \dot{y}_2$):

$$\begin{aligned} d\Theta(x_0)\dot{x} &= \frac{\partial\varphi}{\partial t}(\tau, x_0)\left(\dot{t}(x_0)\begin{bmatrix} 0 \\ \dot{x} \end{bmatrix}\right) + \partial_x\varphi(\tau, x_0)\begin{bmatrix} 0 \\ \dot{x} \end{bmatrix} \\ &= \dot{\gamma}(0)\dot{t}(x_0)\begin{bmatrix} 0 \\ \dot{x} \end{bmatrix} + \begin{bmatrix} 1 & \alpha \\ 0 & A \end{bmatrix}\begin{bmatrix} 0 \\ \dot{x} \end{bmatrix} \\ &= \dot{\gamma}(0)a(\dot{x}) + \begin{bmatrix} \alpha\dot{x} \\ A\dot{x} \end{bmatrix}, \end{aligned}$$

where $a(\dot{x}) = \dot{t}(x_0)\begin{bmatrix} 0 \\ \dot{x} \end{bmatrix}$. The resulting vector has to belong to $T_{x_0}(S)$. Hence $a(\dot{x}) = -\alpha\dot{x}$ and $A\dot{x} = d\Theta(x_0)\dot{x}$. Our assertion is now immediate.

Examples

1.9.1. Let $M = \boldsymbol{R}^2$, $\Phi(x) = [-x^2, x^1]$. The integral curves are (in polar coordinates r, ω):

$$\varphi_t(r_0, \omega_0) = \big(r_0\cos(t+\omega_0), r_0\sin(t+\omega_0)\big).$$

For any closed orbit, the Poincaré map Θ is the identity (Fig. 1.9.2).

1.9.2. Let $M = \boldsymbol{R}^2$. Consider the flow induced by the following system of differential equations (in polar coordinates):

$$\frac{dr}{dt} = \sin r, \qquad \frac{d\omega}{dt} = \text{const} = c > 0.$$

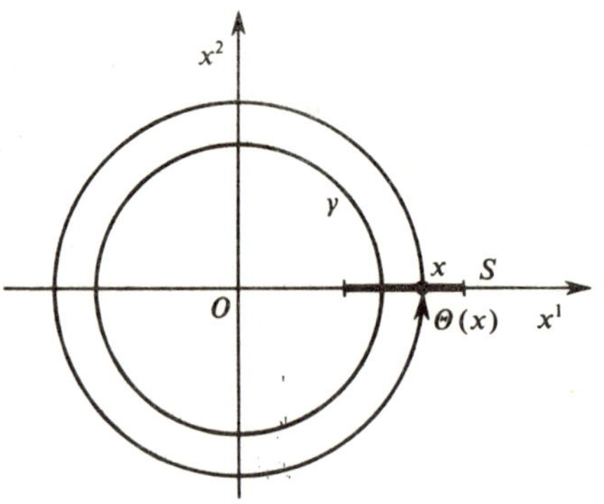

FIG. 1.9.2

The integral curves are

$$\log|\tan\tfrac{1}{2}r| = t + \log|\tan\tfrac{1}{2}r_0|, \qquad \omega = ct + \omega_0,$$

where (r_0, ω_0) is the initial position, $r_0 \neq n\pi$, $n \in Z$. For $r_0 = 0$ we have $r(t) \equiv 0$; for $r_0 = n\pi$, $n \in Z$, $n \neq 0$, the integral curves are circles γ_n. In particular, γ_1 is the circle $r = \pi$.

Consider the point $r_0 = \pi$, $\omega_0 = 0$ and the cross-section S defined as the segment perpendicular to γ_1. The corresponding Poincaré map Θ is, locally, a contraction. Indeed, after simple computation we get

$$\left.\frac{d}{dr}\Theta(r)\right|_{r=\pi} = \frac{1}{e^{2\pi c}} < 1,$$

which implies (see Exercise 1.3.5 on p. 30) that Θ is contractive (on S) at $r = \pi$. This means that close to γ_1, the integral curves approach γ_1; speaking intuitively, they "wind" towards γ_1.

The situation is similar in the vicinity of any other γ_n; the trajectories approach γ_n as $t \to +\infty$ (for n odd) or as $t \to -\infty$ (for n even). We leave to the reader the examination of the behaviour of the flow in the vicinity of the origin.

Definition 1.9.2 (1) A closed orbit γ of a vector field $\Phi \in \mathscr{F}^r(M)$, with prime period τ, is called *transversal* (or: the period τ is *transversal*) iff, for any $x_0 \in \hat{\gamma}$, the unity is an eigenvalue of multiplicity 1 for the operator $d\varphi_\tau(x_0)$.

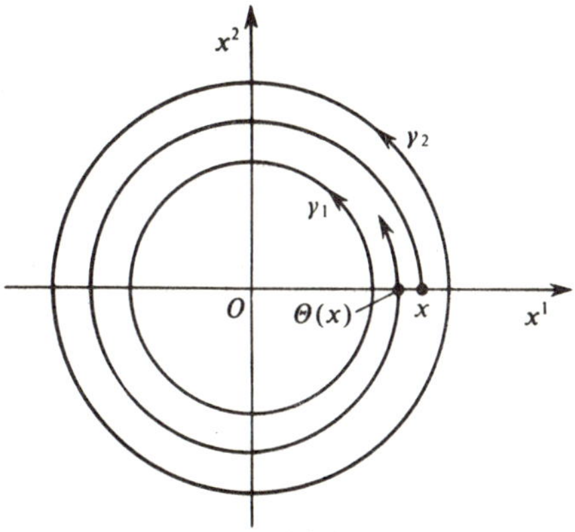

FIG. 1.9.3

(2) A closed orbit γ is called *hyperbolic* (or *elementary*) iff γ is transversal and, for any $x_0 \in \hat{\gamma}$, the operator $d\varphi_\tau(x_0)$ has no eigenvalue λ with $\lambda \neq 1 = |\lambda|$.

As an immediate consequence of Remark 1.9.1(2) we obtain:

Corollary 1.9.1 A closed orbit γ is hyperbolic iff, for any $x_0 \in \hat{\gamma}$ and any cross-section S at x_0, the point x_0 is a hyperbolic fixed point for the corresponding Poincaré map Θ.

We now define the stable and the unstable manifolds of a hyperbolic closed orbit γ. Thus let γ be such an orbit of a vector field $\Phi \in \mathscr{F}^r(M)$ and let φ_t be the flow generated by Φ. Take any point $x_0 \in \hat{\gamma}$ and any cross-section S at x_0. According to the last corollary, x_0 is a hyperbolic fixed point for Θ (the Poincaré map). In virtue of the H–P theorem, Θ defines locally two subsets of S: $W^+_{\text{loc}}(x_0)$ and $W^-_{\text{loc}}(x_0)$. Define

$$W^+(\gamma) = \bigcup_{t \in R} \varphi_t\big(W^+_{\text{loc}}(x_0)\big), \qquad W^-(\gamma) = \bigcup_{t \in R} \varphi_t\big(W^-_{\text{loc}}(x_0)\big).$$

It is not difficult to show that the sets $W^+(\gamma)$, $W^-(\gamma)$ depend neither on the choice of $x_0 \in \hat{\gamma}$ nor on the choice of the cross-section S (because all the Poincaré maps are diffeomorphically conjugate).

$W^+(\gamma)$ and $W^-(\gamma)$ are immersed submanifolds of M and we have:

$$\dim W^+(\gamma) = \dim W^+_{loc}(x_0)+1,$$

$$\dim W^-(\gamma) = \dim W^-_{loc}(x_0)+1,$$

$$\dim W^+(\gamma)+\dim W^-(\gamma) = \dim M+1.$$

Definition 1.9.3 The sets $W^+(\gamma)$ and $W^-(\gamma)$ defined by formulas (1.9.1) are called (respectively); the *stable manifold* and the *unstable manifold* of the given hyperbolic closed orbit γ.

Topologically, $W^+(\gamma)$ and $W^-(\gamma)$ are immersion embeddings into M either of the cylinder $S^1 \times \mathbf{R}^k$ or of the generalized Möbius band $S^1 \dot{\times} \mathbf{R}^k$ (the so-called "twisted product")[13].

1.10. Suspension of a Diffeomorphism

In the preceding section we defined the Poincaré map for a cross-section of a closed orbit of a flow. That means, with a given flow and a given submanifold of codimension 1 (cross-section) we associated a certain diffeomorphism on that submanifold.

Now consider a problem that is, in a sense, converse to the above: given a manifold M and a diffeomorphism φ: $M \to M$, can one find a super-manifold N and a flow φ_t on N such that M be a cross-section for φ_t and that $\varphi_1 = \varphi$ on M?

The answer is positive. Consider the manifold $M_1 = M \times \mathbf{R}$ and the map

$$\alpha(x, s) = (\varphi(x), s-1) \qquad \text{for } (x, s) \in M_1.$$

Put: $\Gamma = \{\alpha^n\colon n \in Z\}$. Then Γ is a group of diffeomorphisms on M_1. Define

$$N = M_1/\Gamma,$$

and denote by π the natural projection from M_1 onto N. We have the quotient topology in N: if (x, s) is a point in M_1, U is a neighbourhood of x in M and $\varepsilon > 0$, then the π-image of any of the sets

$$\varphi^n(U) \times (s+n-\varepsilon, s+n+\varepsilon) \subset M_1, \qquad n \in Z,$$

is a neighbourhood of the π-image of the point (x, s). Thus, if M is a manifold of dimension m, N is a manifold of dimension $m+1$.

We now define a flow $\tilde{\varphi}_t$ on M_1 by putting:

$$\tilde{\varphi}_t(x, s) = (x, t+s).$$

This flow induces a flow φ_t on N. The resulting flow φ_t has the required properties: identifying M with $\pi(M \times \{0\})$ in N, we see that M is a cross-section for φ_t. If $\pi((x, s)) \in M$, i.e. $s = 0$, then

$$\varphi_1 \pi((x, 0)) = \pi(\tilde{\varphi}_1(x, 0)) = \pi(x, 1) = \pi(\varphi(x), 0).$$

Definition 1.10.1 Given a cascade (M, φ), the flow (N, φ_t) just constructed is called the *suspension* of the diffeomorphism φ.

The flow φ_t obtained as a result of this procedure has properties similar to the properties of the original diffeomorphism φ. This observation allows one to construct examples of flows sharing properties with a given diffeomorphism.

The operation of suspension is a particular case of a more general procedure, which we now outline.

Let X be a compact metric space and let $\varphi: X \to X$ be a continuous mapping. Further, let $f(x)$ be a real-valued function on $X, f(x) > 0$ for $x \in X$. As before put:

$$X_1 = X \times \boldsymbol{R}^+,$$

$$\alpha(x, s) = (\varphi(x), s - f(x)) \qquad \text{for } (x, s) \in X_1,$$

$$\Gamma = \{\alpha^n : n \in Z^+\},$$

$$X^f = X_1/\Gamma.$$

In other words, X^f results from X_1 by identification of all pairs of the type: (x, s) and $(\varphi(x), s - f(x))$.

Let π be the natural projection of X_1 onto X^f. The formula

$$\tilde{\varphi}_t(x, s) = (x, t+s)$$

defines a flow on X_1 which induces (via π) a flow φ_t on X^f.

Definition 1.10.2 The resulting flow (X^f, φ_t) is called the *special flow* over the cascade (X, φ) with respect to f.

In the case where $f(x) = 1$ for all $x \in X$ and φ is a homeomorphism the special flow over (X, φ) is precisely the suspension of φ.

1.11. Limit Sets

Definition 1.11.1 (1) Let M be a manifold and let $\varphi \in \mathrm{Diff}^r(M)$, $x_0 \in M$. The sets

$$\omega(x_0) := \{x\colon \exists_{(k_n)}\ k_n \to +\infty,\ \lim \varphi^{k_n}(x_0) = x\},$$

$$\alpha(x_0) := \{x\colon \exists_{(k_n)}\ k_n \to -\infty,\ \lim \varphi^{k_n}(x_0) = x\}$$

are called the *limit sets* of x_0 (at $+\infty$ and $-\infty$, respectively).

(2) Let $\Phi \in \mathscr{F}^r(M)$, let φ_t be the flow generated by Φ and let $x_0 \in M$. The sets:

$$\omega(x_0) := \{x\colon \exists_{(t_n)}\ t_n \to +\infty,\ \lim \varphi_{t_n}(x_0) = x\},$$

$$\alpha(x_0) := \{x\colon \exists_{(t_n)}\ t_n \to -\infty,\ \lim \varphi_{t_n}(x_0) = x\}$$

are called the *limit sets* of x_0 (at $+\infty$ and $-\infty$).

Theorem 1.11.1 *If M is compact and $\Phi \in \mathscr{F}^r(M)$* (case (2) in Definition 1.11.1), *then for every $x_0 \in M$ the sets $\omega(x_0)$, $\alpha(x_0)$ are nonempty, closed and connected.*

In the case of a cascade (M, φ), $\varphi \in \mathrm{Diff}^r(M)$ (case (1) in Definition 1.11.1), *the sets $\omega(x_0)$, $\alpha(x_0)$ are nonempty and closed.*

Proof Only connectedness requires a proof (case (2)); other statements are obvious. Thus assume that

$$\omega(x_0) = X_1 \cup X_2,$$

where X_1 and X_2 are nonempty, closed and disjoint. Then

$$\mathrm{dist}(X_1, X_2) = \delta > 0.$$

Let U_1, U_2 be the open sets given by:

$$U_i = \{x \in M\colon \mathrm{dist}(x, X_i) < \tfrac{1}{3}\delta\}, \qquad i = 1, 2.$$

Then

$$\mathrm{dist}(U_1, U_2) \geqslant \tfrac{1}{3}\delta.$$

By the definition of $\omega(x_0)$, the trajectory $\varphi_t(x_0)$ enters and leaves each of the sets U_1, U_2 infinitely many times as $t \to +\infty$. Thus there exists a pair of sequences $(x_n^{(1)})$, $(x_n^{(2)})$ such that $x_n^{(1)} \in \bar{U}_1$, $x_n^{(2)} \in \bar{U}_2$ $(n \in Z^+)$, $x_n^{(1)} = \varphi_{t_n}(x_0)$, $x_n^{(2)} = \varphi_{s_n}(x_0)$ and $\varphi_t(x_0) \notin U_1 \cup U_2$ for $t \in (t_n, s_n)$. The piece of the trajectory $\varphi_t(x_0)$ corresponding to $t \in (t_n, s_n)$ certainly contains a point x_n whose distance to either of the sets U_1, U_2

exceeds $\frac{1}{9}\delta$. By compactness, the sequence (x_n) has a cluster point $\bar{x}$. Thus we arrive at a contradiction; on the one hand, $\bar{x} \in \omega(x_0)$ by the definition of a limit set, while, on the other hand, $\bar{x} \notin U_1 \cup U_2 \supset \omega(x_0)$.

Example

1.11.1. Let γ be a hyperbolic closed orbit for a vector field $\Phi \in \mathscr{F}^r(M)$, with prime period τ. Suppose that for every $x_0 \in \hat{\gamma}$

$$\operatorname{sp}(\mathrm{d}\varphi_\tau(x_0)) - \{1\} \subset \{\lambda\colon |\lambda| < 1\}.$$

Then there is a neighbourhood U of $\hat{\gamma}$ such that

$$\omega(x) = \hat{\gamma} \qquad \text{for all } x \in U.$$

We shall prove this. Choose any cross-section S at a point $x_0 \in \hat{\gamma}$ and let Θ be the corresponding Poincaré map. According to Remark 1.9.1(2), the spectrum of Θ is contained in the unit circle of the complex plane. Hence, by Exercise 1.3.5 on p. 30, there is a neighbourhood U_0 of x_0 in S such that $\lim_n \Theta^n(x) = x_0$ for all $x \in U_0$.

Fix a $\delta > 0$ and write:

$$U = \bigcup_{-\delta < t < \tau + \delta} \varphi_t(U_0).$$

Evidently, U is a neighbourhood of $\hat{\gamma}$, if U_0 is sufficiently small. Each point $x \in U$ is of the form: $x = \varphi_s(x_1)$, where $x_1 \in U_0$ and $s \in (-\delta, \tau + +\delta)$. Let $t_n(x_1)$ be the succesive instants of recurrence of x_1 in S. The time periods $t_{n+1}(x_1) - t_n(x_1)$ are close to τ; we may assume that they are contained between τ/k and $k\tau$, where k is a constant.

Now let $\bar{x} \in \omega(x)$, i.e., $\bar{x} = \lim \varphi_{t_i}(x)$, $t_i \to +\infty$. Select indices n_i so as to have $t_{n_i}(x_1) \leqslant t_i < t_{n_i+1}(x_1)$, $i = 1, 2, \ldots$ Since $\varphi_{t_n(x_1)}(x_1) = \Theta^n(x_1)$, we have

$$\varphi_{t_i}(x) = \varphi_{s+t_i-t_{n_i}(x_1)}(\varphi_{t_{n_i}(x_1)}(x_1)) = \varphi_{s+t_i-t_{n_i}(x_1)}(\Theta^{n_i}(x_1)).$$

The sequence $(s + t_i - t_{n_i}(x))$ is bounded; we may assume, passing to a subsequence, that this sequence is convergent. Denote by t_0 the limit. Letting $i \to +\infty$ in the last equality, we obtain $\bar{x} = \varphi_{t_0}(x_0) \in \hat{\gamma}$. Thus $\omega(x) = \hat{\gamma}$.

Definition 1.11.2 Let γ be a closed orbit and suppose that $\omega(x) = \hat{\gamma}$ or

$\alpha(x) = \hat{\gamma}$, for any x in some neighbourhood $U \supset \hat{\gamma}$. Then γ is called a *limit cycle*.

For example the curve γ_n in Example 1.9.2 on p. 74 is a limit cycle.

Exercises

1.11.1.A. Let $M = S^1$ and let $\varphi\colon S^1 \to S^1$ be a homeomorphism without periodic points. Show that:

(a) for any $x, y \in S^1$ we have

$$\omega(x) = \omega(y) = \alpha(x) = \alpha(y)$$

(denote this set by Δ);

(b) Δ is a perfect set, either nowhere dense, or the whole of S^1.

Hint. Let $z \in \omega(x)$; then there are $x_n = \varphi^n(x)$, $x_m = \varphi^m(x)$ arbitrarily close to z. The arc joining x_n with x_m is carried by the iterates of φ into a family of arcs on S^1; finitely many of them cover S^1.

1.11.2.B. Let $\varphi\colon S^1 \to S^1$ be a homeomorphism and suppose that $\operatorname{Per}\varphi$ (the set of periodic points) is nonempty. Show that for any $x \in S^1$ we have

$$\alpha(x) \subset \operatorname{Per}\varphi \qquad \text{and} \qquad \omega(x) \subset \operatorname{Per}\varphi.$$

Hint. Reduce the problem to the examination of $\alpha(x)$, $\omega(x)$ for transformations of a segment.

In the situation described in Exercise 1.11.1, the set $\Delta = \alpha(x) = \omega(x)$ (independently of $x \in S^1$) has the property that the trajectory of each $x \in \Delta$ is dense in Δ. Sets with this property deserve more careful examination; they will play a significant part in our further considerations.

Let (X, φ^α) be a dynamical system, $\alpha \in \mathfrak{A}$.

Definition 1.11.3 A set $S \subset X$ is said to be *minimal* iff it is closed, nonempty, invariant under φ^α (i.e., $\varphi^\alpha(S) = S$ for all $\alpha \in \mathfrak{A}$) and such that no proper subset of S shares these properties. If the entire X is a minimal set, we call (X, φ^α) a *minimal* dynamical system.

Examples

1.11.2. Let φ: $X \to X$ be a homeomorphism and let x_0 be a periodic point, with prime period p: $\varphi^p(x_0) = x_0$. Then the set $S = \{x_0, \varphi(x_0), \ldots, \varphi^{p-1}(x_0)\}$ is minimal.

1.11.3. Let φ: $S^1 \to S^1$ be a homeomorphism without periodic points. The set Δ defined in Exercise 1.11.1 is closed, nonempty and φ-invariant. The trajectory of any $x \in \Delta$ is dense in Δ and so Δ contains no proper subsets which are closed and φ-invariant. Thus Δ is a minimal set.

1.11.4. Let $\Phi \in \mathscr{F}^r(M)$. If γ is a closed orbit, then $\hat{\gamma}$ is a minimal set.

Theorem 1.11.2 If X is compact, then every dynamical system (X, φ^α) possesses a minimal set.

Proof It suffices to apply the Zorn Lemma to the family of all closed nonempty invariant subsets of X, ordered by inclusion.

Corollary 1.11.1 In every flow or cascade on a compact space X, each of the sets $\alpha(x)$ and $\omega(x)$ (for any $x \in X$) contains at least one minimal set.

Exercises

1.11.3.A. Let $S \subset X$. Show that S is a minimal set for a dynamical system (X, φ^α) iff the trajectory of every point $x \in S$ is dense in S.

1.11.4.A. Show that if S_1, S_2 are two minimal sets for a dynamical system (X, φ^α), then either $S_1 \cap S_2 = \emptyset$ or $S_1 = S_2$.

1.11.5.A. Let (M, φ_t) be a flow on a manifold M. Show that every minimal set S is connected.

Hint. Apply Exercise 1.11.3.

1.11.6.A. Let (X, φ^α) be a dynamical system and let each φ^α be a homeomorphism. Show that if two sets $S_1, S_2 \subset X$ are invariant, then so are the sets $S_1 \cup S_2$, $S_1 \cap S_2$, $S_1 - S_2$.

1.11.7.B. Give an example of a dynamical system such that for some x the set $\alpha(x)$ is not minimal.

Hint. E.g. a flow on a two-dimensional manifold with trajectories running as shown in Fig. 3.8.2 on p. 180.

We conclude this section with one more theorem on minimal sets.

Theorem 1.11.3 *Let* (X, φ^α) *be a dynamical system. Suppose that* X *is connected and that each* φ^α, $\alpha \in \mathfrak{A}$, *is a homeomorphism. Then every minimal set* S *either is nowhere dense in* X *or is the entire* X.

Proof Assume that $\operatorname{int} S \neq \emptyset$. Since the φ^α are homeomorphisms, we have $\varphi^\alpha(\operatorname{int} S) = \operatorname{int} S$. According to Exercise 1.11.6, the boundary of S, i.e. the set $S - \operatorname{int} S$, is invariant under all the φ^α. Clearly, this set is closed and contained in S – hence it is empty, by the minimality of S. Consequently, S is a closed-open set in X and thus, X being connected, we have $S = X$.

2

Dynamical Systems on Manifolds of Dimension 1 and 2

2.1. Survey of Basic Facts of Differential Topology

In several points of this chapter we shall have to resort to some facts of differential topology. Therefore we devote the first section to a brief discussion of those concepts and theorems of this theory which will be needed in the sequel. Proofs are omitted; they can be found in Milnor (1965).

Let M and N be oriented manifolds [12] of equal dimension. Assume that M is compact, N is connected. Let $f: M \to N$ be a mapping of class C^1.

Definition 2.1.1 A point $x \in M$ is called *regular* iff $\mathrm{d}f(x): T_x(M) \to T_{f(x)}(M)$ is a surjection.

A value $y \in f(M)$ is called *regular* iff every point $x \in f^{-1}(y)$ is regular.

A point [a value] which is not regular is called *critical.*

It follows immediately that the set of regular points is open.

Let x be a regular point for $f: M \to N$. We define the symbol $\operatorname{sgn}\mathrm{d}f(x)$ as $+1$ iff $\mathrm{d}f(x)$ preserves orientation, and as -1 iff $\mathrm{d}f(x)$ alters orientation. For any regular value $y \in N$ we define:

$$\deg(f;\, y) := \sum_{x \in f^{-1}(y)} \operatorname{sgn}\mathrm{d}f(x). \tag{2.1.1}$$

In virtue of the Inverse Mapping Theorem, the points in $f^{-1}(y)$ are isolated and so, by the compactness of M, the set $f^{-1}(y)$ is finite; formula (2.1.1) thus defines correctly a certain integer.

Theorem 2.1.1 *The number* $\deg(f;\, y)$ *depends only on* f *and does not depend on the choice of the regular value* y (Milnor, 1965, Section 5, Theorem A).

This theorem justifies the following definition:

Definition 2.1.2 The number $\deg f = \deg(f;\, y)$, y being an arbitrary regular value, is called the *degree* of f.

Theorem 2.1.2 *If* $f, g\colon M \to N$ *are smoothly homotopic,*[9] *then* $\deg f = \deg g$ (Milnor, 1965, Section 5, Theorem B).

It follows that the concept of degree makes sense not only for a particular mapping f but also for the whole equivalence class of mappings pairwise smoothly homotopic. Since any C^1-map is smoothly homotopic to one having regular values, the definition extends to all C^1-mappings.

Examples

2.1.1. If $M = N = S^1, f(z) = z^m$, then $\deg f = m$.
2.1.2. If $M = N = S^1, f(x) = \lambda \sin 2\pi x \,(\mathrm{mod}\, 1)$, then $\deg f = 0$.
2.1.3. If $f \in \mathrm{Diff}^1(M)$, M any compact manifold, then $\deg f = \pm 1$.
2.1.4. If $M = N = T^2, f = \begin{bmatrix} a & b \\ c & d \end{bmatrix}$, $a, b, c, d \in Z$, $\det f \neq 0$, then

$\deg f = \det f$.

Let $\Phi \in \mathscr{F}^1(U)$, U being an open subset of $\boldsymbol{R}^m$, and let $x_0 \in U$ be an isolated critical point of Φ. For any sufficiently small sphere B with centre at x_0 we define a mapping F of B into S^{m-1}, the unit sphere in $\boldsymbol{R}^m$, by:

$$F(x) = \frac{\Phi(x)}{\|\Phi(x)\|} \quad \text{for } x \in B. \tag{2.1.2}$$

If B_1 and B_2 are two such spheres and neither of the corresponding balls contains critical points other than x_0, then the corresponding mappings F_1 and F_2 are smoothly homotopic. Therefore, according to Theorem 2.1.2, $\deg F_1 = \deg F_2$. This observation allows us to define the index of an isolated critical point:

Definition 2.1.3 Let x_0 be an isolated critical point of a vector field $\Phi \in \mathscr{F}^1(U)$. Let F be given by (2.1.2), B denoting any admissible sphere. The number

$$\iota_\Phi(x_0) := \deg F$$

is termed the *index* of Φ at x_0 (or: the index of x_0 with respect to Φ).

Examples

2.1.5. Let $\Phi(x) = [-x^2, x^1]$, $x = (x^1, x^2) \in \boldsymbol{R}^2$. The point (0, 0) is an isolated critical point of Φ (Fig. 2.1.1). Here F is the rotation of the unit circle by the angle $\frac{1}{2}\pi$, and so

$$\iota_\Phi(0) = \deg F = 1.$$

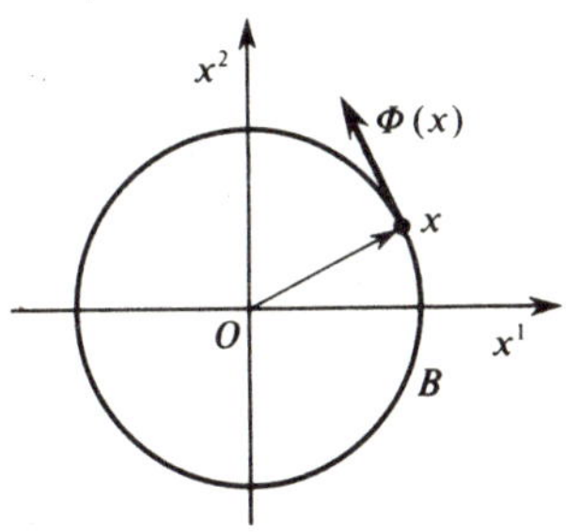

FIG. 2.1.1

2.1.6. Let $U = \boldsymbol{R}^2$, $\Phi(x) = \text{const} \neq 0$. Then for any $x \in \boldsymbol{R}^2$

$$\iota_\Phi(x) = \deg F = 0.$$

Now, let f: $U_1 \to U_2$ be a diffeomorphism between two open sets in $\boldsymbol{R}^m$ and let $\Phi_1 \in \mathscr{F}^1(U_1)$. Then f transports the vector field Φ_1 to U_2; let Φ_2 denote the resulting vector field

$$\Phi_2(y) = \mathrm{d}f(f^{-1}(y)) \circ \Phi_1(f^{-1}(y)) \qquad \text{for } y \in U_2.$$

Theorem 2.1.3 *If x_0 is an isolated critical point of Φ_1, then $y_0 = f(x_0)$ is an isolated critical point of Φ_2 and we have*

$$\iota_{\Phi_1}(x_0) = \iota_{\Phi_2}(y_0).$$

In view of this theorem, it is legitimate to introduce the following definition:

Definition 2.1.4 Let M be a manifold and let $\Phi \in \mathscr{F}^1(M)$. By the *index* of an isolated critical point $x_0 \in M$ we mean the index of its image in any coordinate system (with respect to the vector field in $\boldsymbol{R}^m$ induced by Φ via the chart).

To conclude the section, we formulate one more important theorem (Milnor, 1965).

Theorem 2.1.4 (Poincaré–Hopf) Let M be a compact manifold and let $\Phi \in \mathscr{F}^1(M)$ be a vector field with finitely many critical points: $x_1, \ldots, x_n$. Then

$$\sum_{k=1}^{n} \iota_{\Phi}(x_k) = \chi(M) = \sum_{i=0}^{m} (-1)^i \operatorname{rank} H_i(M),$$

where $\chi(M)$ is the Euler characteristic of M and $H_i(M)$ denotes the i-th homology group of M (for the definitions see e.g. Spanier, 1965).

For $M = S^2$ (2-sphere) we have $\chi(S^2) = 2$; for $M = T^2$ (2-torus) or $M = K^2$ (Klein bottle) we have $\chi(M) = 0$; for other 2-dimensional manifolds we have $\chi(M) < 0$ (see Wallace, 1968).

2.2. Limit Sets for Flows on Two-Dimensional Manifolds

We have introduced the notion of a minimal set for a flow (see Definition 1.11.3). In the case of two-dimensional manifold such sets admit an elegant characterization; a minimal set is either a critical point or a closed orbit or the entire manifold. This fact has been proved by A. J. Schwartz (1963).

Theorem 2.2.1 Let M be a two-dimensional connected manifold and let $\Phi \in \mathscr{F}^2(M)$. If S is a minimal set for the flow φ_t generated by Φ, then either S is a critical element or $S = M$.

Proof We proceed in two steps.

Step I. If S contains a critical element, then S reduces to that element, by the definition of a minimal set. If S has interior points, then $S = M$ by Theorem 1.11.3.

Thus assume that S contains no critical element and that $\operatorname{int} S = \emptyset$, i.e., S is nowhere dense in M. Pick a point $x_0 \in S$ and denote by Γ a cross-section of the orbit passing through x_0. We may assume that M

is a part of $\boldsymbol{R}^2$ and $x_0 = (0, 0)$. Take a rectifying chart for a neighbourhood of x_0. Let (s, u) denote the coordinates in that chart; let this coordinate system be chosen so as to have Γ contained in the s-axis; this means that

$$\Gamma = \{(s, u): u = 0, |s| < a\},$$

a being a constant. Further, let

$$\varphi^1(t, x), \quad \varphi^2(t, x)$$

denote the coordinates of $\varphi_t(x)$ in the coordinate system (s, u). By the transversality of Γ and $\Phi(0)$, we have

$$\frac{\partial \varphi^2}{\partial u}(0, x) \neq 0 \qquad \text{for } x = (s, 0) \in \Gamma.$$

Consider the set I of those points $x = (s, 0) \in \Gamma$ for which there exists an instant $t = t(x) > 0$ such that $\varphi_t(x) \in \Gamma$, i.e., $\varphi^2(t(x), x) = 0$. By the Implicit Function Theorem, I is open in Γ. Let us introduce the function

$$f(s) = \varphi^1(t(s, 0), (s, 0)) \qquad \text{for } (s, 0) \in I.$$

This function maps I into Γ. We shall show that $f'(s) \neq 0$ for all s. Each φ_t is a diffeomorphism and so we have

$$\begin{vmatrix} \dfrac{\partial \varphi^1}{\partial s} & \dfrac{\partial \varphi^1}{\partial u} \\ \dfrac{\partial \varphi^2}{\partial s} & \dfrac{\partial \varphi^2}{\partial u} \end{vmatrix} \neq 0$$

at each point. Obviously,

$$f'(s) = \frac{\partial \varphi^1}{\partial t}\frac{dt}{ds} + \frac{\partial \varphi^1}{\partial x}\frac{\partial x}{\partial s} = \frac{\partial \varphi^1}{\partial t}\frac{dt}{ds} + \frac{\partial \varphi^1}{\partial s}. \tag{2.2.1}$$

The equation $\varphi^2(t(x), x) = 0$ is satisfied identically for $x = (s, 0) \in I$. Differentation with respect to s yields

$$0 = \frac{\partial \varphi^2}{\partial t}\frac{dt}{ds} + \frac{\partial \varphi^2}{\partial x}\frac{\partial x}{\partial s} = \frac{\partial \varphi^2}{\partial t}\frac{dt}{ds} + \frac{\partial \varphi^2}{\partial s}. \tag{2.2.2}$$

Equations (2.2.1) and (2.2.2) form a system with a non-zero determinant. The vector $[f'(s), 0]$ on the left side forces the non-vanishing solution $[dt/ds, 1]$. Consequently $f'(s) \neq 0$ for any s, as claimed.

Since Φ is of class C^2, so are $t(x)$ and $f(s)$.

Write $K = S \cap \Gamma$. The set K is closed and nowhere dense in Γ: if K were dense in some piece of Γ, S would be dense in an open subset of M, contrary to the assumption. We may also assume that the endpoints of Γ do not belong to S.

The properties of minimal sets imply that every point $x \in S$ is a limit point (as $t \to +\infty$) of the trajectory passing through x. Therefore, for every $x \in K$ there is a $t(x) > 0$ such that $\varphi(t(x), x) \in \Gamma$. This means that $K \subset I$ and $f(K) = K$. Let V be an open subset of I such that

$$K \subset V \subset \bar{V} \subset I.$$

Summing up, we see that the function f fulfils the following conditions: there are positive constants σ, ϑ and λ such that

1° $\sigma \leqslant |f'(s)| \leqslant \vartheta$ for $s \in V$,
2° $|f''(s)| \leqslant \lambda$ for $s \in V$;

moreover,

3° $f(K) = K$;

and finally,

4° $f^n(s) \neq s$ for all $(s, 0) \in K$ and $n \neq 0$

(otherwise the point $(s, 0)$ would belong to a closed orbit, contrary to the assumption that S does not contain a critical element).

Step II. We shall show that there is no function $f(s)$ satisfying 1°–4°.

II(a). Suppose that such an f exists. The set $\Gamma - K$ is open and thus can be represented as the union

$$\Gamma - K = \bigcup_{k=1}^{\infty} (\alpha_k, \beta_k).$$

Denote by ε the distance from K to $\Gamma - V$. Then for $\beta_k - \alpha_k < \varepsilon$ we have $\langle \alpha_k, \beta_k \rangle \subset V$.

We claim that at least one interval $\langle \alpha_{k_0}, \beta_{k_0} \rangle$ satisfies the condition:

$$f^n(\langle \alpha_{k_0}, \beta_{k_0} \rangle) \subset V \qquad \text{for } n = 0, 1, 2, \ldots$$

There are only finitely many intervals (α_i, β_i) with $\beta_i - \alpha_i \geqslant \varepsilon$. Thus, according to 4°, there exists an index n_0 such that, for any $n \geqslant n_0$, neither of the numbers $f^n(\alpha_i), f^n(\beta_i)$ is an endpoint of such a "long" interval. Hence the interval whose one endpoint is $f^{n_0}(\alpha_i)$ has the following property: any of its f^n-images has length less than ε. (Here we use condition 3°). Consequently all those f^n-images are contained in V. This proves our claim.

II(b). Let $I_0 = (\alpha, \beta)$ be an interval with the property established in step II(a). Denote $f^n(I_0)$ by I_n, $n = 0, 1, \ldots$ Since f has no periodic points in K, the intervals I_n are pairwise disjoint. Hence

$$\sum_{n=0}^{\infty} |I_n| < +\infty.$$

It can be proved that

$$\sum_{n=1}^{\infty} \left| \frac{\mathrm{d}f^n(\alpha)}{\mathrm{d}x} \right| = \sum_{n=1}^{\infty} \prod_{j=0}^{n-1} |f'(\alpha_j)| < +\infty, \tag{2.2.3}$$

where $\alpha_j = f^j(\alpha)$. We omit the proof; an argument nearly identical to that needed in the proof of (2.2.3) will be applied in the proof of Theorem 2.4.1, step III, p. 99.

II(c). We now prove the existence of a number $d_0 > 0$ such that the interval $H = (\alpha - d_0, \alpha + d_0)$ satisfies:

$$\sum_{n=0}^{\infty} |f^n(H)| < +\infty.$$

First we will try to find an $\alpha' < \alpha$ such that

$$\sum_{n=0}^{\infty} |f^n((\alpha', \alpha))| < +\infty.$$

To this end, denote $f^n(\alpha)$ by α_n, $f^n(\alpha')$ by α'_n, $f'(\alpha_n)$ by u_n and $|\alpha'_n - \alpha_n|$ by d_n.

Using the Taylor formula $f'(\xi) = f'(\alpha_n) + f''(\eta)(\xi - \alpha_n)$ and keeping 2° in mind, we get:

$$\begin{aligned} |\alpha'_{n+1} - \alpha_{n+1}| &= |f(\alpha'_n) - f(\alpha_n)| = |\alpha'_n - \alpha_n|\,|f'(\xi)| \\ &\leqslant |\alpha'_n - \alpha_n|\,(|f'(\alpha_n)| + |f''(\eta)|\,|\alpha'_n - \alpha_n|) \\ &\leqslant |\alpha'_n - \alpha_n|\,(u_n + \lambda\,|\alpha'_n - \alpha_n|), \end{aligned}$$

where ξ and η are some points in $\langle \alpha'_n, \alpha_n \rangle$. Hence

$$d_{n+1} \leqslant d_n(u_n + \lambda d_n).$$

Consider the recurrent sequence

$$\bar{d}_{n+1} = \bar{d}_n(u_n + \lambda \bar{d}_n). \tag{2.2.4}$$

If we can find $\bar{d}_0 > 0$ such that $\sum \bar{d}_n < +\infty$, then putting $d_0 = \bar{d}_0$ we get $\sum d_n < +\infty$ (clearly, $d_n \leqslant \bar{d}_n$). Such a $\bar{d}_0$ can indeed be found; we leave this to the reader (Exercise 2.2.1 below).

Hence follows the existence of an α' with the desired property. In a similar way we find an $\alpha'' > \alpha'$ satisfying:

$$\sum_{n=0}^{\infty} |f^n((\alpha, \alpha''))| < +\infty.$$

Now it suffices to define d_0 as $\min(\alpha-\alpha', \alpha''-\alpha)$.

II(d). Since $\alpha \in K$, there is a positive integer r such that

$$f^r(\alpha) \in H$$

(the points $f^n(\alpha)$ lie densely in K). Thus

$$f^{nr}(\alpha), f^{(n+1)r}(\alpha) \in f^{nr}(H),$$

and so, by II(c),

$$\sum_{n=0}^{\infty} |f^{(n+1)r}(\alpha) - f^{nr}(\alpha)| \leqslant \sum_{n=0}^{\infty} |f^{nr}(H)| < +\infty.$$

It follows that the sequence $(f^{nr}(\alpha))$ converges, the limit is a fixed point of f^r and belongs to K. This contradicts condition 4°. The proof is finished.

Exercise

2.2.1.A. Consider the recurrent sequence

$$\bar{d}_{n+1} = \bar{d}_n(u_n + \lambda \bar{d}_n), \qquad \text{where } \lambda > 0 \text{ and } \sum_{n=0}^{\infty} \prod_{j=0}^{n-1} u_j < +\infty.$$

Show that $\bar{d}_0 > 0$ can be chosen so as to have

$$\sum_{n=1}^{\infty} \bar{d}_n < +\infty.$$

Hint. Postulate $\bar{d}_n = v_n \prod_{j=0}^{n-1} u_j$.

Theorem 2.2.2 (Kneser, 1924) Let M be a compact two-dimensional manifold. If there exists a vector field $\Phi \in \mathscr{F}^1(M)$ such that M is a minimal set for the flow φ_t generated by Φ, then M is the torus T^2.

Proof Since M is minimal, the field has no critical points. Hence $\sum \iota_\Phi(x) = 0$ and by Theorem 2.1.4

$$\sum \iota_\Phi(x) = 0 = \chi(M) = \operatorname{rank} H_0(M) - \operatorname{rank} H_1(M) + \operatorname{rank} H_2(M).$$

The only compact connected 2-manifolds with the Euler characteristic equal to zero are T^2 (2-torus) and K^2 (Klein bottle). It can easily be shown that if a minimal flow existed on K^2, it would necessarily have a closed orbit – a contradiction. Thus $M = T^2$.

Definition 2.2.1 Let $\Phi \in \mathscr{F}^1(M)$ be a complete vector field on a manifold M and let γ_0 be a closed orbit of Φ. We say that a trajectory $\gamma_1^+(t) = \varphi(t, x)$, $\varphi(0, \bar{x}) = \bar{x}$, $t \geqslant 0$, *winds toward* γ_0 iff for every cross-section S of γ_0 at a point $x_0 \in \hat{\gamma}_0$ we have

(1) $\hat{\gamma}_1^+ \cap S \neq \emptyset$,

(2) $\Theta^n(x) \to x_0$ for all $x \in \hat{\gamma}_1^+ \cap S$

(Θ denotes the Poincaré map) (see Fig. 2.2.1).

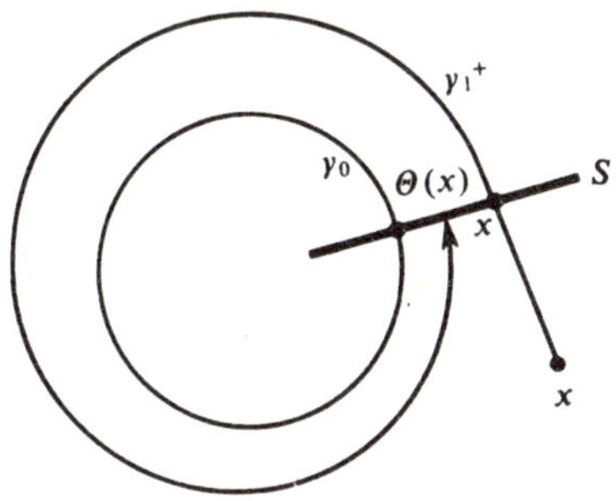

FIG. 2.2.1

The theorem which we now present arose from a number of independent results, proved by several authors at different times and under different assumptions. Putting those results into a single statement, we obtain

Theorem 2.2.3 (Poincaré–Bendixon–Schwartz) Let M be a compact oriented two-dimensional manifold. Let $\Phi \in \mathscr{F}^2(M)$ and denote by $\varphi(t, x)$ the flow generated by Φ. Suppose that for a certain $x_0 \in M$ the set $\omega(x_0)$ contains no critical points and that $\omega(x_0) \neq M$. Then $\omega(x_0)$ is a closed orbit and $\varphi(t, x_0)$ winds toward $\omega(x_0)$.

Proof By Theorem 1.11.2 the set $\omega(x_0)$ contains a minimal subset. According to Theorem 2.2.1, this subset is a closed orbit. Denote it by γ_0. Since M is oriented, a sufficiently small neighbourhood of $\hat{\gamma}_0$ is diffeomorphic to a cylinder $\{(x^1)^2+(x^2)^2 = 1, |x^3| < \delta\}$, the section $x^3 = 0$ being the image of γ_0. Our flow can now be transferred to that cylinder. Without loss of generality we may regard the cylinder as an annulus containing the circle $(x^1)^2+(x^2)^2 = 1$. Clearly, the quantity $\|\Phi(x)\|$ is separated from zero in that annulus. Taking into account that Φ is of class C^2, we easily show, by geometrical arguments, that $\varphi_t(x)$ winds toward γ_0; we leave the details to the reader (see Fig. 2.2.2 and Exercise 2.2.4).

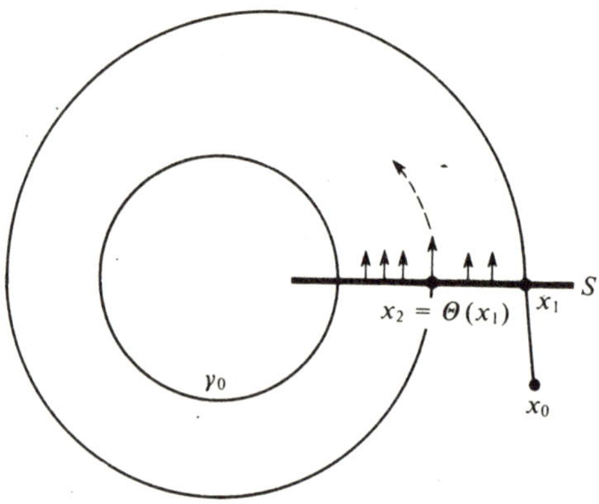

FIG. 2.2.2

Remark 2.2.1 Theorem 2.2.3 remains true also for non-orientable manifolds: if M is non-orientable, we lift the flow to $\tilde{M}$, the double covering[12] of M; $\tilde{M}$ is orientable; it remains to observe that the operation of lifting carries closed orbits in M into closed orbits in $\tilde{M}$.

Exercises

2.2.2.A. Prove formula (2.2.3) on p. 89.
Hint. Compare the proof of Theorem 2.4.1, step III.

2.2.3.B. Prove Theorem 2.2.2 without using Theorem 2.1.4 (Poincaré–Hopf).
Hint. Consider a field orthogonal to the field in question.

2.2.4.B. Let Φ be a continuous vector field in $\boldsymbol{R}^n$ and let $x(t)$ denote a solution of the equation $\mathrm{d}x/\mathrm{d}t = \Phi(x)$, $t > 0$, such that $\omega(x(t))$ is a compact set without critical points. Show that $\omega(x(t))$ is a closed Jordan curve $y(t)$, which is a solution of the same equation.

Hint. By Theorem 2.2.1 $\omega(x)$ contains a closed orbit. Consider that orbit together with a cross-section. The set $\omega(x)$ coincides with the orbit in question (Coddington and Levinson, 1955; Hartman, 1964).

2.2.5.B. Let f and g be two isotopic[9] diffeomorphisms and let $\tilde{M}$ and $\tilde{N}$ be two manifolds admitting suspension of f and g. Prove that $\tilde{M}$ and $\tilde{N}$ are homeomorphic.

2.3. Index of a Critical Point of a Flow on $\boldsymbol{R}^2$

We shall now examine the case of $M = \boldsymbol{R}^2$. The index of an isolated critical point of a vector field $\Phi \in \mathscr{F}^1(\boldsymbol{R}^2)$ can be defined as follows.

Let J be a closed Jordan curve of class C^1. Assume that $\Phi(x) \neq 0$ for $x \in J$. Fix a non-zero vector a and denote by $\Theta(x)$ the signed angle between the vectors a and $\Phi(x)$. Clearly, $\Theta(x)$ is a function of class C^1, because Φ is so. Let $\gamma(t)$, $t \in I$, be a parametrization of J. Define:

$$\Delta := \int_I \frac{\mathrm{d}}{\mathrm{d}t} \Theta(\gamma(t))\, \mathrm{d}t.$$

We call this number the *total change of angle* between $\Phi(x)$ and the direction of a, as x runs over J.

By the definition of the integral, if (π_n) is a normal sequence of partitions of J,

$$\pi_n = \{x_{n,k} \in J \colon k = 1, \ldots, k_n\},$$

then

$$\Delta = \lim_{n \to \infty} \sum_{k=1}^{k_n} \left(\Theta(x_{n,k+1}) - \Theta(x_{n,k})\right).$$

The index of Φ along J is the number

$$\iota_\Phi(J) := \frac{\Delta}{2\pi};$$

this number is independent of parametrization and of the choice of a.

Evidently, $\iota_\Phi(J)$ is an integer. It is not difficult to check that if the region encompassed by J contains exactly one critical point x_0, then $\iota_\Phi(J)$ is equal to $\iota_\Phi(x_0)$ (cf. Definition 2.1.3); we leave the details to the reader.

Thus, Theorem 2.1.2 implies

Theorem 2.3.1 *If J is a closed Jordan curve in $\mathbf{R}^2$ and $\Phi \in \mathscr{F}^1(\mathbf{R}^2)$ is a field without critical points in the closure of the region D surrounded by J, then $\iota_\Phi(J) = 0$.*

As a consequence we obtain further theorems:

Theorem 2.3.2 *If the region D surrounded by J contains a finite number of critical points $x_1, \ldots, x_n$ of a vector field Φ, then*

$$\iota_\Phi(J) = \sum_{k=1}^{n} \iota_\Phi(x_k).$$

Theorem 2.3.3 *If $J = \hat{\gamma}$, where $\gamma(t)$ is a closed trajectory of a vector field Φ, then*

$$\iota_\Phi(J) = 1.$$

Proof If J is a closed orbit, then the total change of angle between $\Phi(x)$ and any fixed direction (as x runs over J) is 2π. Hence $\iota_\Phi(J) = 1$.

Corollary 2.3.1 *Every domain encompassed by a closed orbit of Φ contains at least one critical point of Φ.*

Corollary 2.3.2 *If such a domain contains finitely many critical points, the sum of their indices is equal to 1.*

Lemma 2.3.1 *Let $\Phi_0, \Phi_1 \in \mathscr{F}^1(\mathbf{R}^2)$ be two vector fields and let J be a closed Jordan curve. If the vectors $\Phi_0(x), \Phi_1(x)$ have opposite directions at no point $x \in J$, then*

$$\iota_{\Phi_0}(J) = \iota_{\Phi_1}(J).$$

Proof For any $s \in \langle 0, 1 \rangle$ consider the vector field

$$\Phi_s = (1-s)\Phi_0 + s\Phi_1.$$

The condition of the lemma implies:

$$\Phi_s(x) \neq 0 \qquad \text{for all } x \in J,\ s \in \langle 0, 1 \rangle.$$

Hence the index $\iota_{\Phi_s}(J)$ is well defined and depends continuously on s (this is easy to show). A continuous integer-valued function must be constant. Hence the assertion.

Theorem 2.3.4 Let x_0 *be a non-degenerate critical point of a vector field* $\Phi \in \mathscr{F}^1(\boldsymbol{R}^2)$. *Then*

$$\iota_\Phi(x_0) = \iota_\Psi(x_0),$$

where

$$\Psi(x) = \dot{\Phi}(x_0)(x - x_0).$$

($\dot{\Phi}(x_0)$ *denotes, as usual, the Hessian of* Φ *at* x_0).

Proof We may assume that $x_0 = 0$. Denote the Hessian $\dot{\Phi}(x_0)$ by A, for brevity. Then

$$\Phi(x) = Ax + G(x),$$

where $G(0) = 0$, $\mathrm{d}G(0) = 0$. This means that for every $\varepsilon > 0$ there is an $r > 0$ such that

$$\|G(x)\| \leqslant \varepsilon \|x\| \qquad \text{for } \|x\| \leqslant r.$$

Fix ε. Let J be the circle with centre 0 and radius r. In virtue of the preceding lemma, we shall get what we want if we show that the vectors $\Phi(x)$ and Ax never have opposite directions, as $x \in J$. Assume the contrary:

$$\Phi(x_1) + sAx_1 = 0$$

for some $x_1 \in J$ and some $s \geqslant 0$. Then

$$(1+s)Ax_1 = -G(x_1).$$

Since 0 is a non-degenerate critical point, the operator A is invertible, and so there exists a $\delta > 0$ such that

$$\|Ax\| \geqslant \delta \|x\| \qquad \text{for all } x.$$

Consequently

$$\delta \|x_1\| \leqslant (1+s)\|Ax_1\| = \|G(x_1)\| \leqslant \varepsilon \|x_1\|.$$

The choice of $\varepsilon > 0$ was arbitrary. Choosing $\varepsilon < \delta$, we get a contradiction.

Theorem 2.3.5 *Let x_0 be a non-degenerate critical point of a vector field Φ. The index $\iota_\Phi(x_0)$ is equal to -1 iff x_0 is a saddle* (see p. 27) *for the vector field $\dot{\Phi}(x_0)(x-x_0)$; in other cases $\iota_\Phi(x_0) = 1$.*

Proof We may again assume that $x_0 = 0$ and write $A = \dot{\Phi}(0)$. By Theorem 2.3.4 it suffices to prove our assertion for the vector field Ax. Let

$$A = \begin{bmatrix} a & b \\ c & d \end{bmatrix}, \qquad \det A \neq 0.$$

If $\det A > 0$, then A is isotopic to the identity (see Exercise 2.3.1). Thus, by Theorem 2.1.2,

$$\iota_{Ax}(0) = \iota_{Ix}(0) = 1.$$

If $\det A < 0$, then A is isotopic to $I_1 = \begin{bmatrix} 1 & 0 \\ 0 & -1 \end{bmatrix}$. Thus

$$\iota_{Ax}(0) = \iota_{I_1 x}(0) = -1.$$

Evidently, $\det A < 0$ iff 0 is a saddle for A (see Section 2.3).

Exercises

2.3.1.A. Show that if A: $\boldsymbol{R}^2 \to \boldsymbol{R}^2$ is a non-singular linear operator then A is isotopic either to $I = \begin{bmatrix} 1 & 0 \\ 0 & 1 \end{bmatrix}$ or to $I_1 = \begin{bmatrix} 1 & 0 \\ 0 & -1 \end{bmatrix}$, according as $\det A > 0$ or $\det A < 0$. Write the isotopy explicitly.

Hint. Use the Jordan form of a matrix.

2.3.2.B. Consider the following system of differential equations on the plane:

$$\dot{x}^1 = x^2,$$
$$\dot{x}^2 = -b \sin x^1 - a x^2,$$

where $a, b > 0$, $a - 4b < 0$. Show that for every $x \in \boldsymbol{R}^2$ the set $\omega(x)$ reduces to the point $(k\pi, 0)$ for a certain $k \in Z$. Draw the orbits of the system.

Hint. First show that the set $\omega(x)$ is bounded for every x. Then show that there are no closed orbits.

Remark The system of equations considered in Exercise 2.3.2 describes the damped oscillation of a flat pendulum.

2.3.3.B. Examine the system

$$\dot{x}^1 = x^2,$$
$$\dot{x}^2 = -x^1 + (1 - (x^1)^2 - (x^2)^2)x^2.$$

Show that the only periodic solutions are of the form

$$x^1(t) = \sin(t+c), \qquad x^2(t) = \cos(t+c),$$

c denoting a constant.

Hint. Examine $\dfrac{\mathrm{d}}{\mathrm{d}t}((x^1)^2 + (x^2)^2)$.

2.4. Homeomorphisms of the Circle

Our objective in this section is a classification of homeomorphisms of the circle with regard to topological conjugacy.

The theory which we present is due to Denjoy (1932).

Theorem 2.4.1 *Let* $\varphi \in \mathrm{Diff}^1(S^1)$. *Suppose that* φ *has no periodic points and preserves orientation, and that* φ' *(the derivative of* φ*) is of bounded variation. Let* Δ *denote the minimal set defined in Exercise 1.11.1. Then* $\Delta = S^1$.

Proof The proof will be carried out in 3 steps.

Step I. Fix $x_0 \in S^1$ and write $x_k = \varphi^k(x_0)$, $k \in Z$. Fix a positive integer n and consider the open arc(x_n, x_0). Suppose that

$$(x_n, x_0) \cap \{x_k\colon |k| \leqslant n\} = \emptyset. \tag{2.4.1}$$

Then the points $x_0, x_1, \ldots, x_{n-1}$ and $x_{-n}, x_{-n+1}, \ldots, x_{-1}$ alternate on S^1. This means that every arc (x_k, x_{k-n}), $k = 0, 1, \ldots n$, contains no other points of the set $\{x_k\colon |k| \leqslant n\}$. Indeed, assume that a certain $x_j (|j| \leqslant n$; say, $j > 0)$ lies in some (x_k, x_{k-n}). Let k be the greatest index with this property. We have two possibilities:

1° $(x_j, x_{j-n}) \subset (x_k, x_{k-n})$; then $\varphi^{j-k}(x_k, x_{k-n}) \subset (x_k, x_{k-n})$ and consequently φ^{j-k} has a fixed point, contrary to the assumption.

2° $x_{k-n} \in (x_j, x_{j-n})$; then, k being maximal, we get $j < k$. Hence $\varphi^{n-k}(x_j) \in \varphi^{n-k}(x_k, x_{k-n})$. This yields $x_{n-k+j} \in (x_n, x_0)$, where $0 < n-k+j < n$, contradicting (2.4.1).

The case of $j \leqslant 0$ easily reduces to that of $j > 0$.

Observe that if $x_0 \in \Delta$ then there are infinitely many n satisfying (2.4.1). Indeed, Δ is minimal and so the sequence $\{x_n\}$ is dense in Δ.

Step II. By assumption, φ preserves orientation. Thus $\varphi'(x) > 0$. Then, φ' being continuous, there is an $a > 0$ such that

$$\varphi'(x) \geqslant a > 0 \qquad \text{for all } x \in S^1. \tag{2.4.2}$$

Define:

$$f(x) = \log \varphi'(x). \tag{2.4.3}$$

The function $f(x)$ is of bounded variation. To see this, take any partition $\xi_0, \xi_1, \ldots, \xi_\nu$ of S^1; $\xi_\nu = \xi_0$. Applying the Mean Value Theorem to the function $\log u$ (for $u \geqslant a$) we get

$$\begin{aligned}\sum_{i=0}^{\nu-1} |f(\xi_{i+1}) - f(\xi_i)| &= \sum_{i=0}^{\nu-1} |\log \varphi'(\xi_{i+1}) - \log \varphi'(\xi_i)| \\ &\leqslant \sum_{i=0}^{\nu-1} \frac{1}{a} |\varphi'(\xi_{i+1}) - \varphi'(\xi_i)| \\ &\leqslant \frac{1}{a} \operatorname{Var} \varphi' < +\infty.\end{aligned}$$

Put $\operatorname{Var} f = V$. Let x_0, n and x_k, $|k| \leqslant n$, be as in Step I. By the conclusion of Step I, the arcs (x_k, x_{k-n}), $k = 0, 1, \ldots, n$, are pairwise disjoint. Hence

$$\sum_{k=0}^{n-1} |f(x_k) - f(x_{k-n})| \leqslant V. \tag{2.4.4}$$

On the other hand, we have

$$\sum_{k=0}^{n-1} f(x_k) = \log \prod_{k=0}^{n-1} \varphi'(x_k) = \log \frac{\mathrm{d}\varphi^n}{\mathrm{d}x}(x_0)$$

and

$$\begin{aligned}\sum_{k=0}^{n-1} f(x_{k-n}) &= \log \prod_{k=0}^{n-1} \varphi'(x_{k-n}) = \log \prod_{j=-n}^{-1} \varphi'(x_j) = \log \frac{\mathrm{d}\varphi^n}{\mathrm{d}x}(x_{-n}) \\ &= \log \frac{\mathrm{d}\varphi^{-n}}{\mathrm{d}x}(x_0).\end{aligned}$$

The last two equalities together with (2.4.4) result in:

$$V \geqslant \left| \log \frac{d\varphi^n}{dx}(x_0) + \log \frac{d\varphi^{-n}}{dx}(x_0) \right| = \left| \log\left(\frac{d\varphi^n}{dx}(x_0) \cdot \frac{d\varphi^{-n}}{dx}(x_0) \right) \right|.$$

Hence

$$\frac{d\varphi^n}{dx}(x_0) \cdot \frac{d\varphi^{-n}}{dx}(x_0) \geqslant e^{-V}. \tag{2.4.5}$$

Step III. Assume that $\Delta \neq S^1$. The set $S^1 - \Delta$ is open (see Exercise 1.11.1) and so

$$S^1 - \Delta = \bigcup_{n=0}^{\infty} I_n,$$

where I_n, $n = 0, 1, \ldots$, are a sequence of pairwise disjoint intervals (arcs) on S^1. The set $S^1 - \Delta$ is φ-invariant. Thus $\varphi^k(I_0) = I_{m_k}$ for some m_k; for different k the numbers m_k are also different; otherwise φ would have a periodic point. Hence

$$\sum_{k=1}^{\infty} |I_{m_k}| < +\infty. \tag{2.4.6}$$

The length of I_{m_k} can be estimated from below:

$$|I_{m_k}| = \int_{I_0} \frac{d\varphi^k(x)}{dx} dx \geqslant |I_0| \min_{x \in I_0} \frac{d\varphi^k(x)}{dx} = |I_0| \prod_{i=0}^{k-1} \varphi'(\xi_i), \tag{2.4.7}$$

where $\xi_i \in I_{m_i}$.

Denote by v_i and u_i the minimum and the maximum (respectively) of φ' on I_{m_i}. On account of (2.4.6) and (2.4.7) we have

$$\sum_{k=1}^{\infty} \prod_{i=0}^{k-1} v_i < +\infty. \tag{2.4.8}$$

Note that, by (2.4.2),

$$\frac{u_i}{v_i} = 1 + \left(\frac{u_i}{v_i} - 1 \right) \leqslant 1 + \frac{u_i - v_i}{a} \leqslant e^{(u_i - v_i)/a}.$$

Therefore

$$\frac{\prod_{i=0}^{k-1} u_i}{\prod_{i=0}^{k-1} v_i} \leqslant \exp\left(\frac{1}{a}\sum_{i=0}^{k-1}(u_i - v_i)\right) \leqslant \exp\left(\frac{1}{a}\operatorname{Var}\varphi'\right),$$

and the convergence of the series in (2.4.8) yields

$$\lim_{k\to\infty}\frac{\mathrm{d}\varphi^k}{\mathrm{d}x}(x) = 0 \qquad \text{for every } x\in\bar{I}_0.$$

The same argument shows that

$$\lim_{k\to\infty}\frac{\mathrm{d}\varphi^{-k}}{\mathrm{d}x}(x) = 0 \qquad \text{for every } x\in\bar{I}_0\cdot$$

Thus

$$\lim_{k\to\infty}\frac{\mathrm{d}\varphi^k}{\mathrm{d}x}(x)\frac{\mathrm{d}\varphi^{-k}}{\mathrm{d}x}(x) = 0 \qquad \text{for every } x\in\bar{I}_0.$$

This holds in particular for $x = x_0$, where x_0 denotes an endpoint of I_0. In view of Exercise 1.11.3, p. 81, there is a sequence $n_k \to +\infty$ such that (2.4.1) holds with $n = n_k$. According to what has been shown in Step II, inequality (2.4.5) is satisfied at this point for all $n = n_k$. Thus we arrive at a contradiction.

Remark 2.4.1 The assumption of φ being of bounded variation is essential. There exist diffeomorphisms $\varphi\in\operatorname{Diff}^1(S^1)$ for which $\Delta \neq S^1$ (see Denjoy, 1932).

In our further discussion we shall need the concept of the so-called rotation number of a circle homeomorphism. Let us now work out the definition of this concept.

Let $\varphi\colon S^1 \to S^1$ be an orientation preserving homeomorphism. Denote by $f(x)$ the lifting of φ to the real line $\boldsymbol{R}$:

$$f\colon \boldsymbol{R}\to\boldsymbol{R}; \qquad \varphi(z) = \mathrm{e}^{2\pi \mathrm{i} f(x)} \qquad \text{for } z = \mathrm{e}^{2\pi \mathrm{i} x}.$$

Clearly, there are infinitely many such continuous functions $f(x)$, but any two of them differ by an integer. Since φ preserves orientation, f is monotonically increasing and fulfils

$$f(x+1) = f(x)+1. \tag{2.4.9}$$

If we assumed that φ alters orientation, (2.4.9) ought to be replaced by

$$f(x+1)=f(x)-1, \tag{2.4.10}$$

f being monotonically decreasing.

Theorem 2.4.2 For every $x_0 \in \boldsymbol{R}$ the limit

$$\lim_{|n|\to\infty} \frac{1}{n} f^n(x_0) = \varrho \tag{2.4.11}$$

exists and does not depend on the initial point x_0. This limit is a rational number iff φ has a periodic point.

Proof The proof is carried out in 4 steps.

Step I. We first show that if the limit in (2.4.11) exists for some value of x_0, then it exists for any other value and is the same. Fix $x_0, y_0 \in \boldsymbol{R}$ and write $x_n = f^n(x_0)$, $y_n = f^n(y_0)$. Let k and m be integers such that

$$m \leqslant y_0 - (x_0 + k) < m+1.$$

Since f is increasing, we have

$$f^n\big(m+(x_0+k)\big) \leqslant f^n(y_0) < f^n\big((m+1)+(x_0+k)\big).$$

Hence, by (2.4.9),

$$f^n(x_0)+m+k \leqslant f^n(y_0) < f^n(x_0)+m+k+1,$$

and so

$$\frac{x_n}{n} + \frac{m+k}{n} \leqslant \frac{y_n}{n} < \frac{x_n}{n} + \frac{m+k+1}{n}.$$

Thus, if $\lim x_n/n = \varrho$ exists, then also $\lim y_n/n = \varrho$.

Step II. Assume that φ has a periodic point $z_0 = e^{2\pi i x_0}$ and write $x_n = f^n(x_0)$. Then $\varphi^p(z_0) = z_0$ for a certain $p \in Z^+$; in terms of f this means that $x_p - x_0 = r$ is an integer. By easy induction we get $x_{np} = x_0 + nr$ for $n \in Z$.

Let k be an integer. Writing k in the form

$$k = np+s, \qquad 0 \leqslant s < p,$$

we obtain

$$x_k = x_{np+s} = f^s(x_{np}) = x_s + nr,$$

and consequently

$$\frac{x_k}{k} = \frac{x_s}{k} + \frac{rn}{k} = \frac{x_s}{k} + \frac{rn}{np+s}.$$

If $k \to \pm\infty$, then also $n \to \pm\infty$, and the last expression tends to $\varrho = r/p$.

Thus in this case the limit ϱ in (2.4.11) exists and is rational, as claimed.

Step III. We now show that ϱ exists also if φ has no periodic points. Fix $m \in Z^+$ and consider the point $f^m(0)$. We have

$$r < f^m(0) < r+1$$

for a certain integer r; the inequalities are strict since otherwise 0 would be a periodic point. Applying the mapping f^m to the above inequalities, we get

$$f^m(r) < f^{2m}(0) < f^m(r+1).$$

Now, we have

$$f^m(r) = f^m(0+r) = f^m(0) + r > r + r = 2r$$

and similarly

$$f^m(r+1) < 2(r+1).$$

Thus

$$2r < f^{2m}(0) < 2(r+1).$$

Repeating this argument n times ($n \in Z^+$), we obtain

$$nr < f^{nm}(0) < n(r+1),$$

i.e.,

$$\frac{r}{m} < \frac{1}{nm} f^{nm}(0) < \frac{r+1}{m}.$$

Taking $n = 1$, we get

$$\frac{r}{m} < \frac{1}{m} f^m(0) < \frac{r+1}{m}.$$

The last two inequalities jointly yield

$$\left| \frac{1}{nm} f^{nm}(0) - \frac{1}{m} f^m(0) \right| < \frac{2}{m}.$$

Interchanging n with m, we see that also

$$\left| \frac{1}{nm} f^{nm}(0) - \frac{1}{n} f^{n}(0) \right| < \frac{2}{n}.$$

Hence

$$\left| \frac{1}{n} f^{n}(0) - \frac{1}{m} f^{n}(0) \right| < \frac{2}{m} + \frac{2}{n}.$$

The above argument retains validity also in the case of n or m (or both) negative; one has to use the absolute value of n and m. We are then led to the estimate

$$\left| \frac{1}{n} f^{n}(0) - \frac{1}{m} f^{m}(0) \right| < \frac{2}{|m|} + \frac{2}{|n|},$$

which shows that the limit

$$\varrho = \lim_{|n| \to \infty} \frac{1}{n} f^{n}(0)$$

exists.

Step IV. It remains to show that φ has a periodic point whenever ϱ is rational. Thus let ϱ be rational. Then there exist integers m and k such that $m\varrho + k = 0$. Write

$$g(x) = f^{m}(x) + k.$$

It is not difficult to see that g is a lift of φ^{m} to the real line. For every n we have

$$g^{n}(x) = f^{mn}(x) + nk,$$

whence

$$\frac{g^{n}(x)}{n} = \frac{f^{mn}(x)}{n} + k = m\,\frac{f^{mn}(x)}{mn} + k,$$

and so

$$\lim_{n \to \infty} \frac{g^{n}(x)}{n} = m\varrho + k = 0. \tag{2.4.12}$$

Now assume that φ^{m} has no fixed points. Then

$$g(x) \neq x \qquad \text{for all } x; \tag{2.4.13}$$

suppose that $x < g(x)$ (if the opposite inequality holds, the proof is similar). Starting from $x = 0$, we obtain an increasing sequence:

$$0 < g(0) < \dots < g^n(0) < \dots$$

This sequence is upper-bounded; we have $g^n(0) \leqslant 1$ for all $n \in Z^+$. To see this, note that if we had $g^{n_0}(0) > 1$ for some n_0, then

$$g^{2n_0}(0) = g^{n_0}(g^{n_0}(0)) > g^{n_0}(1) = f^{mn_0}(1) + kn_0 = f^{mn_0}(0) + 1 + kn_0$$
$$= g^{n_0}(0) + 1 > 2$$

and by induction $g^{rn_0}(0) > r$, i.e.,

$$\frac{g^{rn_0}(0)}{rn_0} > \frac{1}{n_0} \qquad \text{for } r = 1, 2, \dots,$$

which contradicts (2.4.12).

It follows that the sequence $g^n(0)$ converges. Denoting the limit by a, we easily obtain $g(a) = a$, contrary to (2.4.13). Consequently, φ has a periodic point, as asserted.

This theorem justifies the following definition.

Definition 2.4.1 The number ϱ given by (2.4.11) is called the *rotation number* of the mapping $f(x)$.

However, we wish to assign a "rotation number" to a homeomorphism of the circle, not of the line. Let $\{f_n\colon n \in Z\}$ be the family of all lifts of φ to $\boldsymbol{R}$. Since any two of them differ by an integer, we may write $f_n = f_0 + n$, f_0 denoting an arbitrarily chosen member of that family. Each f_n has a well defined rotation number ϱ_n, and we have $\varrho_n = \varrho_0 + n$, as is easy to see. Hence:

Definition 2.4.2 The *rotation number* of an orientation preserving homeomorphism $\varphi\colon S^1 \to S^1$ is the number $\varrho \in \langle 0, 1)$ defined by:

$$\varrho = \varrho_n (\text{mod } 1),$$

where ϱ_n are the rotation numbers of the lifts of φ to the real line.

Geometrically, the rotation number can be viewed as the medium angle (divided by 2π) covered by a point of the circle under the action of φ. More formally: we pick a point, apply n iterates of φ, find how many times the point has made the round of the circle, divide by n and pass to the limit.

Theorem 2.4.3 *Let φ and ψ be homeomorphisms of the circle, with rotation numbers ϱ_φ and ϱ_ψ. Suppose that φ and ψ are topologically conjugate by a homeomorphism $\omega\colon S^1 \to S^1$ (i.e., $\varphi \circ \omega = \omega \circ \psi$). If ω preserves orientation then $\varrho_\varphi = \varrho_\psi$; if ω alters orientation, then $\varrho_\varphi + \varrho_\psi = 1$.*

Proof The homeomorphisms φ and ψ either preserve or alter orientation. There is no loss of generality in assuming that they are orientation preserving. Let f, g, h be the respective lifts of φ, ψ, ω to $\boldsymbol{R}$. We may assume that $f \circ h = h \circ g$; moreover,

$$f(x+1) = f(x)+1, \qquad g(x+1) = g(x)+1, \qquad h(x+1) = h(x) \pm 1;$$

in the last equality we have the plus sign if ω preserves orientation and the minus sign otherwise (cf. (2.4.9) and (2.4.10)).

By the definition of the rotation number we have

$$\varrho_f = \lim_{|n| \to \infty} \frac{f^n(y)}{n}, \qquad \varrho_g = \lim_{|n| \to \infty} \frac{g^n(x)}{n} \qquad (y = h(x)).$$

Then either $\varrho_f = \varrho_g = 0$ and there remains nothing to prove, or at least one of these numbers is different from zero, say $\varrho_g \neq 0$. By an appropriate choice of the lift g we may have $\varrho_g > 0$, whence $\lim_{|n| \to \infty} g^n(x) = +\infty$. Since $\lim_{x \to +\infty} h(x)/x = \pm 1$, we obtain, using the identity $f^n = h \circ g^n \circ h^{-1}$,

$$\frac{\varrho_f}{\varrho_g} = \lim_{|n| \to +\infty} \frac{f^n(y)}{g^n(x)} = \lim_{|n| \to +\infty} \frac{h \circ g^n(x)}{g^n(x)} = \pm 1.$$

To see what this means in terms of φ and ψ, write $\varrho_f = \varrho_\varphi + k$, $\varrho_g = \varrho_\psi + l$, $\varrho_\varphi, \varrho_\psi \in \langle 0, 1)$, $k, l \in Z$. If ω is orientation preserving (the plus sign in the last formula), we get $\varrho_\varphi + k = \varrho_\psi + l$, whence $\varrho_\varphi = \varrho_\psi$. In the opposite case, $\varrho_\varphi + k = -\varrho_\psi - l$, whence $\varrho_\varphi + \varrho_\psi = 1$.

Corollary 2.4.1 *Two rotations of the circle are topologically conjugate iff they are either identical or mutually inverse.*

Proof Let φ and ψ be two rotations of S^1, the first by an angle $2\pi\alpha$, the second by an angle $2\pi\beta$, $\alpha, \beta \in \langle 0, 1)$ (we may assume this, because the rotations by the angles ξ and $\xi + 2k\pi$ are, in fact, identical). From the definition of the rotation number it follows that $\alpha = \varrho_\varphi$, $\beta = \varrho_\psi$. Assume that φ and ψ are topologically conjugate. Theorem 2.4.2 then yields either $\alpha = \beta$ or $\alpha + \beta = 1$, and this means exactly that φ and ψ are either equal or inverse rotations.

The proof of the fact that if $\alpha + \beta = 1$ then φ and ψ are topologically conjugate is left to the reader (Exercise 2.4.2). The case where $\alpha = \beta$ requires no proof.

Lemma 2.4.1 *Let φ be an orientation preserving homeomorphism of S^1 and let f be its lift to the real line.*

Consider two subsets of $\boldsymbol{R}$:

$$A = \{n\varrho + m\colon\ n, m \in Z\}, \qquad B = \{f^n(0) + m\colon\ n, m \in Z\}.$$

where ϱ is the rotation number of f. Suppose that ϱ is irrational. Then the assignment

$$n\varrho + m \leftrightarrow f^n(0) + m$$

is a monotone bijection between A and B:

$$n\varrho + m < k\varrho + l \quad \text{iff} \quad f^n(0) + m < f^k(0) + l.$$

Proof Let $p, q, r \in Z$ fulfil the condition

$$p < f^q(0) < r. \tag{2.4.14}$$

We shall show that this implies

$$p < q\varrho < r. \tag{2.4.15}$$

Applying the mapping f^q to the numbers occurring in (2.4.14), we get

$$f^q(p) < f^{2q}(0) < f^q(r),$$

i.e.,

$$f^q(0) + p < f^{2q}(0) < f^q(0) + r.$$

Repeating this argument, we obtain

$$f^q(0) + (k-1)p < f^{kq}(0) < f^q(0) + (k-1)r,$$

whence

$$\frac{1}{k} f^q(0) + \left(1 - \frac{1}{k}\right) p < q \frac{1}{kq} f^{kq}(0) < \frac{1}{k} f^q(0) + \left(1 - \frac{1}{k}\right) r,$$

and so, in the limit $(k \to +\infty)$,

$$p \leqslant q\varrho \leqslant r.$$

Since ϱ is irrational, the inequalities must be strict. This settles (2.4.15).

Now, we have to prove that

$$f^n(0)+m<f^k(0)+l \qquad \text{iff} \qquad n\varrho+m<k\varrho+l.$$

The first of these inequalities is equivalent, via the mapping f^{-k}, to

$$f^{n-k}(0)<l-m,$$

and the second can be rewritten as

$$(n-k)\varrho<l-m.$$

The equivalence between the last two inequalities is an easy consequence of the fact that (2.4.14) implies (2.4.15).

The assignment under consideration is indeed bijective, since ϱ is irrational (i.e., φ has no periodic points).

Theorem 2.4.4 If $\varphi\colon S^1\to S^1$ is an orientation preserving homeomorphism, for which $\Delta=S^1$ (see Theorem 2.4.1 or Exercise 1.11.1), *then φ is topologically conjugate to the rotation by $2\pi\varrho$, where ϱ is the rotation number of φ.*

Proof Let A and B be as in Lemma 2.4.1. Denote by h_0 the bijection considered in the lemma:

$$h_0\colon B\to A, \qquad h_0(f^n(0)+m)=n\varrho+m.$$

The condition: $\Delta=S^1$ implies that every trajectory $\{\varphi^n(z)\colon n\in Z\}$ is dense in S^1 and consequently the set $\{f^n(0) \pmod 1)\colon n\in Z\}$ is dense in $\langle 0, 1)$. Hence B is dense in $\boldsymbol{R}$. As shown in the lemma, h_0 is strictly increasing.

Write

$$r(y)=y+\varrho \qquad \text{for } y\in A.$$

It is easy to verify that

$$h_0(x+1)=h_0(x)+1 \qquad \text{for } x\in B, \tag{2.4.16}$$

$$h_0\circ f(x)=r\circ h_0(x) \qquad \text{for } x\in B. \tag{2.4.17}$$

We define a function $h\colon \boldsymbol{R}\to\boldsymbol{R}$ by

$$h(x_0)=\lim_{\substack{x\to x_0-\\ x\in B}} h_0(x_0).$$

This is correct, since B is dense in $\boldsymbol{R}$ and h_0 is monotone on B. Clearly, h is monotone; it is also continuous, since otherwise it would have jumps, which is impossible because $h(B) = A$ and A is dense in $\boldsymbol{R}$. Consequently, equalities (2.4.16) and (2.4.17) remain valid, with h in place of h_0, for all $x \in \boldsymbol{R}$. Thus h is a lift of a certain homeomorphism of the circle; moreover, h conjugates f, the lift of φ, with r, the lift of the rotation by $2\pi\varrho$. Hence φ and the rotation are topologically conjugate.

Corollary 2.4.2 Every orientation preserving diffeomorphism φ of the circle which has no periodic points and satisfies the condition $\operatorname{Var}\varphi' < +\infty$ *is topologically conjugate to a rotation.*

Proof In virtue of Theorem 2.4.1, the limit set for φ is the entire circle. Hence, by Theorem 2.4.4, φ is topologically conjugate to rotation by the angle $2\pi\varrho$, where ϱ is the rotation number of φ.

Exercises

2.4.1.B. Let $\varphi\colon S^1 \to S^1$ be a homeomorphism for which $\Delta = S^1$, let ψ be the rotation by $2\pi\varrho$, where ϱ is the rotation number of φ, and let f and r denote the respective lifts of φ and ψ to the real line. Show that if φ is not a rotation then there exists exactly one continuous function $h\colon \boldsymbol{R} \to \boldsymbol{R}$ such that $h \circ f = r \circ h$, $h(x+1) = h(x)+1$ and $h(0) = 0$.

Hint. Use the fact that $\{f^n(0) (\operatorname{mod} 1)\colon n \in Z\}$ is a dense subset of $\langle 0, 1)$.

2.4.2.A. Show that the rotations of S^1 by the angles α and $-\alpha$ are topologically conjugate.

Hint. Take $h(x) = -x$ $(x \in \boldsymbol{R})$.

2.5. Differential Equations and Flows on the Torus

The theory of homeomorphisms of the circle is a fundamental tool in the study of differential equations on the torus T^2.

Suppose that we are given a differential equation on the real line

$$\frac{dx}{dt} = f(t, x), \qquad x \in \boldsymbol{R}. \tag{2.5.1}$$

the following conditions being imposed on the right hand side:

1° $f(t, x)$ is a function of class C^1 defined for all $(t, x) \in \boldsymbol{R}^2$,

2° $f(t+1, x) = f(t, x+1) = f(t, x)$ for all $(t, x) \in \boldsymbol{R}^2$.

We will examine equation (2.5.1) jointly with the second (auxiliary) equation

$$\frac{\mathrm{d}x}{\mathrm{d}t} = g(t, y) \equiv 1, \qquad y \in \boldsymbol{R}. \tag{2.5.2}$$

Clearly, g also satisfies conditions 1° and 2°. Thus the vector field (f, g) defined on $\boldsymbol{R}^2$ can be naturally projected onto T^2. Hence the system of equations (2.5.1), (2.5.2) induces a certain differential equation on T^2 – an equation, in fact, of a very special type, because $y = t + \text{const}$. Consequently, we may identify the second variable y with time and regard (2.5.1) as an equation on the torus.

Let $\varphi(t, w)$ be a solution of (2.5.1) satisfying the initial condition $\varphi(0, w) = w$. It is easily checked that

$$\varphi(t, w+1) = \varphi(t, w)+1, \qquad w \in \boldsymbol{R}. \tag{2.5.3}$$

Indeed, the function

$$\psi(t, w) = \varphi(t, w+1) - \varphi(t, w)$$

satisfies the equation

$$\frac{\mathrm{d}\psi(t, w)}{\mathrm{d}t} = f(t, w+1) - f(t, w) = 0,$$

and hence is constant in t; setting $t = 0$, we get $\psi(0, w) = 1$, and (2.5.3) follows.

Write

$$g_t(w) = \varphi(t, w), \qquad t \in \langle 0, 1 \rangle, \ w \in \boldsymbol{R}.$$

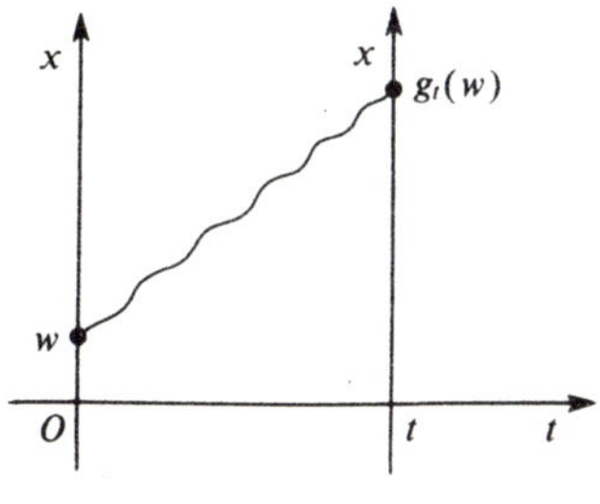

FIG. 2.5.1

The function g_t is continuous, monotone and homotopic to the identity. Thus it is an orientation preserving homeomorphism of the real line onto itself (see Fig. 2.5.1). Since

$$g_t(w+1) = \varphi(t, w+1) = \varphi(t, w)+1 = g_t(w)+1,$$

the map

$$\tilde{\varphi}_t(z) = \exp(2\pi i g_t(w)) \qquad \text{for } z = \exp(2\pi i w)$$

is an orientation preserving homeomorphism of the circle.

Theorem 2.5.1 If the right-hand side of (2.5.1) *satisfies conditions* 1° *and* 2°, *the functions* $\dfrac{\partial f}{\partial x}(t, \cdot)$ *are of uniformly* (*with respect to* t) *bounded variation on* $\langle 0, 1\rangle$ *and if the mappings* $\tilde{\varphi}_t$ (*defined above*) *have no periodic points, then the limit set* Δ *of each* $\tilde{\varphi}_t$ *is the entire circle*: $\Delta = S^1$.

Proof The function $g_t'(w) = \dfrac{\partial}{\partial w}\varphi(t, w)$ satisfies the equation

$$\frac{\mathrm{d}}{\mathrm{d}t}\frac{\partial\varphi(t, w)}{\partial w} = \frac{\partial f}{\partial x}(t, \varphi(t, w))\frac{\partial\varphi}{\partial w}(t, w)$$

with the initial condition $g_0'(w) = 1$. By elementary calculation (which we omit here) we verify that the function

$$\exp\int_0^t \frac{\partial f}{\partial x}(s, \varphi(s, w))\,\mathrm{d}s$$

satisfies the same equation with the same initial condition. Hence

$$g_t'(w) = \exp\int_0^t \frac{\partial f}{\partial x}(s, \varphi(s, w))\,\mathrm{d}s.$$

On account of Theorem 2.4.1, all we need to show is that g_t' is of bounded variation on $\langle 0, 1\rangle$. Define

$$u(w) = \int_0^t \frac{\partial f}{\partial x}(s, \varphi(s, w))\,\mathrm{d}s.$$

If we prove that $u(w)$ is of bounded variation, then the Mean Value Theorem immediately shows that $g_t'(w) = \exp u(w)$ is also of bounded variation.

Let $0 = w_0 < w_1 < \ldots < w_n = 1$ be any partition of $\langle 0, 1\rangle$. Then

$$\sum_{k=0}^{n-1} |u(w_{k+1}) - u(w_k)|$$

$$\leqslant \int_0^t \sum_{k=0}^{n-1} \left| \frac{\partial f}{\partial x}(s, \varphi(s, w_{k+1})) - \frac{\partial f}{\partial x}(s, \varphi(s, w_k)) \right| \mathrm{d}s.$$

In view of the properties of $\varphi(t, w)$, the points $\varphi(s, w_0), \ldots, \varphi(s, w_n)$ define a partition of a certain interval of length 1. By the periodicity of $f(t, x)$ and by the assumption that

$$V := \sup_t \operatorname*{Var}_{0 \leqslant x \leqslant 1} \frac{\partial f}{\partial x}(t, x) < +\infty,$$

we obtain

$$\sum_{k=0}^{n-1} \left| \frac{\partial f}{\partial x}(s, \varphi(s, w_{k+1})) - \frac{\partial f}{\partial x}(s, \varphi(s, w_k)) \right| \leqslant V.$$

Hence

$$\operatorname{Var} u \leqslant tV.$$

Theorem 2.5.2 *Let $\tilde{\varphi}_t$: $S^1 \to S^1$ have the same meaning as in the last theorem and let ϱ denote the rotation number of the homeomorphism $\tilde{\varphi}_1$. If ϱ is irrational, then there exists a homeomorphism h: $T^2 \to T^2$, $h(x, t) = (u, v)$, such that the trajectories of (2.5.1), written additively in the coordinates (u, v), are of the form*

$$v(u, v_0) = (u, v_0 + u\varrho),$$

i.e., they are rectilinear in the covering plane $\boldsymbol{R}^2$.

Proof In virtue of Theorem 2.5.1 and of Corollary 1.4.2, there is a homeomorphism of S^1 whose lift H: $\boldsymbol{R} \to \boldsymbol{R}$ fulfils the identity

$$r \circ H = H \circ g_1\,; \tag{2.5.4}$$

here, as before, $g_t(\cdot) = \varphi(t, \cdot)$, and r denotes the lift of the rotation by

$2\pi\varrho$. Now we introduce new coordinates: through any point (t, x) we draw an integral curve of (2.5.1):

$$x = \varphi(t, x_0),$$

where $x_0 = \varphi(-t, x) = g_{-t}(x)$ and we replace the coordinates (t, x) by (t, x_0); clearly enough, this change of coordinates is defined by a homeomorphism.

Let us write

$$h_1(t, x_0) = (t, H(x_0) + t\varrho).$$

The desired homeomorphism h, or rather its lift to $\boldsymbol{R}^2$, is defined as follows:

$$h(t, x) = h_1(t, \varphi(-t, x)) = h_1(t, g_{-t}(x)),$$

i.e.,

$$h(t, x) = (u, v),$$

where

$$u = t,$$
$$v = H(x_0) + t\varrho = H \circ g_{-t}(x) + t\varrho.$$

In order to check that h indeed defines a homeomorphism of the torus, we have to show that

$$h(t+1, x) = h(t, x+1) = h(t, x) (\mathrm{mod}\ 1).$$

We have

$$\begin{aligned} h(t, x+1) &= (t, H(x_0+1) + t\varrho) = (t, H(x_0) + t\varrho) + (0, 1) \\ &= (t, H(x_0) + t\varrho) (\mathrm{mod}\ 1). \end{aligned}$$

As regards the other equality, it suffices to verify that

$$h(0, x) = h(1, x) (\mathrm{mod}\ 1).$$

Writing $x_0 = g_{-1}(x) = g_1^{-1}(x)$ we get, in view of (2.5.4),

$$\begin{aligned} h(1, x) &= (1, H(x_0) + \varrho) = (1, r \circ H(x_0)) = (1, r \circ H \circ g_1^{-1}(x)) \\ &= (1, H(x)) = (0, H(x)) + (1, 0) = (0, H(x)) \\ &= h(0, x) (\mathrm{mod}\ 1). \end{aligned}$$

Finally, observe that the trajectory $\varphi(t, x_0)$, written in coordinates

(u, v), takes the form

$$h(t, \varphi(t, x_0)) = (u, H(x_0)+u\varrho) = (u, v_0+u\varrho),$$

and this ends the proof.

Remark 2.5.1 If the function f in equation (2.5.1) is of class C^r, $r \geqslant 2$, and if H (in the proof above) is a C^r-diffeomorphism, then also h (the rectifying homeomorphism in Theorem 2.5.2) is a C^r-diffeomorphism.

Remark 2.5.2 In general, little can be said about the regularity of the homeomorphism h. E.g., it is not known whether h must be of class C^1 when f is analytic.

Theorem 2.5.3 (Siegel, 1945, Siegel and Moser, 1971) If $\Phi \in \mathscr{F}^r(T^2)$, $r \geqslant 2$, is a vector field without critical points or closed orbits, then there exists a closed curve $s(t)$ on T^2 such that

$$\dot{s}(t) \upuparrows \Phi(s(t)) \qquad \textit{for every } t;$$

moreover, this curve intersects every trajectory of Φ and does not disconnect the torus.

The curve $s(t)$ is called *Siegel's curve*.

Proof Let $\Phi(x) = (f^1(x), f^2(x))$. Consider the field Ψ orthogonal to Φ at each point, given by

$$\Psi(x) = (-f^2(x), f^1(x)).$$

Ψ is also of class C^r. We will examine the system of equations

$$\frac{d\psi(t, x)}{dt} = \Psi(\psi(t, x)), \qquad \psi(0, x) = x. \tag{2.5.5}$$

The integral curves of (2.5.5) are orthogonal to the trajectories of Φ.

Let $\psi(t, x_0)$ be a fixed solution of (2.5.5). Since T^2 is compact, the set $\omega(x_0)$ is non-empty. Thus there are time instants t_1 and t_2, $t_1 < t_2$, such that the points $\psi(t_1, x_0)$ and $\psi(t_2, x_0)$ lie arbitrarily close to each other.

Since $\Psi(x) \neq 0$ for $x \notin T^2$, we can choose a rectifying chart in which the field trajectories are as in Fig. 2.5.2. The instants t_1 and t_2 can be chosen so that, within the chart, there is no portion of trajectory $\psi(t, x_0)$ corresponding to a time interval in (t_1, t_2) and situated between the two pieces marked in Fig. 2.5.2. It is not difficult to construct

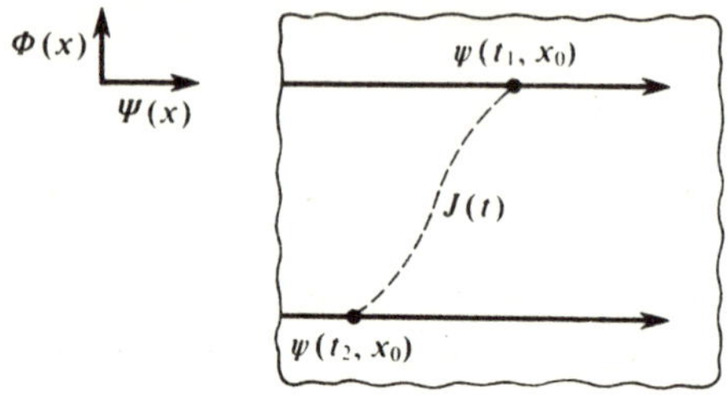

FIG. 2.5.2

a C^r-curve $J(t')$, $0 \leqslant t' \leqslant 1$, (dashed line in Fig. 2.5.2) connecting $\psi(t_1, x_0)$ with $\psi(t_2, x_0)$, tangent to $\psi(t, x_0)$ at these points, and such that the vector $\dot{J}(t')$ is not parallel to $\Phi(J(t'))$ at any point $t' \in (0, 1)$.

We define the curve $s(t)$:

$$s(t) = \begin{cases} \psi(t_1 + t, x_0) & \text{for } 0 \leqslant t \leqslant t_2 - t_1, \\ J(t - t_2 + t_1) & \text{for } t_2 - t_1 \leqslant t \leqslant t_2 - t_1 + 1. \end{cases}$$

Obviously, $s(t)$ is of class C^r; the property

$$\dot{s}(t) \nparallel \Phi(s(t)) \qquad \text{for } 0 \leqslant t \leqslant t_2 - t_1 + 1$$

does not require a proof, either.

We now show that $s(t)$ does not disconnect the torus. Assume the contrary: then one of the components into which $s(t)$ splits T^2 is a disc; denote it by D. On the boundary of D the vector field Φ is directed either permanently inside or permanently outside D. Hence, we have for $x \in D$ either $\omega(x) \subset D$ or $\alpha(x) \subset D$. In virtue of Theorem 2.2.3 (Poincaré–Bendixon–Schwartz) one of the sets $\omega(x)$, $\alpha(x)$ contains a critical point of Φ or is a closed orbit. Both these conclusions contradict the assumption imposed on Φ.

According to Theorem 2.2.1 (Schwartz) the flow φ_t is minimal (has no critical elements). Thus every trajectory approaches the curve $s(t)$ arbitrarily close. Since $s(t)$ is transversal to the field trajectories, every trajectory necessarily intersects $s(t)$ – and this is all that remained to be proved.

Exercises

2.5.1.B. Prove Theorem 2.5.3 for fields $\Phi \in \mathscr{F}^1(T^2)$.

Hint. The proof given above has to be modified as follows: firstly, choose x_0 to be an element of some set $\omega(y)$; secondly, use the fact that if φ is a homeomorphism of the circle then its limit set $\Delta \subset S^1$ is a minimal set (cf. Exercise 1.11.1).

2.5.2.B. Show that there exists a field $\Phi \in \mathscr{F}^r(T^2)$, $r \geqslant 2$, without critical points, admitting no Siegel's curve.

Hint. Consider a field on the torus having two closed orbits γ_1, γ_2, sensed reciprocally and such that for any point $x \notin \hat{\gamma}_1 \cup \hat{\gamma}_2$ the sets $\alpha(x)$ and $\omega(x)$ coincide with $\hat{\gamma}_1$ and $\hat{\gamma}_2$, respectively. Use the fact that each curve on T^2 has two "sides".

Now suppose that we are given a differential equation

$$\frac{dx}{dt} = \Phi(x)$$

on the torus:

$$\Phi = (f^1, f^2) \in \mathscr{F}^r(T^2), \qquad r \geqslant 2,\ x = (x^1, x^2) \in T^2;$$

suppose that Φ has no closed orbits or critical points. Written coordinatewise, this equation turns into the system of two equations

$$\frac{dx^1}{dt} = f^1(x^1, x^2), \qquad \frac{dx^2}{dt} = f^2(x^1, x^2). \tag{2.5.6}$$

Lemma 2.5.1 There exists a coordinate system $y = (y^1, y^2)$ on T^2, C^r-diffeomorphic with $x = (x^1, x^2)$, in which equations (2.5.6) *take the form*

$$\frac{dy^1}{dt} = g^1(y^1, y^2), \qquad \frac{dy^2}{dt} = g^2(y^1, y^2),$$

where

$$g^1(y^1, y^2) > 0 \qquad \textit{for every } y = (y^1, y^2) \in T^2.$$

Proof Let s be Siegel's curve for the system (2.5.6). We will investigate the system in the covering plane $\boldsymbol{R}^2$. The lift $\pi^{-1}(s)$ contains no closed curve because s does not disconnect T^2. Let s_0 and s_1 denote two

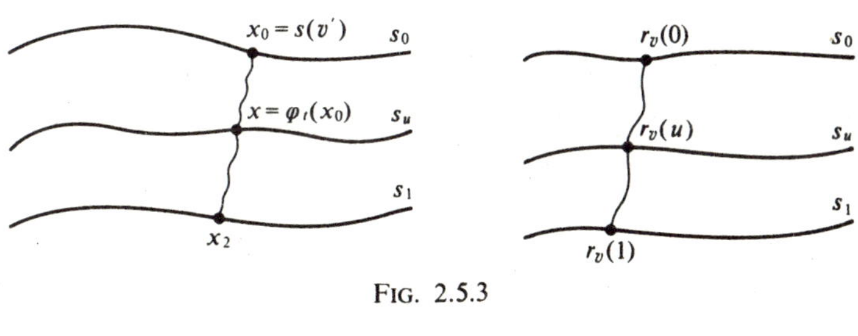

FIG. 2.5.3

adjacent components of $\pi^{-1}(s)$, i.e., such that the area D limited by s_0 and s_1 contains no points of $\pi^{-1}(s)$ (see Fig. 2.5.3). The curves s_0 and s_1 are unbounded and result from each other by translation. Without loss of generality we may assume that on s_0 the field Φ is directed inside D. The closure of D projected onto T^2 covers the entire torus.

Let $x \in D$. Denote by x_0 the point of the curve s_0 such that $x = \varphi(t, x_0)$ for some t; then $x_0 = s_0(v')$. In this way we define a mapping

$$x \mapsto (t, v')$$

of class C^r. Now, let $l(v')$ be the length of the portion of the trajectory $\varphi(t, x)$ contained between x_0 and x_1, the earliest point of intersection with s_1. Define

$$u(x) = \frac{1}{l(v')} \text{length}\{\varphi(\bar{t}, v'):\ 0 \leqslant \bar{t} \leqslant t\}$$

and

$$s_u = \{x \in D:\ u(x) = u\}, \qquad 0 \leqslant u \leqslant 1.$$

Further, we construct curves $r_v(u)$, $0 \leqslant u \leqslant 1$, with the property that each of them crosses every s_u at exactly one point and connects the points on s_0 and s_1 corresponding to the same point on the torus: $\pi(r_v(0)) = \pi(r_v(1))$. The geometrical idea of the construction is the following: by a suitable C^r-diffeomorphism H we transform the curves s_u into straight lines (i.e., the area D into a rectilinear strip), we join the corresponding points of the lines $H(s_0)$ and $H(s_1)$ by straight line segments, and finally we apply H^{-1}: the images of those segments are the desired curves $r_v(u)$. For a fixed x let $v(x)$ be the value of parameter v for which x lies on r_v; then $x = r_{v(x)}(u(x))$.

Write $y^1(x) = u(x)$, $y^2(x) = v(x)$. The change of variables $x \mapsto y$ is of class C^r and fulfils the condition of the lemma. We omit detailed calculation.

Theorem 2.5.4 *Let $\Phi \in \mathscr{F}^r(T^2)$, $r \geqslant 2$, be a vector field without critical points or closed orbits and let φ_t be the flow generated by Φ. Then there exists a coordinate system in which the trajectories of φ_t are rectilinear.*

Proof The flow φ_t is given by the system of equations (2.5.6). In view of Lemma 2.5.1 we may assume that $f^1(x^1, x^2) > 0$. Dividing the second equation of (2.5.6) by the first we obtain

$$\frac{dx^2}{dx^1} = \frac{f^2(x^1, x^2)}{f^1(x^1, x^2)}; \tag{2.5.7}$$

the quotient on the right is of class C^r, since so is the variable change in Lemma 2.5.1. Then by Theorem 2.5.2 there is a coordinate system on T^2 in which the trajectories of (2.5.7), lifted to $\boldsymbol{R}^2$, are straight lines. This proves our theorem, since the trajectories of (2.5.6) and (2.5.7) coincide.

Corollary 2.5.1 The phase portrait of a field $\Phi \in \mathscr{F}^r(T^2)$, $r \geqslant 2$, without critical elements is equivalent to the phase portrait of a certain constant vector field on the torus.

Corollary 2.5.2 If the homeomorphism h in Theorem 2.5.2 is of class C^r, $r \geqslant 2$, then the system of equations (2.5.6) can be reduced by a suitable change of variables (of class C^r) to the system

$$\frac{dy^1}{dt} = G(y^1, y^2), \qquad \frac{dy^2}{dt} = \varrho G(y^1, y^2), \tag{2.5.8}$$

where ϱ is a constant.

This is an immediate consequence of Remark 2.5.1 and the proof of Theorem 2.5.4. The rectilinearity of the trajectories of (2.5.7) means just that $f^2(x^1, x^2)/f^1(x^1, x^2) = \text{const}$. Hence the form (2.5.8).

Exercise

2.5.3.B. Let $U \subset \boldsymbol{R}^2$ be a simply connected closed domain and let $\Phi = (f^1, f^2) \in \mathscr{F}^r(U)$, $r \geqslant 1$, $\Phi(x) \neq 0$ for $x \in U$. Consider the orthogonal vector field $\Psi = (-f^2, f^1)$. Show that there is a coordinate system in U in which the trajectories of Φ and Ψ are mutually perpendicular straight lines.

Hint. The trajectories themselves can be taken as the coordinate axes (and their translates); the coordinate values are to be defined as the transition times along the trajectories.

2.6. Algebraic Automorphisms of the Torus

In Chapter I (Exercise 1.2.4) we introduced the concept of an algebraic automorphism of the torus T^2. In this section we study the properties of such an automorphism in more detail.

The automorphism $\varphi\colon T^2 \to T^2$ given by a matrix

$$\varphi = \begin{bmatrix} a & b \\ c & d \end{bmatrix},$$

where $a, b, c, d \in Z$, $\det\varphi = \pm 1$, is said to be *ergodic* iff neither of the eigenvalues of φ is a root of unity.

Let λ and μ be the eigenvalues of φ. These values are real and we have $\lambda\mu = \pm 1$ (see Exercise 1.2.5). E.g., let $|\lambda| > 1$, $|\mu| < 1$, and let u and v be the corresponding eigenvectors:

$$\varphi(u) = \lambda u, \qquad \varphi(v) = \mu v.$$

The lines: $\boldsymbol{a} = \{tu\colon t \in \boldsymbol{R}\}$, $\boldsymbol{b} = \{tv\colon t \in \boldsymbol{R}\}$ decompose the torus into two parallelograms, which will be denoted by U and V. Fig. 2.6.1 shows how U and V are obtained; U is the shaded part of the unit square, V is the remaining part.

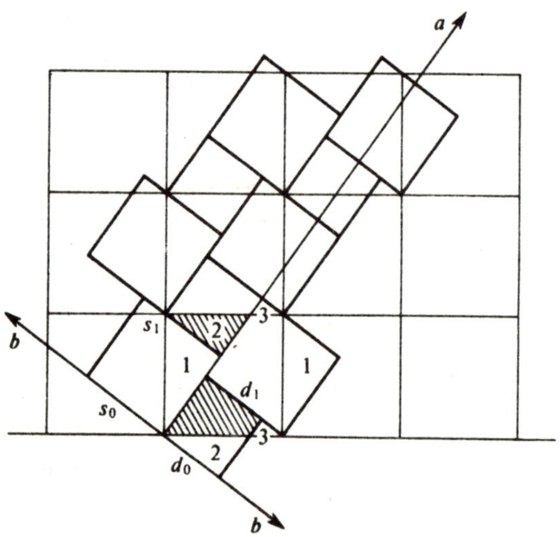

FIG. 2.6.1

Shifting U and V by integer vectors we cover $\boldsymbol{R}^2$ by a network Ξ of parallelograms isometric to U or V. The sides of U and V parallel to v (sides s_1, s_0, d_1, d_0 in Fig. 2.6.1) are called *contracted* sides; those parallel to u are called *expanded* sides.

Let us look at $\varphi(U)$ and $\varphi(V)$. We see that:

1° $\varphi(d_1) \subset d_1 \cup s_1$, $\varphi(s_1) \subset d_1 \cup s_1$, because φ takes integer points into integer points;

2° $\varphi(U)$ and $\varphi(V)$ do not contain any vertex of the network Ξ – otherwise $\varphi(U)$ and $\varphi(V)$ would have some interior points in common.

Denote by A_i, $i = 1, \ldots, k_1$, the components of $U \cap \varphi(U)$; by B_i, $i = 1, \ldots, k_2$, the components of $U \cap \varphi(V)$; by C_i, $i = 1, \ldots, k_3$, the components of $V \cap \varphi(U)$; by D_i, $i = 1, \ldots, k_4$, the components of $V \cap \varphi(V)$. All the sets A_i, B_i, C_i, D_i are parallelograms with sides parallel to u and v; the sides parallel to v are contained in the contracted sides of U and V and their translates (see Fig. 2.6.2). Let

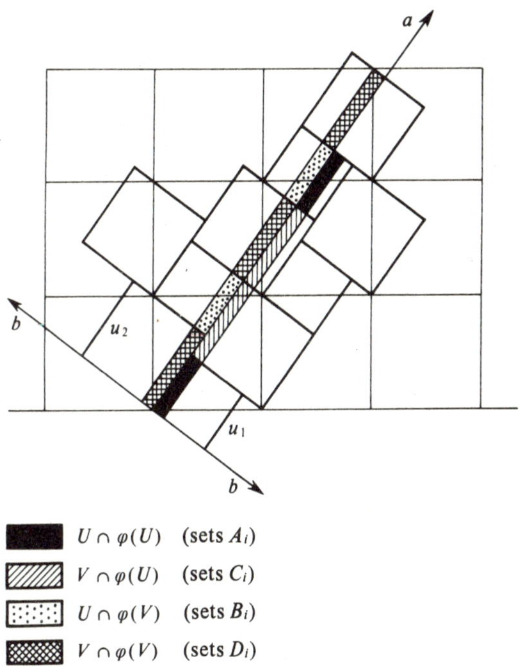

FIG. 2.6.2

X_i, Y_j be any two members of the set families $\{\text{int } A_i\}$, $\{\text{int } B_i\}$, $\{\text{int } C_i\}$, $\{\text{int } D_i\}$. The interrelation between those sets is put down in the form of the following "table of transitions", in which we write the symbol 1 iff $X_i \cap \varphi(Y_j) \neq \emptyset$ and 0 otherwise:

	A_i	B_i	C_i	D_i
$\varphi(A_i)$	1	0	1	0
$\varphi(B_i)$	1	0	1	0
$\varphi(C_i)$	0	1	0	1
$\varphi(D_i)$	0	1	0	1

Definition 2.6.1 The decomposition of T^2 into the sets A_i, B_i, C_i, D_i is called the *Markov partition* (cf. Section 4.12).

The construction described above has been invented by Adler and Weiss (1967).

It is not at all obvious why the matrix of transitions should be the one given here. We invite the reader to reason it out.

We see from the table that

$$\operatorname{int} A_i \cap \varphi(\operatorname{int} A_i) \neq \emptyset$$

for every i. On account of 1° and 2° this leads to a situation as shown in Fig 2.6.3. Since the contracted sides of $\varphi^n(A_i)$ shorten to zero, the set

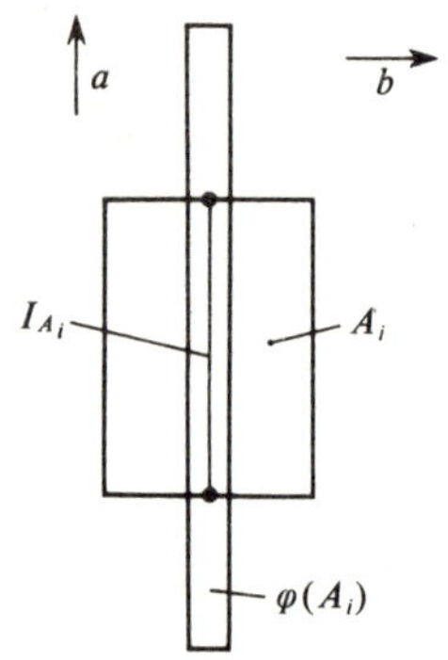

FIG. 2.6.3

$\bigcap_{n=0}^{\infty} \varphi^n(A_i)$ is a segment parallel to u. Denote it by I_{A_i}. Evidently,

$$\varphi^{-1}(I_{A_i}) = \bigcap_{n=-1}^{\infty} \varphi^n(A_i) \subset I_{A_i},$$

and thus there is at least one fixed point in I_{A_i}. Since φ^{-1} is a contraction on I_{A_i} (in the ratio $|\lambda|^{-1}$), the fixed point is unique. It is an interior point of A_i, for the sides of A_i lie on the lines $\boldsymbol{a} = \{tu\}$ and $\boldsymbol{b} = \{tv\}$, and neither of these lines contains any fixed point except zero.

The same argument shows that also every D_i contains exactly one fixed point of φ in its interior. Clearly, in the sets B_i or C_i there are no fixed points.

We now show that

$$k_1 + k_4 = |a+d| = |\operatorname{tr} \varphi|. \tag{2.6.1}$$

(Recall that a and d are two of the entries of the integer matrix defining φ; tr is an abbreviation for: trace.)

The proof is carried out for the case of $\lambda > 0$; if $\lambda < 0$, the reasoning is similar. Let u_1 and u_2 denote the lengths of the expanded sides of U and V. Then (see Fig. 2.6.2) we have

$$\lambda u_1 = k_1 u_1 + k_3 u_2, \qquad \lambda u_2 = k_2 u_1 + k_4 u_2.$$

The homogeneous system of linear equations

$$u_1(k_1 - \lambda) + u_2 k_3 = 0, \qquad u_1 k_2 + u_2(k_4 - \lambda) = 0 \tag{2.6.2}$$

has a non-zero solution (u_1, u_2), and so

$$\det\begin{bmatrix} k_1 - \lambda & k_3 \\ k_2 & k_4 - \lambda \end{bmatrix} = 0,$$

i.e.,

$$\lambda^2 - (k_1 + k_4)\lambda + \det\begin{bmatrix} k_1 & k_3 \\ k_2 & k_4 \end{bmatrix} = 0. \tag{2.6.3}$$

On the other hand, λ is an eigenvalue of φ; thus

$$\lambda^2 - (a+d)\lambda + \det\varphi = 0. \tag{2.6.4}$$

λ is irrational (else $\lambda = \pm 1$); hence the coefficients in equations (2.6.3) and (2.6.4) are equal, and so we have $a + d = k_1 + k_4$.

Corollary 2.6.1 Let $C(\varphi^n)$ denote the number of fixed points of the mapping φ^n. Then

$$\lim_{n\to\infty} \frac{1}{n} \log C(\varphi^n) = \log|\lambda|.$$

Proof It follows from the foregoing considerations that each of the parallelograms A_i and D_i contains exactly one fixed point in its interior; there is one more fixed point, namely, the point 0. Thus

$$C(\varphi) = 1 + k_1 + k_4 = 1 + |\mathrm{tr}\,\varphi| = 1 + |\lambda + \mu|.$$

Repeating this procedure with φ^n in place of φ we obtain:

$$C(\varphi^n) = 1 + |\mathrm{tr}\,\varphi^n| = 1 + |\lambda^n + \mu^n|,$$

whence

$$\lim_{n\to\infty} \frac{1}{n} \log C(\varphi^n) = \lim_{n\to\infty} \frac{1}{n} \log(1 + |\lambda^n + \mu^n|) = \log|\lambda|.$$

Lemma 2.6.1 *Let W denote the set of all points $x \in T^2$ such that for every $n \in Z$*

$$\varphi^n(x) \in \bigcup_i \operatorname{int} A_i \cup \bigcup_i \operatorname{int} B_i \cup \bigcup_i \operatorname{int} C_i \cup \bigcup_i \operatorname{int} D_i.$$

Let $x, y \in W$ and suppose that, for every n, the points $\varphi^n(x)$ and $\varphi^n(y)$ are in the same parallelogram X_{i_n} (one of the A_i, B_i, C_i or D_i). Then $x = y$.

Proof We will take the eigenvectors u and v as the basis of a new coordinate system. For any point $x \in T^2$ denote by x_u, x_v the coordinates of x relative to that basis.

Let x and y fulfil the condition of the lemma and suppose that $x_u \neq y_u$. Let r denote the greatest of the diameters of all the sets A_i, B_i, C_i, D_i. Choose n_0 such that

$$|x_u - y_u| > \frac{r}{|\lambda|^{n_0}}. \tag{2.6.5}$$

By assumption, $\varphi^{n_0}(x)$ and $\varphi^{n_0}(y)$ are in the same parallelogram $X_{i_{n_0}}$; thus

$$|(\varphi^{n_0}(x))_u - (\varphi^{n_0}(y))_u| < r.$$

Hence

$$|(\varphi^{n_0-1}(x))_u - (\varphi^{n_0-1}(y))_u| < \frac{r}{|\lambda|},$$

and, by induction,

$$|x_u - y_u| = |(\varphi^0(x))_u - \varphi^0(y))_u| < \frac{r}{|\lambda|^{n_0}},$$

contrary to (2.6.5). Thus $x_u = y_u$.

Similarly, supposing that $x_v \neq y_v$, we can repeat the argument with φ replaced by φ^{-1} and λ by μ.

Now let $X_1, \ldots, X_m$ be an enumeration of the sets $\operatorname{int} A_i$, $\operatorname{int} B_i$, $\operatorname{int} C_i$, $\operatorname{int} D_i$ and let $\mathfrak{M}$ be the index set $\{1, \ldots, m\}$.

Further, let t denote the mapping

$$t\colon W \to \prod_{n=-\infty}^{+\infty} \mathfrak{M}_n, \qquad \mathfrak{M}_n = \mathfrak{M} \qquad \text{for all } n \in Z,$$

which assigns to each point $x \in W$ the two-tailed sequence

$$t(x) = (k_n)_{n=-\infty}^{+\infty},$$

whose n-th entry k_n is an index such that $\varphi^n(x) \in X_{k_n}$. As a consequence of Lemma 2.6.1 we obtain:

Corollary 2.6.2 *The mapping t is an injection of W into* $\prod_{n=-\infty}^{+\infty} \mathfrak{M}_n$.

We will give a probabilistic interpretation of the above considerations. Call the sets $X_1, \ldots, X_m$ states and regard the normalized Lebesgue measure on T^2 as probability. Denote by $P(Y_j \mid X_i)$ the probability of transition from state X_i to state Y_j; these transition probabilities are given by the following matrix (cf. Fig. 2.6.2):

$X_i \backslash Y_j$	A_j	B_j	C_j	D_j
A_i	$\frac{1}{\lvert\lambda\rvert}$	0	$\frac{u_2}{\lvert\lambda\rvert u_1}$	0
B_i	$\frac{1}{\lvert\lambda\rvert}$	0	$\frac{u_2}{\lvert\lambda\rvert u_1}$	0
C_i	0	$\frac{u_1}{\lvert\lambda\rvert u_2}$	0	$\frac{1}{\lvert\lambda\rvert}$
D_i	0	$\frac{u_1}{\lvert\lambda\rvert u_2}$	0	$\frac{1}{\lvert\lambda\rvert}$

The probability of transition from X_i to Y_j depends only on the state in which x is at the moment, and not on the previous history of x. Thus the dynamical system (W, φ) can be viewed as a Markov chain with states X_i and the transition matrix $(P(Y_j \mid X_i))$, with the initial distribution $|X_1|, \ldots, |X_m|$ ($|\cdot|$ denoting Lebesgue measure). Hence comes the name: *Markov partition.*

Finally, we present a theorem on topological conjugacy of torus automorphisms.

Theorem 2.6.1 (*Arov, 1963*) *Let φ and ψ be algebraic automorphisms of T^2. Suppose that φ and ψ are topologically conjugate. Then there exists an algebraic automorphism A: $T^2 \to T^2$ such that*

$$\varphi \circ A = A \circ \psi.$$

Proof The assumption that φ and ψ are topologically conjugate on T^2 implies that they are also topologically conjugate on $\boldsymbol{R}^2$, up to an integer shift; that is to say, there are a homeomorphism h: $\boldsymbol{R}^2 \to \boldsymbol{R}^2$

and an integer vector $r \in Z^2 = Z \times Z$ such that

$$\varphi \circ h(x) = h \circ \psi(x) + r \qquad \text{for } x \in \boldsymbol{R}^2.$$

(See Fig. 2.6.4; Q denotes the unit square, which we identify with the torus.)

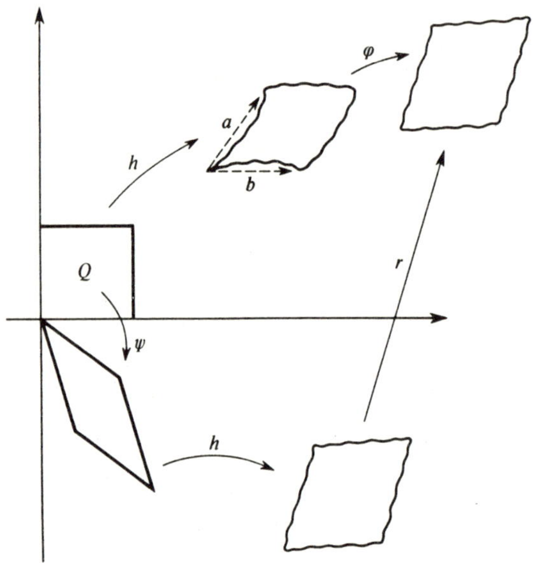

FIG. 2.6.4

Let k be any integer vector. Then

$$h(x+k) = h(x) + A(k, x),$$

where $A(k, x) \in Z^2$ and $A(k, x)$ depends continuously on x. Hence

$$A(k, x) = A(k) = \text{const} \qquad \text{with respect to } x.$$

Since h is a lift of a torus homeomorphism, we have

$$h(x \oplus Z^2) = h(x) \oplus Z^2,$$

and so $A\colon Z^2 \to Z^2$ is surjective.

We will show that A is linear in k. Let $k, l \in Z^2$. Since

$$h(x+k+l) = h(x+k) + A(l) = h(x) + A(k) + A(l)$$

and

$$h(x+k+l) = h(x) + A(k+l),$$

it follows that

$$A(k+l) = A(k)+A(l).$$

Consequently A is given by a matrix with integer entries. As observed, A is a surjection of Z^2; thus the determinant of that matrix is equal to ± 1. This means that A is an algebraic automorphism of T^2.

It remains to show that A conjugates φ with ψ.

Choose $x \in \boldsymbol{R}^2$ and $k \in Z^2$ arbitrarily. The following equalities are fulfilled:

$$\varphi\big(h(x)+k\big) = h\big(\psi(x+k)\big)+r = h\big(\psi(x)+\psi(k)\big)+r,$$
$$\varphi\big(h(x)+A(k)\big) = \varphi \circ h(x)+\varphi \circ A(k) = \varphi\big(h(x+k)\big),$$
$$h \circ \psi(x)+A \circ \psi(k)+r = h\big(\psi(x)+\psi(k)\big)+r.$$

Hence

$$\varphi \circ A(k) = A \circ \psi(k) \qquad \text{for all } k \in Z^2.$$

The two mappings: $\varphi \circ A$ and $A \circ \psi$ are linear; the last equality shows that they coincide, in particular, on the elements of a basis for $\boldsymbol{R}^2$. Thus they are identical; the proof is complete.

3

Transversality Theory and Generic Properties

3.1. Survey of Facts and Notions of Functional Analysis

The study of generic properties requires the use of several non-trivial facts of functional analysis. Customarily, those facts are dealt with in standard courses. To acquaint the reader with them is the aim of the first section in this chapter.

We omit proofs; they can be found in Dieudonné (1960).

Let X and Y be Banach spaces. This notation is kept throughout the section. The space of all continuous linear mappings from X to Y is denoted by $\mathscr{L}(X, Y)$. This is also a Banach space with the usual operator norm.

Definition 3.1.1 A mapping $f\colon X \to Y$ is said to be *differentiable* at a point $x_0 \in X$ iff there exists an operator $A_{x_0} \in \mathscr{L}(X, Y)$ such that

$$\lim_{h \to 0} \frac{\|f(x_0 + h) - f(x_0) - A_{x_0}(h)\|}{\|h\|} = 0.$$

The operator A_{x_0} is called the *derivative* of f at x_0 and is denoted by the symbol $\mathrm{d}f(x_0)$, and its evaluation by $\mathrm{d}f(x_0)h$ (for an increment $h \in X$).

The definition makes sense also if f is defined on some neighbourhood of x_0.

Examples

3.1.1. Let $X = l^2$, $Y = \boldsymbol{R}$, $f(x) = \sum_{n=1}^{\infty} \xi_n^3$ for $x = (\xi_n) \in l^2$. Then we have for $x_0 = (\xi_n)$, $h = (h_n)$

$$\begin{aligned} f(x_0+h)-f(x_0) &= \sum_{n=1}^{\infty} ((\xi_n+h_n)^3-\xi_n^3) \\ &= 3\sum_{n=1}^{\infty} \xi_n^2 h_n + \sum_{n=1}^{\infty} (3\xi_n h_n^2 + h_n^3) \\ &= \mathrm{d}f(x_0)h + R(h). \end{aligned}$$

It is easy to see that $\|R(h)\|/\|h\| \to 0$ as $h \to 0$.

3.1.2. Let $A \in \mathscr{L}(X, Y)$. Then $\mathrm{d}A(x_0) = A$ for every $x_0 \in X$.

3.1.3. Let $X = C\langle 0, 1\rangle = Y$, $f(x)(t) = \int_0^1 (t+s)\,x^2(s)\,\mathrm{d}s$ for $x \in C\langle 0, 1\rangle$. Then

$$(\mathrm{d}f(x)\,h)(t) = 2\int_0^1 (t+s)\,x(s)\,h(s)\,\mathrm{d}s \qquad \text{for } h \in C\langle 0, 1\rangle.$$

3.1.4. Let F be a continuous symmetric bilinear map of $X \times X$ into Y. Write $f(x) = F(x, x)$. Then

$$\mathrm{d}f(x_0)h = 2F(x_0, h).$$

Indeed,

$$f(x_0+h)-f(x_0) = F(x_0+h, x_0+h) - F(x_0, x_0) = 2F(x_0, h) + F(h, h)$$

and [33]

$$\frac{\|F(h, h)\|}{\|h\|} \leqslant \frac{C\,\|h\|^2}{\|h\|} \to 0 \qquad \text{as } h \to 0.$$

3.1.5. Let F be a continuous symmetric n-linear map of $\underbrace{X \times \dots \times X}_{n \text{ times}}$ to Y. Write $f(x) = F(x, \dots, x)$. (Any function of this shape is termed an *n-homogeneous polynomial*; cf. Definition 3.1.2). As in Example 3.1.4, it can be shown that

$$\mathrm{d}f(x)h = nF(\underbrace{x, \dots, x}_{n-1 \text{ times}}, h).$$

Exercises

3.1.1.B. Show that the function $f(x) = \|x\|, f\colon C\langle 0, 1\rangle \to \boldsymbol{R}$, is not differentiable at any point $x \in C\langle 0, 1\rangle$.

Hint. The directional derivatives do not exist.

3.1.2.B. Examine the differentiability of $f\colon U \to \boldsymbol{R}$,

$$U = \{x\colon 0 < x(t) < 1\} \subset C\langle 0, 1\rangle, f(x) = x(x(1/2)).$$

Hint. Assume that x is of class C^1. The directional derivatives exist; nevertheless, the derivative in the sense of Definition 3.1.1 does not exist.

3.1.3.B. Calculate the derivative of the mapping $f\colon C\langle 0, 1\rangle \to C\langle 0, 1\rangle$ given by

$$f(x)(t) = \int_0^1 (t+s)e^{x(s)}ds.$$

Result: $\big(df(x)h\big)(t) = \int_0^1 (t+s)\,h(s)\,ds.$

Theorem 3.1.1 Let X, Y, W be Banach spaces, U an open subset of X, $f\colon U \to Y$, $x_0 \in U$, $y_0 = f(x_0)$, V a neighbourhood of y_0 in Y, $g\colon V \to W$. If f is differentiable at x_0 and g is differentiable at y_0, then the map $h = g \circ f\colon U \to W$ is differentiable at x_0 and

$$dh(x_0) = dg(y_0) \circ df(x_0).$$

Corollary 3.1.1 In notation as above, if g is linear, then

$$dh(x_0) = g \circ df(x_0).$$

Let $U \subset X$ be an open set and let $f\colon U \to Y$ be a mapping differentiable at each point of U. Assigning to every point $x \in U$ the derivative $df(x)$, we obtain a mapping from U to $\mathscr{L}(X, Y)$ (called the *derivative* of f). If this mapping is differentiable at $x_0 \in U$, we say that f is twice differentiable at x_0. The derivative of df at x_0,

$$\big(d(df)\big)(x_0)$$

is denoted by $d^2f(x_0)$. Clearly,

$$d^2f(x_0) \in \mathscr{L}\big(X, \mathscr{L}(X, Y)\big) \cong \mathscr{L}^2(X, Y)$$

(see the definition after Exercise 3.1.4), and thus $\mathrm{d}^2 f(x_0)$ can be naturally identified with a bilinear mapping from $X \times X$ into Y.

Theorem 3.1.2 *If f is twice differentiable at x_0, the bilinear mapping $\mathrm{d}^2 f(x_0)$ is symmetric:*

$$\mathrm{d}^2 f(x_0)(h, k) = \mathrm{d}^2 f(x_0)(k, h) \qquad \text{for } h, k \in X.$$

Theorem 3.1.3 (*Mean Value Theorem*) *If f: $U \to Y$ is differentiable at each point of an open convex set $U \subset X$, then for any $a, b \in U$ we have*

$$\|f(b) - f(a)\| \leqslant \|b - a\| \sup_{\xi \in I} \| \mathrm{d} f(\xi)\|,$$

I denoting the segment joining a with b.

Definition 3.1.2 A mapping f: $X \to Y$ is called an *n-homogeneous polynomial* iff there is an n-linear continuous mapping

$$F\colon \underbrace{X \times \ldots \times X}_{n \text{ times}} \to Y$$

such that

$$f(x) = F(x, \ldots, x).$$

We may always assume that F is symmetric; if it is not, we replace it by

$$\tilde{F}(x_1, \ldots, x_n) := \frac{1}{n!} \sum F(x_{p_1}, \ldots, x_{p_n}),$$

summation ranging over all permutations $(p_1, \ldots, p_n)$ of $(1, \ldots, n)$.

Examples

3.1.6. Let $X = C\langle 0, 1\rangle = Y$. The map f given by: $f(x)(t) = x^n(t)$ for $x \in C\langle 0, 1\rangle$ is an n-homogeneous polynomial.

3.1.7. Let $X = \boldsymbol{R}^m$, $Y = \boldsymbol{R}$. The function

$$f(x) = \sum_{i,j=1}^{m} a_{ij} x^i x^j \qquad \text{for } x = (x^1, \ldots, x^m)$$

is a 2-homogeneous polynomial.

The norm of an n-homogeneous polynomial f is defined by

$$\|f\| = \sup_{\|x\| \leqslant 1} \|f(x)\|.$$

Obviously, this number is not larger than

$$\|F\| = \sup_{\substack{\|x_1\| \leqslant 1 \\ \cdots\cdots \\ \|x_n\| \leqslant 1}} \|F(x_1, \ldots, x_n)\|.$$

Exercise

3.1.4*.B. Let f and F be as above, F symmetric. Show that

$$\|F\| \leqslant \frac{n^n}{n!} \|f\|$$

and prove that the constant $n^n/n!$ is the best possible (Kopeć and Musielak, 1956).

Hint. Put $x = \varepsilon_1 x_1 + \ldots + \varepsilon_n x_n$, where x_j are any vectors of norm 1 and ε_j are any combination of the numbers ± 1 and $\pm \mathrm{i}$.

Now we introduce the regularity classes C^k, $k \geqslant 1$. We will denote by $\mathscr{L}^k(X, Y)$ the space of all continuous k-linear mappings from $X \times \ldots \times X$ (k times) into Y and by $\mathscr{L}^k_s(X, Y)$ the subspace of $\mathscr{L}^k(X, Y)$ consisting of all symmetric maps.

Definition 3.1.3 A mapping f: $U \to Y$, $U \subset X$ open, is of class C^0 iff f is continuous at each point. Suppose that we have already defined mappings of class C^k and the k-th order derivative $\mathrm{d}^k f$. We say that f is of class C^{k+1} iff:

1° f is of class C^k,

2° the map $\mathrm{d}^k f$: $U \to \mathscr{L}^k_s(X, Y)$ is differentiable at every point $x \in U$ and its derivative

$$\mathrm{d}^{k+1} f := \mathrm{d}(\mathrm{d}^k f)$$

is continuous.

Just as in the case of the second order derivative $\mathrm{d}^2 f$, the k-th derivative $\mathrm{d}^k f(x)$ can be regarded as a k-linear operation from $\underbrace{X \times \ldots \times X}_{k \text{ times}}$ into Y:

$$\mathrm{d}^k f(x)(h_1, \ldots, h_k) \in Y \qquad \text{for } h_i \in X,\ i = 1, \ldots, k.$$

If $h_1 = \ldots = h_k = h$, we simply write $\mathrm{d}^k f(x) h^k$ in place of $\mathrm{d}^k f(x)(h, \ldots, h)$.

Example

3.1.8. Let $f\colon X \to Y$ be an n-homogeneous polynomial, $f(x) = F(x, \ldots, x)$, F being symmetric n-linear. Then

$$\mathrm{d}^k f(x)(h_1, \ldots, h_k) = n(n-1)\ldots(n-k+1)F(\underbrace{x, \ldots, x}_{n-k \text{ times}}, h_1, \ldots, h_k). \tag{3.1.1}$$

This formula is proved by induction on k (for $k = 1$, see Example 3.1.5 on p. 127).

Corollary 3.1.2 If f and F are as above, then

$$\mathrm{d}^n f(x)(h_1, \ldots, h_n) = n!\,F(h_1, \ldots, h_n),$$
$$\mathrm{d}^{n+k} f(x)(h_1, \ldots, h_n) = 0 \qquad \textit{for } h_1, \ldots, h_n \in X.$$

We now define partial derivatives. Let X_1, X_2, Y be Banach spaces and let $f\colon U_1 \times U_2 \to Y, f(x_1, x_2) = y$, be a differentiable mapping, U_i being an open subset of X_i, $i = 1, 2$. By fixing elements $a \in U_2$, $b \in U_1$ we obtain mappings $f(x_1, a)$ and $f(b, x_2)$ leading, respectively, from U_1 and U_2 into Y. Denote these mappings by $f(\cdot, a)$ and $f(b, \cdot)$.

Definition 3.1.4 The derivatives of $f(\cdot, a)$ and $f(b, \cdot)$ (for a and b fixed) are called *partial derivatives* of the map f and are denoted by $\partial_1 f$ and $\partial_2 f$, respectively.

Clearly, for $x_1 \in U_1$, $x_2 \in U_2$ we have $\partial_i f(x_1, x_2) \in \mathscr{L}(X_i, Y)$ for $i = 1, 2$.

Theorem 3.1.4 Let $f\colon U \to Y$ be a continuous mapping defined on an open set $U \subset X_1 \times X_2$. A necessary and sufficient condition for f to be of class C^1 on U is the existence and continuity of both partial derivatives $\partial_1 f$ and $\partial_2 f$ at each point $(x_1, x_2) \in U$. In this case the derivative of f at (x_1, x_2) is given by:

$$\mathrm{d}f(x_1, x_2)(h_1, h_2) = \partial_1 f(x_1, x_2)h_1 + \partial_2 f(x_1, x_2)h_2, \quad h_1 \in X_1,\ h_2 \in X_2.$$

Theorem 3.1.5 (Taylor Formula) Let X and Y be Banach spaces and let $f\colon U \to Y$ be a mapping defined on a convex open set $U \subset X$. If f is of class C^r on U and if the points x, $x+h$ are in U, then

$$f(x+h) = \sum_{k=0}^{r} \frac{1}{k!}\mathrm{d}^k f(x)h^k + \underbrace{R(x, h)h^r}_{\varrho(x,h)},$$

where

$$R(x, h) = \int_0^1 \frac{(1-t)^{r-1}}{(r-1)!}\left(\mathrm{d}^r f(x+th) - \mathrm{d}^r f(x)\right)\mathrm{d}t \in \mathscr{L}_s^r(X, Y).$$

The next theorem is, in a sense, converse to Taylor's.

Theorem 3.1.6 *Let X and Y be Banach spaces and let f: $U \to Y$ be a mapping defined on a convex open set $U \subset X$. Suppose that there exist mappings φ_k: $U \to \mathscr{L}_s^k(X, Y)$, $k = 0, \ldots, r$, with the following properties:*

1° $\varphi_0 = f$,
2° φ_k *are continuous,*
3° *writing, for any* x, $x+h \in U$,

$$\varrho(x, h) := f(x+h) - \sum_{k=0}^{r} \frac{1}{k!}\varphi_k(x)h^k,$$

we have

$$\lim_{(x,h)\to(x_0,0)} \frac{\|\varrho(x, h)\|}{\|h\|^r} = 0 \qquad \textit{for all } x_0 \in U.$$

Then f is of class C^r and

$$\mathrm{d}^k f(x) = \varphi_k(x) \qquad \textit{for } x \in U,\ k = 0, \ldots, r.$$

For $X = Y = \boldsymbol{R}$ this theorem was proved by Marcinkiewicz and Zygmund (1936), and for the general case by Glaeser (1958).

Theorem 3.1.7 (Implicit Mapping Theorem) *Let X, Y, W be Banach spaces and let f: $U \to W$ be a C^r-mapping, $r \geqslant 1$, defined on an open set $U \subset X \times Y$. Suppose that $f(x_0, y_0) = 0$ at some point $(x_0, y_0) \in U$ and that the partial derivative (at that point) $\partial_2 f(x_0, y_0)$ is an isomorphism between Y and W. Then there exists a neighbourhood V_0 of x_0 in X such that, for each neighbourhood $V \subset V_0$, there is a unique C^r-map h: $V \to Y$ with the properties: $h(x_0) = y_0$, $(x, h(x)) \in U$ for all $x \in V$ and*

$$f(x, h(x)) = 0 \qquad \textit{for all } x \in V.$$

Theorem 3.1.8 (Inverse Mapping Theorem) *Let X, Y be Banach spaces and let f: $U \to Y$ be a C^r-mapping, $r \geqslant 1$, defined on an open set $U \subset X$. If at some point $x_0 \in U$ the derivative $\mathrm{d}f(x_0)$ is an isomorphism between X and Y, then there exists a neighbourhood V of x_0 in X such that f*

restricted to V is a C^r-diffeomorphism of V onto a neighbourhood of $y_0 = f(x_0)$ in Y.

We now specify a certain subclass of all C^r-mappings on a set U. Thus let $U \subset X$ be an open convex set and let $C^r(U, Y)$ denote the space of all C^r-mappings $f\colon U \to Y$ (as usual, X and Y are Banach spaces). We define:

$$B^r(U, Y) := \{f \in C^r(U, Y)\colon \sup_{x \in U} \|d^k f(x)\| < +\infty,\ k = 0, \ldots, r\}.$$

Evidently, $B^r(U, Y)$ is a linear subspace of $C^r(U, Y)$.

Exercise

3.1.5.A. Show that $B^r(U, Y)$ becomes a Banach space if the norm is given by

$$\|f\|_r = \sum_{k=0}^{r} \sup_{x \in U} \|d^k f(x)\|.$$

Consider the map

$$\text{ev}\colon B^r(U, Y) \times U \to Y$$

defined as follows:

$$\text{ev}(f, x) = f(x) \qquad \text{for } f \in B^r(U, Y),\ x \in U;$$

this is the so-called *evaluation map.*

Theorem 3.1.9 The evaluation map ev: $B^r(U, Y) \times U \to Y$ *is of class C^r and we have*

$$d^k \text{ev}(f, x)(\xi, h)^k = k d^{k-1}\xi(x)h^{k-1} + d^k f(x)h^k$$

for any $f, \xi \in B^r(U, Y)$ and $x, h \in X$ such that $x, x+h \in U$.

Proof We prove this for $r = 2$. Let f, ξ, x, h satisfy the conditions of the theorem. Applying the Taylor formula, we see that

$$\begin{aligned} \text{ev}(f+\xi, x+h) &= (f+\xi)(x+h) = f(x+h) + \xi(x+h) \\ &= [f(x) + df(x)h + \tfrac{1}{2}d^2 f(x)h^2] + \\ &\quad + [\xi(x) + d\xi(x)h + \tfrac{1}{2}d^2\xi(x)h^2] + \\ &\quad + R_f(x, h)h^2 + R_\xi(x, h)h^2. \end{aligned}$$

Define

$$\varphi_0(f, x) = f(x),$$
$$\varphi_1(f, x)(\xi, h) = \mathrm{d}f(x)h + \xi(x),$$
$$\varphi_2(f, x)(\xi_1, \xi_2; h_1, h_2) = \mathrm{d}^2 f(x)(h_1, h_2) + \mathrm{d}\xi_1(x)h_2 + \mathrm{d}\xi_2(x)h_1,$$
$$\varrho((f, x); (\xi, h)) = R_f(x, h)h^2 + R_\xi(x, h)h^2 + \tfrac{1}{2}\mathrm{d}^2\xi(x)h^2.$$

For any couple (ξ, h) we define the norm:

$$\|(\xi, h)\| := \max(\|\xi\|_2, \|h\|).$$

Then $\|(\xi, h)\| \geqslant \|h\|$ and so

$$\begin{aligned}\frac{\|\varrho((f, x); (\xi, h))\|}{\|(\xi, h)\|^2} &\leqslant \frac{\|R_f(x, h)\| \cdot \|h\|^2}{\|(\xi, h)\|^2} + \frac{\|R_\xi(x, h)\| \cdot \|h\|^2}{\|(\xi, h)\|^2} + \frac{\|\mathrm{d}^2\xi(x)h^2\|}{\|(\xi, h)\|^2} \\ &\leqslant \|R_f(x, h)\| + \|R_\xi(x, h)\| + \|\mathrm{d}^2\xi(x)\|.\end{aligned} \tag{3.1.2}$$

Let $\varepsilon > 0$ be arbitrary. For $\|h\|$ and $\|\xi\|_2$ small enough we have:

$$\|R_\xi(x, h)\| = \left\|\int_0^1 (1-t)\left(\mathrm{d}^2\xi(x+th) - \mathrm{d}^2\xi(x)\right)\mathrm{d}t\right\| < \varepsilon \tag{3.1.3}$$

and

$$\|\mathrm{d}^2\xi(x)h^2\| \leqslant \|\mathrm{d}^2\xi(x)\| \cdot \|h\|^2 < \varepsilon\|h\|^2. \tag{3.1.4}$$

Estimates (3.1.2) and (3.1.4) imply

$$\lim_{(\xi, h)\to 0} \frac{\|\varrho((f, x); (\xi, h))\|}{\|(\xi, h)\|^2} = 0$$

and an application of Theorem 3.1.6 ends the proof.

Let us switch over to topics of another type.

Definition 3.1.5 Let T be a topological space. A set $A \subset T$ is called *residual* iff it is the intersection of countably many open dense subsets of T:

$$A = \bigcap_{n=1}^{\infty} G_n, \qquad G_n \text{ dense and open.}$$

By complementation we obtain the alternative description: A is residual iff

$$T - A = \bigcup_{n=1}^{\infty} F_n, \qquad F_n \text{ closed and nowhere dense.}$$

In a complete metric space every residual set is dense.

Theorem 3.1.10 (Sard) *Let $f\colon U \to \mathbf{R}^p$, $U \subset \mathbf{R}^n$ open. Suppose that f is of class C^r where $r > \max(0, n-p)$. Then the set of all critical values of f is of Lebesgue measure zero in $\mathbf{R}^p$* (see Milnor, 1965; Abraham and Robbin, 1967).

Theorem 3.1.11 (second version of Sard's theorem) *Let M and N be finite-dimensional manifolds, $\dim M = n$, $\dim N = p$, and let $f\colon M \to N$ be of class C^r where $r > \max(0, n-p)$. Then the set R_f of all regular values of f is residual (and consequently dense) in N* (see Abraham and Robbin, 1967).

Regular and critical values are defined in Definition 2.1.1.
Theorem 3.1.11 can be derived from Theorem 3.1.10.

Definition 3.1.6 Let $A \in \mathscr{L}(X, Y)$, X, Y being Banach spaces. A is called a *Fredholm operator* iff:

1° $\ker A$, $\operatorname{im} A$ are closed in X, Y, respectively,

2° $\dim \ker A = n < +\infty$,

3° $\operatorname{codim} \operatorname{im} A = p < +\infty$.

The number

$$\dim \ker A - \operatorname{codim} \operatorname{im} A = n - p$$

is called the *index* of A.

Let M and N be Banach manifolds.[36] A C^1-mapping $f\colon M \to N$ is called a *Fredholm map* iff, for every $x \in M$, $df(x)$ is a Fredholm operator.

Theorem 3.1.12 (Smale's Density Theorem). Let M, N be Banach manifolds of class C^r and let $f\colon M \to N$ be a Fredholm map of class C^r. Assume that $r > \max(0, \text{index } df(x))$ for every $x \in M$. Then the set R_f of all regular values of f is residual in N (see Abraham and Robbin, 1967).

Now we turn attention to the class $\mathscr{F}^r(M)$ of all C^r-vector fields on a finite-dimensional compact manifold M. We shall show how $\mathscr{F}^r(M)$ can be made into a Banach space.

Thus let M be compact, $\dim M = m < +\infty$. For every finite atlas $\mathfrak{A} = \{(U_k, \alpha_k): k = 1, \ldots, n\}$ we can find another atlas $\mathfrak{A}_1 = \{(Z_k, \alpha_k): k = 1, \ldots, n\}$ such that

$$Z_k \subset \bar{Z}_k \subset U_k, \qquad \bigcup_{k=1}^{n} Z_k = M.$$

Let $\Phi \in \mathscr{F}^r(M)$, $r \geqslant 1$. Denote by Ψ_k the image of Φ in the chart (Z_k, α_k). Define:

$$\|\Phi\|_{\mathfrak{A}} = \max_k \sup_{y \in \alpha_k(Z_k)} \sum_{i=0}^{r} \|\mathrm{d}^i \Psi_k(y)\|.$$

The two theorems which follow are quite easy. We invite the reader to supply them with proofs.

Theorem 3.1.13 *$\mathscr{F}^r(M)$ is a Banach space with the norm $\|\cdot\|_{\mathfrak{A}}$.*

Theorem 3.1.14 *If $\mathfrak{A}_1$ and $\mathfrak{A}_2$ are two atlases and $\|\cdot\|_{\mathfrak{A}_1}$, $\|\cdot\|_{\mathfrak{A}_2}$ are norms obtained in the way just described, then these norms are equivalent, i.e., there are constants a and b such that*

$$a\|\Phi\|_{\mathfrak{A}_1} \leqslant \|\Phi\|_{\mathfrak{A}_2} \leqslant b\|\Phi\|_{\mathfrak{A}_1} \qquad \textit{for every } \Phi \in \mathscr{F}^r(M).$$

We pass to the definition of the k-jet bundle. Let $\Phi_1, \Phi_2 \in \mathscr{F}^r(M)$, let $x \in M$ and let (U, α) be a chart, $x \in U$, $y = \alpha(x)$. Denote by $\Psi_{1,\alpha}$, $\Psi_{2,\alpha}$ the images of Φ_1, Φ_2 in that chart. We will write

$$\Psi_{1,\alpha}(y) \stackrel{k}{\equiv} \Psi_{2,\alpha}(y)$$

iff

$$\mathrm{d}^i \Psi_{1,\alpha}(y) = \mathrm{d}^i \Psi_{2,\alpha}(y) \qquad \text{for } i = 0, \ldots, k;$$

here k is an integer between 0 and r.

Lemma 3.1.1 *Let (U_0, α_0) be some fixed chart and let $x \in U_0$. If two vector fields $\Phi_1, \Phi_2 \in \mathscr{F}^r(M)$ satisfy the relation*

$$\Psi_{1,\alpha_0}(y_0) \stackrel{k}{\equiv} \Psi_{2,\alpha_0}(y_0), \qquad y_0 = \alpha_0(x),$$

then they satisfy also:

$$\Psi_{1,\alpha}(y) \overset{k}{\equiv} \Psi_{2,\alpha}(y), \qquad y = \alpha(x),$$

for any other chart (U, α), $x \in U$.

We leave the proof to the reader.

This lemma shows that the relation $\overset{k}{\equiv}$ may be interpreted as a relation between vector fields on M (at a point $x \in M$): by definition,

$$\Phi_1(x) \overset{k}{\equiv} \Phi_2(x)$$

iff

$$\Psi_{1,\alpha}(y) \overset{k}{\equiv} \Psi_{2,\alpha}(y), \qquad y = \alpha(x),$$

in some (equivalently: any) chart (U, α), $x \in U$. Clearly, this is an equivalence relation.

Definition 3.1.7 Given a point $x \in M$, a vector field $\Phi \in \mathscr{F}^r(M)$ and an integer k, $0 \leqslant k \leqslant r$, the equivalence class

$$\{\Phi_1 \in \mathscr{F}^r(M): \ \Phi_1(x) \overset{k}{\equiv} \Phi(x)\}$$

is called the *k-jet* of Φ at x and is denoted by the symbol $j^k_\Phi(x)$.

The family of all k-jets $j^k_\Phi(x)$, $\Phi \in \mathscr{F}^r(M)$, is called the *k-jet* at x and is denoted by $j^k(x)$.

The class of all $j^k(x)$, $x \in M$, is called the *k-jet bundle* and is denoted by $J^k(M)$.

Speaking more intuitively, though less precisely, any member of $j^k(x)$ is a $(k+1)$-tuple $(\dot{x}, A_1, \ldots, A_k)$, where $\dot{x} \in T_x(M)$ and $A_i \in \mathscr{L}^i(E, E)$, E denoting the model space for M.

Examples

3.1.9. $M = \boldsymbol{R}$. A vector field $\Phi \in \mathscr{F}^r(M)$ is simply a C^r-function f: $\boldsymbol{R} \to \boldsymbol{R}$. Then, for instance, the triple (0, 0, 2) can be viewed as the 2-jet $j^2_f(0)$, where f is any function such that $f(0) = 0, f'(0) = 0, f''(0) = 2$, e.g. $f(x) = x^2$.

3.1.10. $M = \boldsymbol{R}^2$. The triple

$$\big((1, 0), A_1, A_2\big),$$

where

$$A_1 = \begin{bmatrix} 1 & 1 \\ 0 & 0 \end{bmatrix} \in \mathscr{L}(\boldsymbol{R}^2, \boldsymbol{R}^2),$$

$$A_2 \in \mathscr{L}^2(\boldsymbol{R}^2, \boldsymbol{R}^2): \ A_2(h, k) = (0, hk) \qquad \text{for } h, k \in \boldsymbol{R},$$

is identified with the 2-jet $j_f^2(0, 0)$ of any C^2-map $f\colon \boldsymbol{R}^2 \to \boldsymbol{R}^2$ such that

$$f(0, 0) = (1, 0),$$

$$\mathrm{d}f(0, 0) = \begin{bmatrix} 1 & 1 \\ 0 & 0 \end{bmatrix},$$

$$\mathrm{d}^2 f(0, 0)(h, k) = (0, hk),$$

e.g.

$$f(x) = (x^1 + x^2 + 1, \ x^1 x^2).$$

There is a natural manifold structure on $J^k(M)$. Charts can be defined as follows: if (U_0, α_0) is a chart on M, let

$$\alpha^k\big(j_\Phi^k(x)\big) = \big(\alpha_0(x), \mathrm{d}\Psi(\alpha_0(x)), \ldots, \mathrm{d}^k\Psi(\alpha_0(x))\big),$$

where Ψ is the image of Φ via α_0. Then α^k assigns to every pair (x, Φ) a point of a Euclidean space of a suitably high dimension.

Exercise

3.1.6.B. Find $\dim J^k(M)$ if $\dim M = m$.
Result: $(2 + m + \ldots + m^k)m$.

The last concept which we wish to introduce in this section is the linear map bundle.

Let $T(M)$ and $T(M')$ be the tangent bundles over two Banach manifolds M and M'; let E_x and $E_{x'}$ denote the fibres over $x \in M$ and $x' \in M'$.

Definition 3.1.8 The *linear map bundle* is the set

$$\mathscr{L}\big(T(M), T(M')\big) := \bigcup_{(x,x') \in M \times M'} \mathscr{L}(E_x, E_{x'}),$$

where, as usual, $\mathscr{L}(E_x, E_{x'})$ denotes the space of all continuous linear mappings from E_x to $E_{x'}$.

$\mathscr{L}(T(M), T(M'))$ is also a manifold. Namely, we equip this set with the system of charts defined as follows:

If (U, α) and (U', α') are any charts on M and M', respectively, and if

$$A_{x,x'} \in \mathscr{L}(T(M), T(M')), \qquad x \in U, \ x' \in U',$$

we put

$$\beta(A_{x,x'}) = (\alpha(x), \alpha'(x'); A) \in \alpha(U) \times \alpha'(U') \times \mathscr{L}(E, E'),$$

where E, E' are the model spaces for M, M' and A denotes the operator $A_{x,x'}$ written in the charts (U, α) and (U', α').

Now, let X and Y be Banach spaces and $U \subset X$ an open convex set. Consider the map

$$\mathrm{ev}_k\colon B^r(U, Y) \times U \to U \times \mathscr{L}^1(X, Y) \times \ \dots \ \times \mathscr{L}^k(X, Y)$$

(where r and k are integers, $0 \leqslant k \leqslant r$), defined by:

$$\mathrm{ev}_k(f, x) = (x, \mathrm{d}f(x), \dots, \mathrm{d}^k f(x)) \qquad \text{for } f \in B^r(U, Y), \ x \in U.$$

Theorem 3.1.15 The mapping ev_k *has the following properties*:

(1) *is of class* C^{r-k};

(2) *for* $(f, x) \in B^r(U, Y) \times U$ *the linear map* $\mathrm{d}(\mathrm{ev}_k)(f, x)$ *is surjective; the subspace* $\ker \mathrm{d}(\mathrm{ev}_k)(f, x)$ *has a closed complement in* $B^r(U, Y) \times X$.

We omit the proof; the reader should again resort to Abraham and Robbin (1967).

Definition 3.1.9 A closed subspace X_1 of a Banach space X is said to *split* X iff X_1 has a closed complement in X.[26] A linear operator $A\colon X \to Y$ is said to be *splitting* iff $\ker A$ splits X.

Using this terminology, we can restate assertion (2) of Theorem 3.1.15 as follows: *the derivative of* ev_k *is a splitting surjection.*

3.2. Elements of the Transversality Theory

We have already come across the idea of transversality in our course: in the construction of the Poincaré map we introduced the concept of a *transversal section*, or *cross-section*, of an orbit of a vector field Φ at a point x_0. Recall that such a section was defined as any submanifold S of codimension 1 in M such that $x_0 \in S$ and $\Phi(x_0) \notin T_{x_0}(S)$. The latter condition can be rewritten as

$$T_{x_0}(S) \oplus [\Phi(x_0)] = T_{x_0}(M);$$

of course, here $[\Phi(x_0)]$ denotes the one-dimensional space spanned by the vector $\Phi(x_0)$.

We now extend the notion of transversality to a larger class of objects.

Let M be a Banach manifold of class C^1, having E as its model space. A set $W \subset M$ is called a *submanifold* iff for every $x \in W$ there is an admissible chart (U, α), $\alpha(x) = 0$, and E admits a decomposition

$$E = E_1 \oplus E_2$$

into the direct sum of two subspaces (independently of x), such that

$$\alpha(U) = V_1 \times V_2,$$

V_i denoting a neighbourhood of zero in E_i, $i = 1, 2$, and

$$\alpha(W \cap U) = V_1 \times \{0\}.$$

It follows that $T_x(W)$ is then a complemented subspace of $T_x(M)$.

Definition 3.2.1 (1) Let N and W be two C^1-submanifolds of a Banach C^1-manifold M. We say that N and W are *transversal* at a point $x \in N \cap W$ iff

$$T_x(M) = T_x(N) + T_x(W)$$

(the sum on the right need not be direct). In symbols, this is written as follows:

$$N \pitchfork_x W.$$

(2) Let M and N be Banach C^1-manifolds, W a C^1-submanifold of N, and $f\colon M \to N$ a C^1-map. We say that f is *transversal* to W at $x \in M$ iff either

(2a) $f(x) \notin W$,

or

(2b) if $y = f(x) \in W$, the subspace $(\mathrm{d}f(x))^{-1}(T_y(W))$ splits $T_x(M)$ and the image $\mathrm{d}f(x)(T_x(M))$ contains a closed complement to $T_y(W)$ in $T_y(N)$. In symbols:

$$f \pitchfork_x W.$$

If f is transversal to W at every point $x \in M$, we say that f is transversal to W and we write

$$f \pitchfork W.$$

Theorem 3.2.1 Let M and N be Banach C^1-manifolds, W a C^1-submanifold of N, and f: $M \to N$ a C^1-map. Let $x \in M$ and $y = f(x) \in W$. A necessary and sufficient condition for $f \pitchfork_x W$ is that there exist admissible charts (U, α) and (V, β) on M and N, respectively, $x \in U$, $y \in V$, with the following properties:

1° $\alpha(U) = U_1 \times V_2,\ \ \beta(V) = V_1 \times V_2,\ \ f(U) \subset V,$

$\alpha(x) = (0, 0),\ \ \beta(y) = (0, 0),\ \ \beta(V \cap W) = V_1 \times \{0\};$

2° *writing*

$$f_{\alpha\beta} = \beta \circ f \circ \alpha^{-1}\colon\ U_1 \times U_2 \to V_1 \times V_2$$

we have

$$f_{\alpha\beta}(x_1, x_2) = (\eta(x_1, x_2), x_2),$$

where η is a certain function, η: $\alpha(U) \to V_1$.

Proof The sufficiency is obvious. To prove the necessity suppose that $f \pitchfork_x W$. Let E and F be the model spaces for M and N, respectively. Choose charts (U, α) and (V, β) on M and N so as to have

$$x \in U,\ y \in V, \qquad \alpha(x) = 0, \qquad \beta(y) = 0,$$
$$\alpha(U) \subset E, \qquad \beta(V) = V_1 \times V_2 \subset F,$$
$$\beta(V \cap W) = V_1 \times \{0\}, \qquad f(U) \subset V.$$

Then

$$g = f_{\alpha\beta}\colon\ \alpha(U) \to V_1 \times V_2, \qquad g(0) = 0.$$

Denote by F_1 and F_2 the subspaces of F spanned by V_1 and V_2: $F = F_1 \oplus F_2$. Since $f \pitchfork_x W$, we have

$$g \pitchfork_0 V_1 \times \{0\}.$$

By assumption (and by the definition of transversality), $E_1 := (\mathrm{d}g(0))^{-1}(F_1)$ is a complemented subspace of E:

$$E = E_1 \oplus E_2.$$

The elements of E can thus be written in the form:

$$u = (u_1, u_2) \in E,\ u_1 \in E_1,\ u_2 \in E_2;$$

the mapping g is of the form

$$g(u_1, u_2) = (h(u_1, u_2), k(u_1, u_2)),$$

h and k denoting certain C^1-maps.

We will use the symbols $\dot{u}_1$ and $\dot{u}_2$ to stand for vectors in the tangent spaces to E_1 and E_2 (which are, in fact, identical to E_1 and E_2). Then

$$\mathrm{d}g(\dot{u}_1, \dot{u}_2) = (\mathrm{d}h(\dot{u}_1, \dot{u}_2), \mathrm{d}k(\dot{u}_1, \dot{u}_2)).$$

In this equality, the symbol $\mathrm{d}g$ is used, for brevity, instead of $\mathrm{d}g(0, 0)$; also $\mathrm{d}h$, $\mathrm{d}k$ are used in the same meaning. Thus in a precise notation, the left-hand side should read: $\mathrm{d}g(0, 0)(\dot{u}_1, \dot{u}_2)$, and so on. We keep this convention till the end of the proof: all derivatives are taken at $(0, 0)$.

By the definition of E_1, $\mathrm{d}g$ maps E_1 onto $F_1 \times \{0\}$, and so $\mathrm{d}k(\dot{u}_1, \dot{u}_2) = \partial_2 k(\dot{u}_2)$. By transversality,

$$\mathrm{d}g(E) + (F_1 \times \{0\}) = F,$$

and thus $\partial_2 k$ maps E_2 onto $\{0\} \times F_2$. Since $\mathrm{d}g$ vanishes only for $\dot{u}_2 = 0$ in E_2, the map $\mathrm{d}k$ defines an isomorphism between E_2 and F_2. Therefore we may simply assume that $E_2 = F_2$ (up to an isomorphism).

Consider the mapping δ: $E \to E$,

$$\delta(u_1, u_2) = (u_1', u_2'),$$

$$u_1' = u_1, \qquad u_2' = k(u_1, u_2).$$

The derivative of δ at $(0, 0)$ is of the following form (in coordinates u_1, u_2):

$$\mathrm{d}\delta = \begin{bmatrix} I & 0 \\ 0 & \partial_2 k \end{bmatrix}.$$

Thus $\mathrm{d}\delta$ is surjective. In virtue of the Inverse Mapping Theorem, δ is C^1-invertible.

Let us replace the chart (U, α) by (U, α_1) where $\alpha_1 = \delta \circ \alpha$. Then

$$\beta \circ f \circ \alpha_1^{-1}\colon \alpha_1(U) \to V_1 \times V_2,$$

$$\begin{aligned}\beta \circ f \circ \alpha_1^{-1}(u_1', u_2') &= \beta \circ f \circ \alpha^{-1} \circ \delta^{-1}(u_1', u_2')\\ &= g(u_1, u_2) = (h(u_1, u_2), k(u_1, u_2))\\ &= (h(u_1', \delta_2^{-1}(u_1', u_2')), u_2'),\end{aligned}$$

where δ_2^{-1} denotes the "second coordinate" of δ^{-1}.

The charts (U, α_1), (V, β) and the function

$$\eta(u_1', u_2') = h(u_1', \delta_2^{-1}(u_1', u_2'))$$

satisfy conditions 1° and 2°, as required.

Corollary 3.2.1 *Let M, N, W and f be as in the above theorem. Suppose that $f \pitchfork W$. Then:*

(1) *$f^{-1}(W)$ is a submanifold of M;*
(2) *for every $x \in f^{-1}(W)$ we have*

$$(\mathrm{d}f(x))^{-1} T_y(W) = T_x f^{-1}(W),$$

where $y = f(x)$;

(3) *for every $x \in f^{-1}(W)$, any closed complement to $T_x(f^{-1}(W))$ in $T_x(M)$ is isomorphic to any closed complement to $T_y(W)$ in $T_y(N)$;*

(4) *if W is closed and M is compact, then $f^{-1}(W)$ has finitely many components.*

Now, let $\mathscr{A}$, M, N be Banach manifolds of class C^r, $r \geqslant 0$, and suppose that we are given a mapping

$$\varrho\colon \mathscr{A} \to C^r(M, N).$$

We will write ϱ_a in place of $\varrho(a)$; each ϱ_a is a C^r-map from M to N: $\varrho_a(x) \in N$ for $x \in M$. The *evaluation* map

$$\mathrm{ev}_\varrho\colon \mathscr{A} \times M \to N$$

is defined by the formula:

$$\mathrm{ev}_\varrho(a, x) = \varrho_a(x).$$

If the evaluation map ev_ϱ is of class C^r, we say that ϱ is a *C^r-representation.*

Theorem 3.2.2 *Let $\mathscr{A}$, M, N be C^0-manifolds, let $W \subset N$ be a closed subset (not necessarily a submanifold) and $K \subset M$ a compact subset, and let*

$$\varrho\colon \mathscr{A} \to C^0(M, N)$$

be a C^0-representation. Then

$$A^0_{KW} := \{a \in \mathscr{A}\colon \varrho_a(K) \cap W = \emptyset\}$$

is an open subset of $\mathscr{A}$.

Proof Let $x_0 \in K$, $a_0 \in A^0_{KW}$. Since $\varrho_{a_0}(x_0) \notin W$, there exists a neighbourhood $V_{a_0} \times U_{x_0}$ of the point (a_0, x_0) in $\mathscr{A} \times M$ such that $\varrho_a(x) \notin W$ for any $a \in V_{a_0}$, $x \in U_{x_0}$. This is a consequence of the assumption that ev_ϱ is continuous and W is closed. Keep a_0 fixed and let x_0 vary in K. By compactness, finitely many $U_{x_1}, \ldots, U_{x_n}$ cover K. To each U_{x_i} there corresponds a neighbourhood $V_{a_0,i}$. The set $V = V_{a_0,1} \cap \ldots \cap V_{a_0,n}$ is a neighbourhood of a_0 and $V \subset A^0_{KW}$.

Theorem 3.2.3 Let $\mathscr{A}$, M, N be Banach C^1-manifolds, $\dim M < +\infty$, *let $W \subset N$ be a closed C^1-submanifold and $K \subset M$ a compact subset, and let*

$$\varrho\colon \mathscr{A} \to C^1(M, N)$$

be a C^1-representation. Then

$$A_{KW} := \{a \in \mathscr{A}\colon \varrho_a \pitchfork_x W \text{ for } x \in K\}$$

is an open subset of $\mathscr{A}$.

Proof We define a certain set $Z \subset \mathscr{L}(T(M), T(N))$. By definition, a point $(x, y, A_{x,y})$ $(=$ an element of $\mathscr{L}(T(M), T(N)))$ belongs to Z iff either

1° $y \notin W$,

or

2° $y \in W$ and the image $A_{x,y}(T_x(M))$ contains a closed complement to $T_y(W)$ in $T_y(N)$.

The set Z is open. Indeed, if $y \notin W$, the set Z contains the point $(x, y, A_{x,y})$ together with any neighbourhood

$$U_x \times V_y \times B(A_{x,y}, \varepsilon),$$

where $V_y \cap W = \emptyset$ and $B(A_{x,y}, \varepsilon)$ denotes the ε-ball around $A_{x,y}$ in the space $\mathscr{L}(T_x(M), T_y(N))$; if $y \in W$, the image of $T_x(M)$ under $A_{x,y}$ contains a complement to $T_y(W)$ in $T_y(N)$; thus every operator A sufficiently close to $A_{x,y}$ in $\mathscr{L}(T_x(M), T_y(N))$ shares this property (we leave the details to the reader), and again the set

$$U_x \times V_y \times B(A_{x,y}, \varepsilon)$$

is contained in Z, provided that U_x, V_y and $\varepsilon > 0$ are small enough.

Consider the set

$$W' = \mathscr{L}(T(M), T(N)) - Z.$$

This is a closed subset of $\mathscr{L}(T(M), T(N))$. Since ϱ is a C^1-representation, the mapping

$$\varrho_a' : \mathscr{A} \to C^0(M, \mathscr{L}(T(M), T(N)))$$

given by

$$\varrho_a'(x) = \mathrm{d}\varrho_a(x)$$

is a C^0-representation. By construction,

$$\varrho_a \pitchfork_x W \qquad \text{iff} \qquad \varrho_a'(x) \notin W';$$

the condition (in the definition of transversality) that $T_x(M)$ should be split (by a certain subspace) is here immaterial, because $\dim M < +\infty$. Applying Theorem 3.2.2 to the representation ϱ' and the closed set $W' \subset \mathscr{L}(T(M), T(N))$, we conclude that A_{KW} is an open subset of $\mathscr{A}$.

Theorem 3.2.4 (Transversal Density Theorem) Let $\mathscr{A}$, M, N be Banach C^r-manifolds and $W \subset N$ a C^r-submanifold, and let

$$\varrho : \mathscr{A} \to C^r(M, N)$$

be a C^r-representation. Further, assume that:

1° $\dim M = n < +\infty$, $\operatorname{codim} W = q < +\infty$,
2° $r > \max(0, n-q)$,
3° $\mathrm{ev}_\varrho \pitchfork W$.

Then the set

$$A_W := \{a \in \mathscr{A} : \varrho_a \pitchfork W\}$$

is residual in $\mathscr{A}$.

Proof Consider the set

$$B = \mathrm{ev}_\varrho^{-1}(W) \subset \mathscr{A} \times M.$$

According to Corollary 3.2.1, B is a submanifold of codimension q in $\mathscr{A} \times M$. Denote by π_1 the projection

$$\pi_1 : \mathscr{A} \times M \to \mathscr{A}, \qquad \pi_1(a, x) = a,$$

and write

$$\pi = \pi_1|_B.$$

Then π is a C^r-map. Denote by R_π the set of all regular values of π. We shall show that:

(a) π is a Fredholm mapping with index $n-q$,

(b) $A_W = R_\pi$.

Hence, by Theorem 3.1.12, follows our assertion.

In the proof of (a) we will need the following lemma, which we give without proof, inviting the reader to regard it as an exercise.

Lemma 3.2.1 Let F and G be Banach spaces, $\dim G = n < +\infty$, *and let* $E \subset F \times G$ *be a subspace of codimension q. Denote by* p_1 *the projection of* $F \times G$ *onto the first factor and write* $p = p_1|_E$. *Then p is a Fredholm operator with index* $n-q$.

The proof of the theorem will be carried out in 3 steps.

Step I. Proof of (a). We have to show that

$$d\pi\colon T_{(a,x)}(B) \to T_a(\mathscr{A})$$

is a Fredholm operator. This follows directly from Lemma 3.2.1 applied to $G = T_x(M)$, $F = T_a(\mathscr{A})$, $E = T_{(a,x)}(B)$; the assumptions of the lemma are fulfilled, because $\operatorname{codim} T_{(a,x)}(B) = \operatorname{codim} W = q$, and so (a) is proved.

Step II. We now prove the inclusion $R_\pi \subset A_W$. Let $a \in R_\pi$. We have to show that $\varrho_a \pitchfork W$. Since $T_x(M)$ is finite-dimensional, it suffices to show that

$$T_y(W) + d\varrho_a(T_x(M)) = T_y(N)$$

for $x \in M$, $y = \varrho_a(x) \in W$. Thus let $\dot{y} \in T_y(N)$. All that we need is to find a pair of vectors: $\dot{w} \in T_y(W)$ and $\dot{x} \in T_x(M)$ such that

$$\dot{y} = \dot{w} + d\varrho_a \dot{x}; \tag{3.2.1}$$

as in the proof of Theorem 3.2.1, we write for brevity $d\varrho_a\dot{x}$, instead of $d\varrho_a(x)\dot{x}$; this will cause no confusion, because x remains fixed.

On account of assumption 3° there are: a vector $\dot{w}_1 \in T_y(W)$ and a vector $(\dot{a}, \dot{x}_1) \in T_a(\mathscr{A}) \times T_x(M)$ satisfying the equality

$$\dot{y} = \dot{w}_1 + d(\mathrm{ev}_\varrho)(\dot{a}, \dot{x}_1). \tag{3.2.2}$$

(Here again $d(\mathrm{ev}_\varrho)(\dot{a}, \dot{x})$ stands for $d(\mathrm{ev}_\varrho)(a, x)(\dot{a}, \dot{x}_1)$; an analogous notational convention is used with respect to all derivative symbols occurring in this proof.)

Since $d(\mathrm{ev}_\varrho) = \partial_a \mathrm{ev}_\varrho + \partial_x \mathrm{ev}_\varrho$, equality (3.2.2) can be rewritten as

$$\dot{y} = \dot{w}_1 + \partial_a(\mathrm{ev}_\varrho)\dot{a} + d\varrho_a \dot{x}_1. \tag{3.2.3}$$

By assumption, $a \in R_\pi$; thus there exists a vector $\dot{b} \in T_{(a,x)}(B)$, $\dot{b} = (\dot{a}, \dot{x}_2)$, such that $\mathrm{d}\pi(\dot{a}, \dot{x}_2) = \dot{a}$. Further, by the definition of B, we have

$$\mathrm{d}(\mathrm{ev}_\varrho)(\dot{a}, \dot{x}_2) \in T_y(W),$$

i.e.,

$$\dot{w}_2 := \partial_a(\mathrm{ev}_\varrho)\dot{a} + \mathrm{d}\varrho_a \dot{x}_2 \in T_y(W). \tag{3.2.4}$$

From (3.2.3) and (3.2.4) we obtain, by eliminating the summand $\partial_a(\mathrm{ev}_\varrho)\dot{a}$,

$$\dot{y} = (\dot{w}_1 + \dot{w}_2) - \mathrm{d}\varrho_a \dot{x}_2 + \mathrm{d}\varrho_a \dot{x}_1,$$

which is just (3.2.1) with

$$\dot{w} = \dot{w}_1 + \dot{w}_2 \in T_y(W), \qquad \dot{x} = \dot{x}_1 - \dot{x}_2 \in T_x(M).$$

Step III. Finally, we prove the inclusion $A_W \subset R_\pi$. Let $a \in A_W$ and let $(a, x) \in B$. We have to show that $\mathrm{d}\pi(a, x)$ maps $T_{(a,x)}(B)$ onto $T_a(\mathscr{A})$.

Choose $\dot{a} \in T_a(\mathscr{A})$ arbitrarily. Let $\dot{x}_1 \in T_x(M)$ and $\dot{w}_1 \in T_y(W)$, where $y = \varrho_a(x)$. Clearly, the vector

$$\dot{y} := \dot{w}_1 + \partial_a(\mathrm{ev}_\varrho)\dot{a} + \mathrm{d}\varrho_a \dot{x}_1 \tag{3.2.5}$$

belongs to $T_y(N)$. On the other hand, since $\varrho_a \pitchfork_x W$, this vector can be written in the form

$$\dot{y} = \dot{w}_2 + \mathrm{d}\varrho_a \dot{x}_2, \tag{3.2.6}$$

where $\dot{w}_2$ and $\dot{x}_2$ are some vectors in $T_y(W)$ and $T_x(M)$, respectively. Equalities (3.2.5) and (3.2.6) result in

$$\partial_a(\mathrm{ev}_\varrho)\dot{a} = \mathrm{d}\varrho_a \dot{x} + \dot{w},$$

where $\dot{w} = \dot{w}_2 - \dot{w}_1$, $\dot{x} = \dot{x}_2 - \dot{x}_1$. Consequently,

$$\mathrm{d}(\mathrm{ev}_\varrho)(\dot{a}, -\dot{x}) = \dot{w} \in T_y(W),$$

whence

$$(\dot{a}, -\dot{x}) \in (\mathrm{d}(\mathrm{ev}_\varrho))^{-1} T_y(W) = T_{(a,x)}(B).$$

Thus

$$\dot{a} = \mathrm{d}\pi(a, x)(\dot{a}, -\dot{x})$$

is in the image of $T_{(a,x)}(B)$ under the operator $\mathrm{d}\pi(a, x)$. Since a was chosen arbitrarily, this operator is surjective, and this ends the proof.

Exercises

3.2.1.A. Let M be a finite-dimensional manifold. Every vector field $\Phi \in \mathscr{F}^1(M)$ can be regarded as a mapping of M into $T(M)$. Let $\mathscr{G}_0$ denote the set of vector fields $\Phi \in \mathscr{F}^1(M)$ whose critical points are all non-degenerate. Prove that

$$\mathscr{G}_0 = \mathscr{G}_0^1(M) := \{\Phi \in \mathscr{F}^1(M)\colon \ \Phi \pitchfork (TM)_0\},$$

where $(TM)_0$ denotes the 0-section of $T(M)$ (i.e., $(TM)_0 = \{(x, 0)\colon \ x \in M\}$).

Hint. If $\Phi(x_0) = 0$ and x_0 is non-degenerate, than the matrix $d\Phi(x_0)$ is non-singular.

3.2.2.B. Show that $\mathscr{G}_0^1(M)$ is dense in $\mathscr{F}^1(M)$ (in C^1-topology).

Hint. Apply Theorem 3.2.4.

3.3. Differentiation of a Flow on a Manifold with Respect to the Generating Vector Field

Let M be a finite-dimensional manifold of class C^∞, $\dim M = m$. We may consider the map

$$G\colon \ \mathscr{F}^1(M) \times M \times \boldsymbol{R} \to M$$

given by the formula

$$G(\Phi, x, t) = \varphi_t(x) \tag{3.3.1}$$

for $\Phi \in \mathscr{F}^1(M)$, $x \in M$, $t \in \boldsymbol{R}$; as usual, φ_t denotes the flow generated by Φ.

Theorem 3.3.1 The mapping G defined by (3.3.1) *is of class C^1.*

Proof It suffices to prove that G is of class C^1 in each variable (Φ, x, t) separately. That G is so with respect to x and t requires no proof: it is a direct consequence of basic theorems on ordinary differential equations. We thus focus attention on continuous differentiability of G with respect to the "variable" $\Phi \in \mathscr{F}^1(M)$.

Let us fix a point $x_0 \in M$ and a time-value $t = a$, and let $\Phi \in \mathscr{F}^1(M)$ be a given vector field. Differentiability is a property of a local character; therefore we are only interested in the behaviour of Φ in a subregion of M which contains a neighbourhood of the trajectory

$\varphi_t(x_0)$ for $0 \leqslant t \leqslant a$. We may thus transfer that region to $\boldsymbol{R}^m$, assuming that $x_0 = 0$ and, say,

$$\{\varphi_t(0)\colon 0 \leqslant t \leqslant a\} = \{(0, \ldots, 0, r)\colon 0 \leqslant r \leqslant r_0\},$$

where r_0 is a positive number. (We call this procedure *rectification* of an orbit.)

The set

$$U = \{x \in \boldsymbol{R}^m\colon |x^i| < \delta,\ i = 1, \ldots, m-1,\ -\varepsilon < x^m < r_0 + \varepsilon\}$$

(δ and ε fixed) is an open neighbourhood of the trajectory under consideration (or rather of its fragment passed in the time-period $0 \leqslant t \leqslant a$). Let $\Psi \in \mathscr{F}^1(U)$ be another vector field such that the trajectory of $\Phi + \Psi$ starting from $x = 0$ remains within U up to the instant $t = a$. Denote by $\psi_t(x)$ the trajectories of the field $\Phi + \Psi$. Since the initial point $x = 0$ is fixed, we will write $\varphi(t)$ and $\psi(t)$ instead of $\varphi_t(0)$ and $\psi_t(0)$.

The functions $\varphi(t)$ and $\psi(t)$ satisfy the equations:

$$\frac{d\varphi(t)}{dt} = \Phi(\varphi(t)), \qquad \frac{d\psi(t)}{dt} = \Phi(\psi(t)) + \Psi(\psi(t)).$$

For any t in the interval $0 \leqslant t \leqslant a$ we have

$$\begin{aligned}
&\psi(t) - \varphi(t) \\
&\quad = \int_0^t \big(\Phi(\psi(s)) - \Psi(\psi(s))\big)ds - \int_0^t \Phi(\varphi(s))ds \\
&\quad = \int_0^t \big(\Phi(\psi(s)) - \Phi(\varphi(s))\big)ds + \\
&\qquad + \int_0^t \big(\Psi(\psi(\varepsilon)) - \Psi(\varphi(s))\big)ds + \int_0^t \Psi(\varphi(s))ds. \qquad (3.3.2)
\end{aligned}$$

Using the Taylor formula, we obtain

$$\begin{aligned}
&\Phi(\psi(s)) - \Phi(\varphi(s)) \\
&\quad = d\Phi(\varphi(s))(\psi(s) - \varphi(s)) + R(\varphi(s), \psi(s) - \varphi(s)).
\end{aligned}$$

Hence the first summand in the expression on the right of (3.3.2) is equal to

$$\int_0^t d\Phi(\varphi(s))(\psi(s) - \varphi(s))ds + \int_0^t R(\varphi(s), \psi(s) - \varphi(s))ds.$$

In the first of the above integrals let us replace the difference $\psi(s)-\varphi(s)$ by an integral, just as in (3.3.2). This gives

$$\int_0^t d\Phi(\varphi(s))(\psi(s)-\varphi(s))ds$$

$$= \int_0^t d\Phi(\varphi(s))\Big[\int_0^s (\Phi(\psi(u))-\Phi(\varphi(u)))du +$$

$$+ \int_0^s (\Psi(\psi(u))-\Psi(\varphi(u)))\,du + \int_0^s \Psi(\varphi(u))\,du\Big]ds$$

$$= \int_0^t d\Phi(\varphi(s))\int_0^s (\Phi(\psi(u))-\Phi(\varphi(u)))du\,ds +$$

$$+ \int_0^t d\Phi(\varphi(s))\int_0^s (\Psi(\psi(u))-\Psi(\varphi(u)))du\,ds +$$

$$+ \int_0^t d\Phi(\varphi(s))\int_0^s \Psi(\varphi(u))du\,ds.$$

Continuing this procedure, we observe that the right-hand side of (3.3.2) gives rise to a series $\sum(u_k+v_k+w_k)$ containing three types of terms:

(a) terms linear with respect to Ψ:

$$u_0 = \int_0^t \Psi(\varphi(s))ds,$$

$$u_1 = \int_0^t d\Phi(\varphi(s))\int_0^s \Psi(\varphi(u))du\,ds,$$

. . . .

$$u_k = \int_0^t d\Phi(\varphi(s_1))\int_0^{s_1} d\Phi(\varphi(s_2)) \dots \int_0^{s_k} \Psi(\varphi(u))du\,ds_k \dots ds_1,$$

. . . .

(b) terms of second order in Ψ:

$$v_0 = \int_0^t [\Psi(\psi(s))-\Psi(\varphi(s))]ds,$$

$$v_1 = \int_0^t d\Phi(\varphi(s))\int_0^s [\Psi(\psi(u))-\Psi(\varphi(u))]du\,ds,$$

. . . .

$$v_k = \int_0^t d\Phi(\varphi(s_1))\int_0^{s_1} d\Phi(\varphi(s_2)) \dots \int_0^{s_k} [\Psi(\psi(u))-\Psi(\varphi(u))]du\,ds_k \dots ds_1,$$

. . . .

(c) terms of the "tail" type:

$$w_0 = \int_0^t R(\varphi(s), \psi(s) - \varphi(s))\mathrm{d}s,$$

$$w_1 = \int_0^t \mathrm{d}\Phi(\varphi(s)) \int_0^s R(\varphi(u), \psi(u) - \varphi(u))\mathrm{d}u\,\mathrm{d}s,$$

. . . .

$$w_k = \int_0^t \mathrm{d}\Phi(\varphi(s_1)) \int_0^{s_1} \mathrm{d}\Phi(\varphi(s_2)) \ldots \int_0^{s_k} R(\varphi(u), \psi(u) - \varphi(u))\mathrm{d}u\,\mathrm{d}s_k \ldots \mathrm{d}s_1,$$

. . . .

Each of the three series $\sum u_k, \sum v_k, \sum w_k$ converges. To see this, we estimate their terms as follows:

$$\|u_k\| = \left\|\int_0^t \mathrm{d}\Phi(\varphi(s_1)) \int_0^{s_1} \mathrm{d}\Phi(\varphi(s_2)) \ldots \int_0^{s_k} \Psi(\varphi(u))\mathrm{d}u\,\mathrm{d}s_k \ldots \mathrm{d}s_1\right\|$$

$$\leqslant \|\Phi\|_1^k \|\Psi\|_0 \frac{t^{k+1}}{(k+1)!}, \tag{3.3.3}$$

$$\|v_k\| \leqslant \|\Phi\|_1^k \frac{t^{k+1}}{(k+1)!} \|\Psi\|_1 \max_{0 \leqslant u \leqslant a} \|\psi(u) - \varphi(u)\|, \tag{3.3.4}$$

$$\|w_k\| \leqslant \frac{t^{k+1}}{(k+1)!} \|\Phi\|_1^k \sup_{0 \leqslant u \leqslant a} \|R(\varphi(u), \psi(u) - \varphi(u))\|, \tag{3.3.5}$$

where $\|\cdot\|_1$ denotes the C^1-norm and $\|\cdot\|_0$ the C^0-norm (of the vector field in question).

It is not difficult to show, by reductio ad absurdum, that there exists a constant $c > 0$ such that

$$\sup_{0 \leqslant u \leqslant a} \|\psi(u) - \varphi(u)\| \leqslant c\|\Psi\|_1. \tag{3.3.6}$$

Since Φ is of class C^1, for every $\varepsilon > 0$ there is an $r > 0$ such that

$$\frac{\|R(x, h)\|}{\|h\|} < \varepsilon \qquad \text{for } \|h\| < r,\ x \in U.$$

Hence, applying (3.3.6), we obtain for $\|\Psi\|_1$ small enough:

$$\|R(\varphi(u), \psi(u) - \varphi(u))\| \leqslant \varepsilon \|\psi(u) - \varphi(u)\| \leqslant c\varepsilon \|\Psi\|_1,$$

and finally,

$$\|w_k\| \leqslant \frac{t^{k+1}}{(k+1)!}\|\Phi\|_1^k \varepsilon c \|\Psi\|_1.$$

The series $\sum u_k + \sum v_k + \sum w_k$ can be defined for every vector field $\Psi \in \mathscr{F}^1(U)$ (fulfilling the conditions imposed). Define:

$$A(\Psi) := \sum_{k=0}^{\infty} u_k. \tag{3.3.7}$$

In virtue of the Banach–Steinhaus theorem, A is a continuous linear operator (relative to the $\|\cdot\|_1$-topology for Ψ). We shall show that this operator is precisely the partial derivative of G with respect to the first variable. Indeed,

$$G(\Phi+\Psi, 0, a) - G(\Phi, 0, a) = \psi(a) - \varphi(a)$$
$$= \sum_{k=0}^{\infty} u_k + \sum_{k=0}^{\infty} v_k + \sum_{k=0}^{\infty} w_k,$$

where u_k, v_k, w_k are calculated at $t = a$. Hence, in view of (3.3.4) and (3.3.5) we get (for small $\|\Psi\|_1$)

$$\|G(\Phi+\Psi, 0, a) - G(\Phi, 0, a) - A(\Psi)\|$$
$$\leqslant c(\|\Psi\|_1^2 + \varepsilon\|\Psi\|_1) \sum_{k=0}^{\infty} \|\Phi\|_1^k \frac{a^{k+1}}{(k+1)!}$$
$$\leqslant ace^{a\|\Phi\|_1}\|\Psi\|_1(\varepsilon + \|\Psi\|_1).$$

Thus if $\|\Psi\|_1$ is small enough, we have

$$\frac{\|G(\Phi+\Psi, 0, a) - G(\Phi, 0, a) - A(\Psi)\|}{\|\Psi\|_1} \leqslant ace^{a\|\Phi\|_1}(\varepsilon + \|\Psi\|_1),$$

which shows that

$$A = \partial_\Phi G(\Phi, 0, a),$$

as claimed. In view of estimate (3.3.3) A is a uniform limit of a series of functions continuous in (Φ, x, a); consequently, A is jointly continuous, and this means that G is a C^1-map. This ends the proof.

Theorem 3.3.2 Let $\Phi \in \mathscr{F}^1(M)$, M being finite-dimensional, $x \in M$, and suppose that $\Phi(x) \neq 0$; let G be the C^1-map given by formula (3.3.1)*. Then for every $t_0 > 0$ the derivative $\partial_\Phi G(\Phi, x, t_0)$ is a surjection onto the space $T_{\varphi_{t_0}(x)}(M)$.*

Proof We preserve the notation introduced in the proof of the preceding theorem. As we have shown, the derivative under consideration is the operator given by (3.3.7); i.e., for $0 < t < t_0$ we have

$$\partial_\Phi G(\Phi, x, t)(\Psi) = A(\Psi) = \sum_{k=0}^{\infty} A_k(\Phi) = A_0(\Psi) + \sum_{k=1}^{\infty} A_k(\Psi),$$

where $A_k(\Psi)$ denotes the term u_k (of type (a)), satisfying the estimate (see (3.3.3)):

$$\|A_k(\Psi)\| \leqslant \|\Phi\|_1^k \frac{t^{k+1}}{(k+1)!} \|\Psi\|_0. \tag{3.3.8}$$

Write

$$B(\Psi) := \sum_{k=1}^{\infty} A_k(\Psi);$$

then $A = A_0 + B$. The operator A_0 is a surjection of $\mathscr{F}^1(M)$ onto $T_{\varphi_t(x)}(M)$; this is what we now show. As in the preceding proof, we transfer our considerations to the set U in $\boldsymbol{R}^m$. If e_i denotes the vector $e_i = [0, \ldots, 0, 1, 0, \ldots 0]$ (the unit on the i-th place, $i = 1, \ldots, m$) or the constant vector field $\Psi = e_i$, we have $A_0(e_i) = te_i$. Thus A_0 is indeed surjective.

Now suppose that A is not surjective. Then there exist scalars $\alpha_1, \ldots, \alpha_m$ such that

$$\sum_{i=1}^{m} \alpha_i A(e_i) = 0 \qquad \text{and} \qquad \sum_{i=1}^{m} |\alpha_i|^2 = 1.$$

Then

$$\sum_{i=1}^{m} \alpha_i A_0(e_i) = -\sum_{i=1}^{m} \alpha_i B(e_i),$$

whence, in view of (3.3.8),

$$t\Big\|\sum_{i=1}^{m} \alpha_i e_i\Big\| = \Big\|\sum_{i=1}^{m} \alpha_i B(e_i)\Big\| \leqslant \sum_{i=1}^{m} |\alpha_i| \sum_{k=1}^{\infty} \frac{1}{(k+1)!} \|\Phi\|_1^k t^{k+1} \leqslant c_1 t^2,$$

the constant c_1 depending on Φ only. This yields

$$1 \leqslant c_1 t,$$

which is nonsense for small t. Thus, for small t, the operator $\partial_\Phi G(\Phi, x, t)$ is a surjection onto $T_{\varphi_t(x)}(M)$.

Definition (3.3.7) of $A(\Psi)$ together with the definition of terms u_k of type (a) imply that $A(\Psi)$ satisfies the following differential equation:

$$\frac{\mathrm{d}A(\Psi)}{\mathrm{d}t} = \Psi(\varphi(t)) + \mathrm{d}\Phi(\varphi(t))A(\Psi). \tag{3.3.9}$$

Suppose that s_0 is such that $\partial_\Phi G(\Phi, x, s_0)$ is surjective and that $\varphi_s(x) \neq x$ for $0 < s \leqslant s_0$. Further, suppose that either the trajectory $\varphi_t(x)$ is non-periodic or it is periodic with prime period not less than t_0. We can find an open set V such that $\varphi_s(x) \in V$ for $0 < s < s_0$ and $\varphi_s(x) \notin V$ for $s_0 \leqslant s < t_0$. Consider the following class of vector fields

$$\mathscr{E} = \{\Psi\colon\ \Psi(x) = 0 \text{ for } x \notin V\}.$$

By a slight modification of the previous argument we show that the operator $\partial_\Phi G(\Phi, x, s_0)$, i.e., the operator $A|_{t=s_0}$ (A calculated at $t = s_0$) carries the family $\mathscr{E}$ onto the entire $T_{\varphi_{s_0}}(M)$. Thus, for every $\dot{x} \in T_{\varphi_{s_0}(x)}(M)$ we can find a vector field $\Psi \in \mathscr{E}$ such that $A|_{t=s_0}(\Psi) = \dot{x}$. Now, observe that for $\Psi \in \mathscr{E}$ equation (3.3.9) is reduced to

$$\frac{\mathrm{d}A(\Psi)}{\mathrm{d}t} = \mathrm{d}\Phi(\varphi(t))A(\Psi)$$

in the interval $s_0 \leqslant t < t_0$. We easily conclude that also for every $\dot{x} \in T_{\varphi_{t_0}(x)}(M)$ there is a vector field $\Psi \in \mathscr{E}$ such that $A|_{t=t_0}(\Psi) = \dot{x}$. Thus the operator $\partial_\Phi G(\Phi, x, t_0) = A|_{t=t_0}$ is surjective, as asserted.

It remains to consider the case where the trajectory $\varphi_t(x)$ is periodic with prime period $\tau < t_0$. From what we have already shown it follows that the derivative $\partial_\Phi G$ taken at $t = \tau$ is a surjection, and hence it is easy to deduce that $\partial_\Phi G$ is surjective for arbitrary t_0.

3.4. Critical Elements of Vector Fields on Compact Manifolds

The technique which we have developed, namely, the concept of transversality and theorems of Sections 3.2 and 3.3, will now be used in the study of closed orbits of vector fields on finite-dimensional compact manifolds.

Let M be a compact manifold, $\dim M = m < +\infty$. For every vector field $\Phi \in \mathscr{F}^r(M)$ we define a mapping

$$f_\Phi\colon\ M \times \boldsymbol{R}^+ \to M \times \boldsymbol{R}^+ \times M$$

by the formula

$$f_{\Phi}(x, t) = (x, t, \varphi_t(x)) \qquad \text{for } x \in M,\ t > 0;$$

of course, φ_t is the flow generated by Φ.

We shall denote by Δ the set

$$\Delta := \{(x, t, x) \colon x \in M,\ t \in \boldsymbol{R}^+\}.$$

Evidently, Δ is a submanifold or $M \times \boldsymbol{R}^+ \times M$. It is called the *diagonal.*

The concept of diagonal is extremely useful in the investigation of critical elements of vector fields. For any vector field $\Phi \in \mathscr{F}^r(M)$ and for $x \in M$, $t \in \boldsymbol{R}^+$, the condition

$$f_{\Phi}(x, t) \in \Delta$$

is satisfied if and only if x either is a critical point or belongs to a closed orbit with period t.

Let us recall that we have termed a closed orbit γ *transversal* iff for $x \in \hat{\gamma}$ the operator $d\varphi_\tau(x)$ has eigenvalue 1 with multiplicity 1; τ denotes the prime period.

Definition 3.4.1 (1) A period τ of a closed orbit is called *transversal* iff for $x \in \hat{\gamma}$ the operator $d\varphi_\tau(x)$ has eigenvalue 1 with multiplicity 1.

(2) Let x be a critical point of a field Φ. A number $t \in \boldsymbol{R}^+$ is called a *transversal period* of x iff 1 is not an eigenvalue of $d\varphi_t(x)$.

Theorem 3.4.1 (1) *Let x be a critical point of a vector field $\Phi \in \mathscr{F}^r(M)$. Then*

$$f_{\Phi} \pitchfork_{(x,\tau)} \Delta$$

if and only if τ is a transversal period of x.

(2) *Let γ be a closed orbit of Φ and let τ be a period of γ. Then*

$$f_{\Phi} \pitchfork_{(x,\tau)} \Delta \qquad \textit{for } x \in \hat{\gamma}$$

if and only if τ is a transversal period.

Proof By the definition of transversality the condition

$$f_{\Phi} \pitchfork_{(x,\tau)} \Delta \tag{3.4.1}$$

is equivalent to:

$$df_{\Phi}(x, \tau)(T_{(x,\tau)}(M \times \boldsymbol{R}^+)) + T_{(x,\tau,x)}(\Delta) = T_{(x,\tau,x)}(M \times \boldsymbol{R}^+ \times M)$$

(since M is finite-dimensional, we may disregard the condition involving splitting). This, in turn, means that for every $\dot{y}_1, \dot{y}_2 \in T_x(M)$ and every $\dot{t}_3 \in \mathrm{T}_\tau(\boldsymbol{R}^+)(=\boldsymbol{R})$ the system of equations:

$$\begin{aligned} \dot{y}_1 &= \dot{x}_1 + \dot{x}_2, \\ \dot{t}_3 &= \dot{t}_1 + \dot{t}_2, \\ \dot{y}_2 &= \mathrm{d}\varphi_\tau(x)\dot{x}_1 + \Phi(x)\dot{t}_1 + \dot{x}_2 \end{aligned} \tag{3.4.2}$$

has at least one solution $\dot{x}_1, \dot{x}_2 \in T_x(M)$, $\dot{t}_1, \dot{t}_2 \in \boldsymbol{R}$. Further, the solvability of system (3.4.2) is equivalent to the solvability of the single equation

$$\mathrm{d}\varphi_\tau(x)\dot{x} - \dot{x} + \Phi(x)\dot{t} = \dot{y}, \tag{3.4.3}$$

where $\dot{y}, \dot{x} \in T_x(M)$, $\dot{t} \in \boldsymbol{R}$. Summing up, we see that condition (3.4.1) is equivalent to the following: equation (3.4.3) has a solution $(\dot{x}, \dot{t})$ for every $\dot{y} \in T_x(M)$.

We pass to the proof of the theorem.

(1) Let x be a critical point. Then $\Phi(x) = 0$, and so equation (3.4.3) reduces to:

$$\mathrm{d}\varphi_\tau(x)\dot{x} - \dot{x} = \dot{y}.$$

This equation has a solution $\dot{x}$ for every $\dot{y} \in T_x(M)$ if and only if 1 is not an eigenvalue of $\mathrm{d}\varphi_\tau(x)$. This establishes (1).

(2) Now let γ be a closed orbit and let $x \in \hat{\gamma}$. Then $\Phi(x) \neq 0$. Let us choose a local coordinate system in which

$$\Phi(x) = \begin{bmatrix} 1 \\ 0 \\ \vdots \\ 0 \end{bmatrix}, \qquad \mathrm{d}\varphi_\tau(x) = \begin{bmatrix} 1 & \alpha \\ 0 & A \end{bmatrix},$$

where A is an $(m-1) \times (m-1)$-matrix and α is an $(m-1)$-vector. In this coordinate system equation (3.4.3) takes the form

$$\left[\begin{array}{c|c} 1 & \alpha \\ \hline 0 & A \end{array}\right] \begin{bmatrix} \dot{x}^1 \\ \dot{x}^2 \\ \vdots \\ \dot{x}^m \end{bmatrix} - \begin{bmatrix} \dot{x}^1 \\ \dot{x}^2 \\ \vdots \\ \dot{x}^m \end{bmatrix} + \begin{bmatrix} \dot{t} \\ 0 \\ \vdots \\ 0 \end{bmatrix} = \begin{bmatrix} \dot{y}^1 \\ \dot{y}^2 \\ \vdots \\ \dot{y}^m \end{bmatrix},$$

where

$$\dot{y} = \begin{bmatrix} y^1 \\ \vdots \\ y^m \end{bmatrix}, \qquad \dot{x} = \begin{bmatrix} x^1 \\ \vdots \\ x^m \end{bmatrix}.$$

This equation can be rewritten as:

$$\begin{bmatrix} \alpha \begin{bmatrix} x^2 \\ \vdots \\ x^m \end{bmatrix} + \dot{t} \\ (A-I) \begin{bmatrix} x^2 \\ \vdots \\ x^m \end{bmatrix} \end{bmatrix} = \begin{bmatrix} y^1 \\ y^2 \\ \vdots \\ y^m \end{bmatrix}.$$

The above equation is solvable for any $\dot{y}$ iff 1 is not an eigenvalue of A. This establishes (2).

Theorem 3.4.2 Let $\Phi_0 \in \mathscr{F}^r(M)$, M compact, and let γ_0 be a closed orbit of Φ_0 with prime period τ_0. Suppose that τ_0 is transversal. Then there exist:

(1) *a neighbourhood V of Φ_0 in $\mathscr{F}^r(M)$,*
(2) *an arbitrarily small neighbourhood U of $\hat{\gamma}_0$ in M,*
(3) *an arbitrarily small neighbourhood I of τ_0 in $\mathbf{R}$,*

with the following property:

for every $\Phi \in V$ there is a unique closed orbit γ of Φ such that $\hat{\gamma} \subset U$ and $\tau \in I$, τ denoting the period of γ.

Proof Let $x_0 \in \hat{\gamma}_0$ and let S be a transversal section of γ_0 at x_0. Consider the manifolds $\mathscr{F}^r(M) \times S \times \mathbf{R}$ and $M \times \mathbf{R}^+ \times M$ locally at the points (Φ_0, x_0, τ_0) and (x_0, τ_0, x_0), respectively. Denote by F and E the model spaces for S and M; then clearly $\dim F = m-1$, $\dim E = m$. Passing to charts, we may assume that M and S are (locally) portions of E and F, and thus we may identify the manifolds $\mathscr{F}^r(M) \times S \times \mathbf{R}$ and $M \times \mathbf{R}^+ \times M$ with the spaces $\mathscr{F}^r(M) \times F \times \mathbf{R}$ and $E \times \mathbf{R} \times E$. The diagonal Δ is then a linear subspace of $E \times \mathbf{R} \times E$ of dimension $m+1$. The subspace

$$\Delta^c = \{(0, 0, y) \colon y \in E\}$$

is a complement to Δ in $E \times \mathbf{R} \times E$.

Let π denote the projection of $E \times \boldsymbol{R} \times E$ onto Δ^c given by:

$$\pi(x, t, y) = (0, 0, y - x).$$

We define a mapping

$$G\colon \mathscr{F}^r(M) \times F \times \boldsymbol{R} \to \Delta^c$$

by the formula

$$G(\Phi, x, t) = \pi \circ f_\Phi(x, t) = \big(0, 0, \varphi_t(x) - x\big), \tag{3.4.4}$$

where φ_t is the flow generated by Φ. Of course,

$$G(\Phi_0, x_0, \tau_0) = 0.$$

In virtue of Theorem 3.3.1, G is a C^1-map. Since the period τ_0 is transversal, we have, on account of Theorem 3.4.1,

$$\Delta + \mathrm{d}f_{\Phi_0}(x_0, \tau_0)(E \times \boldsymbol{R}) = E \times \boldsymbol{R} \times E = \Delta \oplus \Delta^c. \tag{3.4.5}$$

It follows that the linear operator

$$\pi \circ \mathrm{d}f_{\Phi_0}(x_0, \tau_0)$$

maps $E \times \boldsymbol{R}$ onto Δ^c.

By assumption, S is a cross-section; thus E admits the decomposition

$$E = [\Phi(x_0)] \oplus F,$$

where F is tangent to S at x_0, $\dim F = m - 1$. This means that every vector $\dot{x} \in E$ is of the form

$$\dot{x} = (\dot{u}, \dot{v}), \qquad \text{where} \qquad \dot{u} \in [\Phi(x_0)],\ \dot{v} \in F.$$

Furthermore, the coordinates u, v can be chosen so as to have

$$\Phi(x_0) = [1, 0, \ldots, 0], \qquad \mathrm{d}\varphi_{\tau_0}(x_0) = \begin{bmatrix} 1 & \alpha \\ 0 & A \end{bmatrix},$$

where A is an operator from F into F, and α is an $(m-1)$-vector. Writing vectors columnwise, we have

$$\begin{aligned} \pi \circ \mathrm{d}f_{\Phi_0}(x_0, \tau_0)(\dot{x}, \dot{t}) &= \pi\big(\dot{x}, \dot{t}, \mathrm{d}\varphi_{\tau_0}(x_0)(\dot{x}, \dot{t})\big) \\ &= \pi\big((\dot{x}, \dot{t}, \dot{x}) + (0, 0, \mathrm{d}\varphi_{\tau_0}(x_0)(\dot{x}, \dot{t}) - \dot{x})\big) \\ &= \left(0, 0, \frac{\partial}{\partial t}\varphi_{\tau_0}(x_0)\dot{t} + \frac{\partial}{\partial x}\varphi_{\tau_0}(x_0)\dot{x} - \dot{x}\right) \end{aligned}$$

$$= \left(0,\ 0,\ \Phi_0(x_0)\dot{t} + \begin{bmatrix} 1 & \alpha \\ 0 & A \end{bmatrix} \begin{bmatrix} \dot{u} \\ \dot{v} \end{bmatrix} - \begin{bmatrix} \dot{u} \\ \dot{v} \end{bmatrix}\right)$$

$$= \left(0,\ 0,\ \begin{bmatrix} \dot{t} + \alpha \dot{v} \\ A\dot{v} - \dot{v} \end{bmatrix}\right).$$

The number 1 is not an eigenvalue of A, and so $A-I$ is a surjection of F onto F. Since $\dot{t}$ can be arbitrary, the operator $\pi \circ \mathrm{d}f_{\Phi_0}(x_0, \tau_0)$ carries $F \times \boldsymbol{R}$ onto Δ^c. Comparing the dimensions

$$\dim F \times \boldsymbol{R} = (m-1)+1 = m = \dim \Delta^c,$$

we infer that the operator

$$\pi \circ \mathrm{d}f_{\Phi_0}(x_0, \tau_0) = \frac{\partial}{\partial(x,\, t)} G(\Phi_0, x_0, \tau_0)$$

is an isomorphism between the spaces $F \times \boldsymbol{R}$ and Δ^c. Hence, by the Implicit Mapping Theorem, there are: a neighbourhood V of Φ_0 in $\mathscr{F}^r(M)$, an arbitrarily small neighbourhood U_1 of x_0 in S and an arbitrarily small interval $I = (\tau_0 - \varepsilon, \tau_0 + \varepsilon)$, for which there exist unique functions $x(\Phi)$ and $t(\Phi)$ (defined for $\Phi \in V$) such that $x(\Phi) \in U_1$, $t(\Phi) \in I$ and

$$G(\Phi, x(\Phi), t(\Phi)) = 0 \qquad \text{for every } \Phi \in V.$$

Thus, taking a sufficiently small neighbourhood U of $\hat{\gamma}_0$ (U has to satisfy: $U \cap S \subset U_1$), we see that, given $\Phi \in V$, there is exactly one closed orbit γ of Φ contained within U and having period $\tau \in I$. The theorem is proved.

Let $\Gamma_\Phi(a)$ denote the set consisting of all critical points of Φ and of all points belonging to closed orbits with periods $\leqslant a$.

Theorem 3.4.3 If M is a compact manifold, then for every $\Phi \in \mathscr{F}^r(M)$, $r \geqslant 1$, the set $\Gamma_\Phi(a)$ is closed.

Proof Suppose that $x_n \in \Gamma_\Phi(a)$ and $x_n \to x_0$. If the sequence $\{x_n\}$ contains infinitely many critical points, then x_0 is critical as well by the continuity of Φ. Let us thus assume that each x_n belongs to a closed orbit γ_n with period $\tau_n \leqslant a$. Periods τ_n can be chosen from the interval $\frac{1}{2}a \leqslant \tau_n \leqslant a$. Without loss of generality we may assume that the sequence τ_n converges to a limit τ_0, $\frac{1}{2}a \leqslant \tau_0 \leqslant a$. Since $\varphi_{\tau_n}(x_n) = x_n$, we obtain, passing to the limit, $\varphi_\tau(x_0) = x_0$. Thus $x_0 \in \Gamma_\Phi(a)$.

Exercise

3.4.1.A. Let M be a compact manifold and let x_0 be a non-degenerate critical point of a vector field $\Phi_0 \in \mathscr{F}^r(M)$. Show that there exist:

(a) a neighbourhood V of Φ_0 in $\mathscr{F}^r(M)$,

(b) a neighbourhood U of x_0 in M,

such that, for every $\Phi \in V$, the neighbourhood U contains exactly one critical point of Φ.

Hint. Apply the Implicit Mapping Theorem.

Definition 3.4.2 Let $\mathscr{G}^r_\Delta(a)$ denote the class of all vector fields $\Phi \in \mathscr{F}^r(M)$ satisfying the condition

$$f_\Phi \pitchfork_{(x,t)} \Delta \qquad \text{for all } x \in M \text{ and } 0 < t \leqslant a.$$

This condition means that every closed orbit of Φ of period $\tau \leqslant a$ is transversal.

It is not difficult to see that

$$\mathscr{G}^r{}_1(a) \subset \mathscr{G}^r_0(M)$$

(the latter symbol is defined analogously to $\mathscr{G}^1_0(M)$, see Exercise 3.2.1 on p. 148).

We now prove:

Theorem 3.4.4 Let $\Phi_0 \in \mathscr{G}^r_0(M)$, M compact. Then there are a neighbourhood V of Φ_0 in $\mathscr{F}^r(M)$ and a number $b > 0$ such that every vector field $\Phi \in V$ has closed orbits with periods $\tau \geqslant b$ only.

Proof Since $\Phi_0 \in \mathscr{G}^r_0(M)$ and M is compact, there are only finitely many critical points of Φ_0; by the definition of $\mathscr{G}^r_0(M)$, they are all non-degenerate. Therefore every sufficiently small number $t > 0$ is a transversal period of all critical points of Φ_0.

Assume that the theorem is false. Then there is a sequence of closed orbits $\gamma_1, \gamma_2, \ldots$, with periods $\tau_n \to 0$. Select a point x_n in each $\hat{\gamma}_n$; by compactness, we may assume that $x_n \to x_0$. If $\Phi_0(x_0) \neq 0$, then there exist numbers $t_0, c > 0$ and a neighbourhood U of x_0 such that

$$\operatorname{dist}(x, \varphi_t(x)) \geqslant tc \qquad \text{for } 0 \leqslant t \leqslant t_0,\ x \in U.$$

For n large enough we have $x_n \in U$, $\tau_n < t_0$, and so

$$0 = \operatorname{dist}(x_n, \varphi_{\tau_n}(x_n)) \geqslant \tau_n c > 0,$$

which is absurd. Thus x_0 is a critical point.

Since $x_n \to x_0$, we have

$$d\varphi_{t_j}(x_n) \to d\varphi_t(x_0)$$

for any sequence $t_j \to t$. Set $t_j = k_j \tau_{n_j} \to t > 0$, where $k_j \in Z^+$. Then 1 is an eigenvalue of $d\varphi_{t_j}(x_{n_j})$; consequently, 1 is also an eigenvalue of $d\varphi_t(x_0)$. However, observe that every number $t > 0$ can be obtained as the limit of a sequence t_j of the type considered. As we have pointed out, any small enough number $t > 0$ is a transversal period of x_0, i.e., 1 is not an eigenvalue of $d\varphi_t(x_0)$. We arrive at a contradiction.

As an easy consequence of the last theorem we obtain

Corollary 3.4.1 *For every vector field $\Phi_0 \in \mathscr{G}^r_\Delta(a)$ on a compact manifold M there are a neighbourhood V of Φ_0 in $\mathscr{F}^r(M)$ and a number $b > 0$ such that any $\Phi \in V$ has no periodic orbit with period $\tau < b$.*

Theorem 3.4.5 *Let $\Phi \in \mathscr{G}^r_\Delta(a)$. Then:*

(1) *critical elements of Φ of periods $\leqslant a$ are isolated;*

(2) *if M is compact, then $\Gamma_\Phi(a)$ contains at most finitely many critical elements;*

(3) *if M is compact, there exists an $\varepsilon > 0$ such that Φ has no closed orbit with period τ belonging to the interval $a < \tau < a + \varepsilon$.*

Proof (1) and (2) follow from Theorems 3.4.2 and 3.4.4. Assume that (3) does not hold. Then there is a sequence of closed orbits γ_n with periods $\tau_n \to a$. Since M is compact, there exists a closed orbit γ_0 such that

$$\operatorname{dist}(\hat{\gamma}_n, \hat{\gamma}_0) \to 0.$$

Then clearly a is a period of γ_0; by the assumption that $\Phi \in \mathscr{G}^r_\Delta(a)$ this period is transversal. Hence, in view of Theorem 3.4.2, there is a neighbourhood of $\hat{\gamma}_0$ containing no other closed orbits of Φ with periods close to a. This is a contradiction, because the orbits γ_n approach γ_0.

Theorem 3.4.6 *If M is compact, then $\mathscr{G}^r_\Delta(a)$ is an open subset of $\mathscr{F}^r(M)$.*

Proof Let $\Phi \in \mathscr{G}^r_\Delta(a)$. That is to say, $f_\Phi \pitchfork_{(x,t)} \Delta$ for all $x \in M$ and $0 < t \leqslant a$. Now, the condition

$$f_\Phi \pitchfork_{(x,t)} \Delta \qquad \text{for all } x \in M,\ 0 < t \leqslant a$$

is equivalent to:

$$f_\Phi \pitchfork_{(x,t)} \Delta \qquad \text{for all } x \in M,\ \tfrac{1}{2}a \leqslant t \leqslant a.$$

Only "$\Leftarrow$" requires a proof. Thus let τ be a period of a closed orbit γ and suppose that $\tau \leqslant \frac{1}{2}a$. We then have $\frac{1}{2}a \leqslant k\tau \leqslant a$ for some positive integer k. Since $\mathrm{d}\varphi_{k\tau}(x) = (\mathrm{d}\varphi_\tau(x))^k$ and, by assumption, the number 1 is an eigenvalue of multiplicity 1 for the operator $\mathrm{d}\varphi_{k\tau}(x)$, $x \in \hat{\gamma}$, it follows that also for $\mathrm{d}\varphi_\tau(x)$ the unity is such an eigenvalue. Thus the two conditions are indeed equivalent.

Write $K = M \times \langle \frac{1}{2}a, a \rangle$. Then

$$\Phi \in \mathscr{G}^r_\Delta(a) \qquad \text{iff} \qquad f_\Phi \pitchfork_{(x,t)} \Delta \qquad \text{for all } (x, t) \in K.$$

The set K is compact. Hence the set $\mathscr{G}^r_\Delta(a)$ is open, in virtue of Theorem 3.2.2.

The theorem that follows seems to be the most important in this section. It will be resorted to several times in the sequel.

Theorem 3.4.7 Let M be a compact manifold and let $\Phi_0 \in \mathscr{G}^r_\Delta(a)$. Let $\gamma_1, \ldots, \gamma_n$ be all the closed orbits with periods $\tau_i \leqslant a$, $i = 1, \ldots, n$, and let $x_1, \ldots, x_k$ be all the critical points of Φ_0. Then there exist:

(1) *a system of pairwise disjoint neighbourhoods $U_1, \ldots, U_n$ of $\hat{\gamma}_1, \ldots, \hat{\gamma}_n$,*

(2) *a system of pairwise disjoint neighbourhoods $Z_1, \ldots, Z_k$ of $x_1, \ldots, x_k$,*

(3) *arbitrarily small numbers $\delta > 0$ and $\varepsilon > 0$,*

(4) *a neighbourhood V of Φ_0 in $\mathscr{F}^r(M)$*

with the following properties:

(a) *Φ_0 has no closed orbits with periods τ in the interval $a < \tau < a + 2\varepsilon$,*

(b) *every vector field $\Phi \in V$ has exactly one critical point x_i' in each Z_i and exactly one closed orbit γ_i' in each U_i, the period τ_i' of γ_i' satisfying the condition: $|\tau_i' - \tau_i| < \delta$,*

(c) *for every $\Phi \in V$ the sets*

$$\Gamma_\Phi(a+\varepsilon), \qquad \Gamma_{\Phi_0}(a+\varepsilon), \qquad \Gamma_{\Phi_0}(a)$$

contain an equal number of critical elements.

Proof. According to Theorems 3.4.6, 3.4.5, 3.4.4 and to Exercise 3.4.2, there exist neighbourhoods $U_1, \ldots, U_n$, $Z_1, \ldots, Z_k$, numbers $\varepsilon > 0$,

$\delta > 0$ (arbitrarily small) and a neighbourhood V_0 of Φ_0 in $\mathscr{F}^r(M)$, satisfying conditions (a) and (b). All that remains to be shown is the existence of a neighbourhood $V \ni \Phi_0$, $V \subset V_0$, for which (c) is fulfilled.

By assumption, M is compact and so the set $Q = M - (\bigcup U_i \cup \bigcup Z_i)$ is compact, as well. Evidently, Q is disjoint with $\Gamma_{\Phi_0}(a)$. Thus there are positive numbers c and $t_0 < \min \tau_i$ such that

$$\operatorname{dist}(x, \varphi_t^0(x)) \geq ct \qquad \text{for } 0 < t \leq t_0,\ x \in Q,$$

φ_t^0 denoting the flow generated by Φ_0. Thus a neighbourhood V_1 of Φ_0 in $\mathscr{F}^r(M)$ can be found such that every Φ in V_1 fulfils the condition

$$\operatorname{dist}(x, \varphi_t(x)) \geq \tfrac{1}{2}ct \qquad \text{for } 0 < t \leq t_0,\ x \in Q,$$

φ_t denoting the flow generated by Φ.

Observe that for $(x, t) \in Q \times \langle t_0, a \rangle$ we have $f_{\Phi_0}(x, t) \notin \Delta$, i.e., $G(\Phi_0, x, t) \notin \Delta$. Hence, by the compactness of $Q \times \langle t_0, a \rangle$, there exists a neighbourhood V_2 of Φ_0 in $\mathscr{F}^r(M)$ such that

$$G(\Phi, x, t) \notin \Delta \qquad \text{for } \Phi \in V_2,\ (x, t) \in Q \times \langle t_0, a \rangle.$$

Write: $V = V_0 \cap V_1 \cap V_2$. By the construction of V_0, V_1, V_2, we see that for every $\Phi \in V$ the sets $\Gamma_\Phi(a)$ and $\Gamma_{\Phi_0}(a)$ have equally many critical elements. In view of property (a), also $\Gamma_{\Phi_0}(a)$ and $\Gamma_{\Phi_0}(a+\varepsilon)$ have equally many critical elements. This completes the proof.

3.5. Generic Properties. Genericity of Property $\mathscr{G}_1$

We will use the capital $\mathscr{G}$ (possibly with subscripts) to denote any property which can be shared by some vector fields $\Phi \in \mathscr{F}^r(M)$; M is a given manifold. If $\mathscr{G}$ is such a property, the symbol $\mathscr{G}^r(M)$ will denote the family of all $\Phi \in \mathscr{F}^r(M)$ having this property:

$$\mathscr{G}^r(M) := \{\Phi \in \mathscr{F}^r(M) \colon \Phi \text{ has property } \mathscr{G}\}.$$

Definition 3.5.1 A property $\mathscr{G}$ is called *generic in the space* $\mathscr{F}^r(M)$ (or shortly: C^r*-generic*) iff $\mathscr{G}^r(M)$ contains a residual subset of $\mathscr{F}^r(M)$.

We shall now distinguish the following specific properties, which will be the objects of our considerations:

1° Property $\mathscr{G}_0$: all critical points (of the vector field under examination) are non-degenerate (cf. Exercise 3.2.1),

2° Property $\mathscr{G}_1$: all critical points are hyperbolic,

3° Property $\mathscr{G}_\Delta(a)$: all periods $\tau \leqslant a$ of periodic orbits are transversal (a denoting a given positive number),

4° Property $\mathscr{G}_{3/2}(a)$: Properties $\mathscr{G}_1$ and $\mathscr{G}_\Delta(a)$ jointly satisfied,

5° Property $\mathscr{G}_2(a)$: Property $\mathscr{G}_1$ is satisfied and all closed orbits with prime periods $\leqslant a$ are hyperbolic,

6° Property $\mathscr{G}_3(a)$: Property $\mathscr{G}_2(a)$ is satisfied and all stable and unstable manifolds of critical elements with prime periods $\leqslant a$ intersect transversally,

7° Property $\mathscr{G}_2$: Property $\mathscr{G}_2(a)$ satisfied for every a,

8° Property $\mathscr{G}_3$: Property $\mathscr{G}_3(a)$ satisfied for every a.

Two further properties $\mathscr{G}_4$ and $\mathscr{G}_5$ will be introduced in Section 3.8.

We are going to formulate and prove the most significant theorems asserting the genericity of the above properties. In all those theorems M denotes a compact manifold.

Theorem 3.5.1 *The set $\mathscr{G}_1^r(M)$ is open and dense in $\mathscr{F}^r(M)$.*

Proof For any $\Phi \in \mathscr{F}^r(M)$ write

$$\varrho_\Phi(x) = j_\Phi^1(x).$$

Then

$$\varrho_\Phi\colon M \to J^1(M)$$

is a map of class C^{r-1}. The evaluation map

$$\mathrm{ev}_\varrho\colon \mathscr{F}^r(M) \times M \to J^1(M)$$

is onto the whole of $J^1(M)$ and, moreover, the derivative $\mathrm{d}(\mathrm{ev}_o)(\Phi, x)$ is a surjection onto $T_{j_\Phi^1(x)}(J^1(M))$ (see Exercise 3.5.1 on p. 168). Consequently, ev_ϱ is transversal to every submanifold of $J^1(M)$.

Let (U, α) be a chart on M; write $y = \alpha(x)$. We define a set $W \subset J^1(M)$ as follows: a triple $w = (x, \Phi, \mathrm{d}\Phi(x))$ is said to belong to W iff, in the chart (U, α), w is of the form $(y, 0, A)$, where $(y, 0, A) \in \alpha(U) \times E \times \mathscr{L}(E, E)$ and A has at least one eigenvalue with real part equal to 0; $E = \boldsymbol{R}^m$ is the model space for M.

We now have to introduce certain auxiliary facts and notions. The proof of Theorem 3.5.1 is continued on page 167.

Lemma 3.5.1 (1) *The set W is closed;*

(2) *W can be written as the union*

$$W = W_1 \cup \ldots \cup W_l,$$

where W_i is a submanifold of $J^1(M)$, $i = 1, \ldots, l$, and $\operatorname{codim} W_i \geqslant m+1\,(m = \dim M)$.

Proof The proof is carried out in two steps.

Step I. We need the concepts of an algebraic map and a semi-algebraic set.

A mapping f: $\boldsymbol{R}^n \to \boldsymbol{R}^m$ is called *algebraic* iff f is of the form

$$f(x^1, \ldots, x^n) = (f_1(x^1, \ldots, x^n), \ldots, f_m(x^1, \ldots, x^n)),$$

where f_i are polynomials in $x^1, \ldots, x^n$.

A set $A \subset \boldsymbol{R}^n$ is called *semi-algebraic* iff it is described by a finite system of conditions

$$f_i(x) = 0,\ i = 1, \ldots, r, \qquad g_j(x) > 0,\ j = 1, \ldots, s,$$

where f_i and g_j are polynomials.

The set $B \subset \boldsymbol{R}^{2m} \approx \boldsymbol{C}^m$ which we now define is an example of a semi-algebraic set:

$$\begin{aligned} B = \{(\lambda_1, \bar{\lambda}_1, \ldots, \lambda_s, \bar{\lambda}_s, \mu_{2s+1}, \ldots, \mu_m)\colon & \\ & \lambda_j = x_j + \mathrm{i}y_j,\ x_j, y_j \in \boldsymbol{R},\ y_j \neq 0,\ j = 1, \ldots, s; \\ & \mu_j \in \boldsymbol{R},\ j = 2s+1, \ldots, m;\ 0 < 2s \leqslant m; \\ & \exists j\colon\ 1 \leqslant j \leqslant s,\ x_j = 0\}. \end{aligned}$$

B is the set of all systems of roots of polynomials of degree n with real coefficients, having at least one purely imaginary root. This means that for every point $(\lambda_1, \ldots, \lambda_m) \in B$ there is such a polynomial P and the coordinates λ_j, $\bar{\lambda}_j$, μ_j of that point are the zeros of P.

We now describe the construction of that polynomial in more detail. Let P_m denote the set of all real polynomials of degree m; the polynomials will be identified with the $(m+1)$-tuples of coefficients. For any point $(c_1, \ldots, c_m) \in B$ write

$$\begin{aligned} & v(c_1 \ldots c_m) \\ & = \Big(\mp c_1 \ldots c_m,\ \pm \sum_{(i_1, \ldots, i_{m-1})} c_{i_1} \ldots c_{i_{m-1}}, \ldots, +\sum_{i,j} c_i c_j, -\sum_j c_j\Big). \end{aligned}$$

By Viète's formulas, the entries of the sequence on the right are the coefficients of a polynomial with zeros at $c_1, \ldots, c_m$. We have thus defined a mapping v of B into P_m.

Let $k\colon \mathscr{L}(\boldsymbol{R}^m, \boldsymbol{R}^m) \to P_m$ be defined as follows:

$$k(A) = \text{characteristic polynomial of } A \in \mathscr{L}(\boldsymbol{R}^m, \boldsymbol{R}^m).$$

It is easy to see that k is an algebraic map from $\boldsymbol{R}^{m^2}$ onto P_m. Let $\mathscr{A}$ denote the set of all linear operators $\boldsymbol{R}^m \to \boldsymbol{R}^m$ which have at least one purely imaginary eigenvalue. Then $\mathscr{A}$ coincides with $k^{-1}(v(B))$.

Now we employ the following two theorems of algebra; their proofs can be found in Abraham and Robbin (1967).

Theorem 3.5.2 The image and the inverse-image of a semi-algebraic set under an algebraic map is also a semi-algebraic set.

Theorem 3.5.3 Every semi-algebraic set in $\boldsymbol{R}^n$ is a finite union of submanifolds.

In virtue of these theorems the set $\mathscr{A} = k^{-1}(v(B)) \subset \mathscr{L}(\boldsymbol{R}^m, \boldsymbol{R}^m)$ is a finite union of submanifolds. Denote them by $S_1, \ldots, S_l$. Every operator $A \in \mathscr{A}$ can be made, by an arbitrarily small perturbation, into an operator that does not belong to $\mathscr{A}$. Therefore the codimension of each S_j is at least 1.

The set $\mathscr{A}$ is closed. To see this, take a sequence $A_n \to A$, $A_n \in \mathscr{A}$. For every A_n there is a vector $z_n \in \boldsymbol{C}^m$, $\|z_n\| = 1$, such that $Az_n = \mathrm{i}\alpha_n z_n$, where $\alpha_n \in \boldsymbol{R}$. We may assume that $z_n \to z_0$ and $\alpha_n \to \alpha_0$ (because $|\alpha_n| \leqslant \|A_n\| \leqslant \text{const}$). Letting $n \to \infty$ we obtain $Az_0 = \mathrm{i}\alpha_0 z_0$, and so $A \in \mathscr{A}$.

Step II. For every chart (U, α) on M, $J^1(M)$ can be locally represented as the product $\alpha(U) \times E \times \mathscr{L}(E, E)$. In view of what has been shown in Step I, the set W is the finite union of submanifolds

$$W_j = \alpha(U) \times \{0\} \times S_j, \qquad j = 1, \ldots, l.$$

Obviously, $\operatorname{codim} W_j \geqslant m+1$, since the second factor is of codimension m and the third is of codimension not less than 1.

W is closed because $\mathscr{A}$ is closed.

Proof of Theorem 3.5.1 (continued) The following conditions are equivalent:

1° $\Phi \in \mathscr{G}_1^r(M)$,

2° $\varrho_\Phi(M) \cap W = \emptyset$,

3° $\varrho_\Phi \pitchfork W_j$ for $j = 1, \ldots, l$.

The implications 1° ⇔ 2° and 2° ⇒ 3° are obvious. Now, assume that 3° holds and that for some $x \in M$

$$\varrho_\Phi(x) \in W,$$

i.e.,

$$\varrho_\Phi(x) \in W_j \text{ for some } j.$$

According to 3° we have

$$\mathrm{d}\varrho_\Phi(x)\big(T_x(M)\big) + T_w(W_j) = T_w\big(J^1(M)\big), \tag{3.5.1}$$

where $w = \varrho_\Phi(x)$. On the other hand,

$$\operatorname{codim} T_w(W_j) \geqslant m+1 \text{ and } \dim T_x(M) = m,$$

and this contradicts (3.5.1). Thus 3° ⇒ 2°.

Since W is closed and M is compact, the set $\mathscr{G}_1^r(M)$ is open, in virtue of Theorem 3.2.2 and the equivalence between conditions 1° and 2°. Using the equivalence between 1° and 3°, we may write $\mathscr{G}_1^r(M)$ as the intersection

$$\mathscr{G}_1^r(M) = \bigcap_{j=1}^{l} H_j, \tag{3.5.2}$$

where H_j is the set of vector fields Φ such that ϱ_Φ is transversal to W_j. If $r \geqslant 2$, then each H_j is residual in $\mathscr{F}^r(M)$ on account of Theorem 3.2.4 (ϱ_Φ is of class C^{r-1} and $r-1 \geqslant 1$). Hence, by (3.5.2), $\mathscr{G}_1^r(M)$ is residual in $\mathscr{F}^r(M)$, and thus is dense in $\mathscr{F}^r(M)$.

It remains to consider the case of $r = 1$. Every open set V in $\mathscr{F}^1(M)$ contains a vector field Φ_0 of class C^2 together with a set V_1 which is a neighbourhood of Φ_0 in $\mathscr{F}^2(M)$. By the conclusion of the preceding case ($r \geqslant 2$), there exists

$$\Phi \in V_1 \cap \mathscr{G}_1^2(M) \subset V \cap \mathscr{G}_1^1(M).$$

This ends the proof.

Corollary 3.5.1 Property $\mathscr{G}_1$ is C^r-generic.

Exercise

3.5.1.A. Let $\varrho_\Phi(x) = j^1_\Phi(x)$, $x \in M$, $\Phi \in \mathscr{F}^r(M)$, $r \geqslant 2$. Show that

$$\mathrm{d}(\mathrm{ev}_\varrho)(\Phi, x)\colon \ \mathscr{F}^r(M) \times T_x(M) \to T_{j^1_\Phi(x)}(J^1(M))$$

is a surjection.

Hint. This follows from the definition of a jet.

3.6. Genericity of Property $\mathscr{G}_2$

In this section we prove the following theorem:

Theorem 3.6.1 *If M is a compact manifold and if* $r \geqslant 1$, *then the set* $\mathscr{G}^r_2(a)$ *is open and dense in* $\mathscr{F}^r(M)$ *for every* $a > 0$.

Proof We first present the general scheme of the proof. The entire procedure is divided into six steps and we now characterize briefly each of them.

In Step I we show that the set $\mathscr{G}^r_2(a)$ is open.

In Step II we show that the set $\mathscr{G}^r_{3/2}(a)$ is open.

In Step III we show that every $\Phi_0 \in \mathscr{G}^r_1(M)$ has a neighbourhood V such that $V \subset \mathscr{G}^r_2(a_0)$ for some a_0.

In Step IV we show that if $V \subset \mathscr{G}^r_1(M)$ is an open set and if $V \cap \mathscr{G}^r_2(a)$ is dense in V, then $V \cap \mathscr{G}^r_{3/2}(\frac{3}{2}a)$ is also dense in V.

In Step V we show that if $V \subset \mathscr{G}^r_1(M)$ is an open set and if $V \cap \mathscr{G}^r_{3/2}(a)$ is dense in V, then $V \cap \mathscr{G}^r_2(a)$ is also dense in V.

Eventually, in Step VI we deduce the assertion of the theorem from the conclusions of Steps I–V.

Step I. We prove that the set $\mathscr{G}^r_2(a)$ is open in $\mathscr{F}^r(M)$. Write

$$G'(\Phi, x, t) = \mathrm{d}\varphi_t(x),$$

where φ_t is the flow generated by the vector field $\Phi \in \mathscr{F}^r(M)$. Then G' maps $\mathscr{F}^r(M) \times M \times \boldsymbol{R}$ into $\mathscr{L}(T(M), T(M))$. Let W be the subset of $\mathscr{L}(T(M), T(M))$ defined as follows: a triple $(x, y, A_{x,y})$ is an element of W iff

1° $x = y$,

2° $A_{x,x}$ has at least two eigenvalues of modulus 1 (counting multiplicities).

It is easy to see that W is a closed subset of $\mathscr{L}(T(M), T(M))$. By the definition of Property $\mathscr{G}_2$, a vector field Φ is in $\mathscr{G}^r_2(a)$ iff $G'(\Phi, x, t) \notin W$

for $(x, t) \in M \times (0, a\rangle$. Assume that $\Phi_0 \in \mathscr{G}_2^r(a)$. Since $\mathscr{G}_2^r(a) \subset \mathscr{G}_\Delta^r(a)$, there exists a neighbourhood V_2 of Φ_0 in $\mathscr{F}^r(M)$ and a number b with $0 < b < a$, such that, for every $\Phi \in V_2$, all closed orbits of Φ have prime periods not less than b (see Corollary 3.4.1). Now, the set $M \times \langle b, a\rangle$ is compact, the mapping G' is continuous and we have

$$G'(\Phi_0, M, \langle b, a\rangle) \cap W = \emptyset. \tag{3.6.1}$$

Consequently there exists a neighbourhood V_3 of Φ_0 such that

$$G'(\Phi, M, \langle b, a\rangle) \cap W = \emptyset$$

for all $\Phi \in V_3$. In virtue of Theorem 3.5.1 there is a neighbourhood V_1 of Φ_0 such that $V_1 \subset \mathscr{G}_1^r(M)$. Thus $V = V_1 \cap V_2 \cap V_3$ is contained in $\mathscr{G}_2^r(a)$, and so the set $\mathscr{G}_2^r(a)$ is open.

Step II. The set $\mathscr{G}_{3/2}^r(a)$ is open in $\mathscr{F}^r(M)$. Indeed, we have

$$\mathscr{G}_{3/2}^r(a) = \mathscr{G}_1^r(a) \cap \mathscr{G}_\Delta^r(a); \tag{3.6.2}$$

$\mathscr{G}_1^r(a)$ is open by the conclusion of Step I, and $\mathscr{G}_\Delta^r(a)$ is open by Theorem 3.4.5.

Step III. Let $\Phi_0 \in \mathscr{G}_1^r(M)$. We show that there are a neighbourhood V of Φ_0 in $\mathscr{G}_1^r(M)$ and a number $a_0 > 0$ such that $\mathscr{G}_2^r(a_0) \cap V = V$ (i.e., $V \subset \mathscr{G}_2^r(a_0)$).

Clearly $\mathscr{G}_0^r(M) \supset \mathscr{G}_1^r(M)$. Hence, by Theorem 3.4.7, there are a neighbourhood V_0 of Φ_0 and a number $b > 0$ such that the prime periods of all closed orbits of any $\Phi \in V_0$ are $\geqslant b$. Now it suffices to put $a_0 = \frac{1}{2}b$ and $V = V_0 \cap \mathscr{G}_1^r(M)$. Obviously $\mathscr{G}_2^r(a_0) \cap V = V$.

Step IV. We will prove that if $V \subset \mathscr{G}_1^r(M)$ is an open set and if $\mathscr{G}_2^r(a) \cap V$ is dense in V, then $\mathscr{G}_{3/2}^r(\frac{3}{2}a) \cap V$ is dense in V.

To this end we first show that the map G given by

$$G(\Phi, x, t) = \big(x, t, \varphi_t(x)\big)$$

is transversal to the diagonal Δ:

$$G \pitchfork_{(\Phi, x, t)} \Delta \tag{3.6.3}$$

at any point (Φ, x, t) with $\Phi \in \mathscr{G}_2^r(a)$, $x \in M$, $t \in (0, \frac{3}{2}a\rangle$. Let us write the derivative of G:

$$\begin{aligned} \mathrm{d}G(\Phi, x, t)(\Psi, \dot{x}, \dot{t}) &= \big(\dot{x}, \dot{t}, \partial_{(x,t)}\varphi_t(x)(\dot{x}, \dot{t}) + \partial_\Phi \varphi_t(x)(\Psi)\big) \\ &= \mathrm{d}f_\Phi(x, t)(\dot{x}, \dot{t}) + \big(0, 0, \partial_\Phi \varphi_t(x)(\Psi)\big), \end{aligned} \tag{3.6.4}$$

where

$$f_{\Phi}(x, t) = G(\Phi, x, t)$$

(cp. notation introduced in Section 3.4).

Suppose that $G(\Phi, x, t) \in \Delta$, i.e., $\varphi_t(x) = x$, $t \in (0, \frac{3}{2}a\rangle$. We have 3 possibilities:

1° x is a critical point,

2° the orbit $\gamma(x)$ passing through x has prime period $\tau \leqslant a$,

3° t is the prime period.

In cases 1° and 2° either x or $\gamma(x)$ is a hyperbolic critical element (since $\Phi \in \mathscr{G}_2^r(a)$), and thus t is a transversal period. This, in view of Theorem 3.4.1, yields

$$f_{\Phi} \pitchfork_{(x,t)} \Delta,$$

whence by (3.6.4) follows (3.6.3) (because $\mathrm{d}f_{\Phi}(x, t)\big(T_x(M) \times \boldsymbol{R}\big)$ contains a complement of Δ to the whole space).

In case 3° $\partial_{\Phi}\varphi_t(x)$ is a surjection, by Theorem 3.3.2. Hence

$$\Delta + \{0\} \times \{0\} \times \partial_{\Phi}\varphi_t(x)\big(\mathscr{F}^r(M)\big) = T_{(x,t,x)}(M \times \boldsymbol{R} \times M).$$

The auxiliary statement (3.6.3) being settled, we may pass to the proof of the assertion of Step IV. According to Theorem 3.2.1, if a mapping is transversal to a submanifold at some point, it remains transversal within a neighbourhood of that point. Hence G is transversal to Δ in some neighbourhood of the set

$$\big(\mathscr{G}_2^r(a) \cap V\big) \times (0, \tfrac{3}{2}a\rangle.$$

By the compactness of M there is an $\varepsilon > 0$ such that

$$G \pitchfork_{(\Phi,x,t)} \Delta$$

for $(\Phi, x, t) \in \big(\mathscr{G}_2^r(a) \cap V\big) \times M \times (0, \frac{3}{2}a + \varepsilon)$.

The set $\mathscr{G}_2^r(a) \cap V$ is open in $\mathscr{F}^r(M)$ (as was shown in Step I), and so it is a Banach manifold. Thus

$$\mathrm{ev} f_{\Phi}(x, t) = G(\Phi, x, t) \pitchfork_{(x,t)} \Delta$$

for $(x, t) \in M \times (0, \frac{3}{2}a + \varepsilon)$ and $\Phi \in \mathscr{A} := \mathscr{G}_2^r(a) \cap V$. In virtue of Theorem 3.2.4 the set A_Δ consisting of all $\Phi \in \mathscr{G}_2^r(a) \cap V$ satisfying

$$f_{\Phi} \pitchfork_{(x,t)} \Delta \qquad \text{for } (x, t) \in M \times (0, \tfrac{3}{2}a + \varepsilon)$$

is residual in $\mathscr{G}_2^r(a) \cap V$; a fortiori, it is dense in $\mathscr{G}_2^r(a) \cap V$. To conclude the proof of the assertion of this step, it remains to observe that

$$A_\Delta \subset \mathscr{G}^r_{3/2}(\tfrac{3}{2}a).$$

Step V. Before formulating the claim of this step we state without proof an auxiliary fact; the proof will be given later on (Theorem 4.3.1 on p. 198):

If $\Phi \in \mathscr{G}^r_{3/2}(a)$ and γ is a closed orbit of Φ with period $\tau \leqslant a$ and if U is a neighbourhood of $\hat{\gamma}$, then there exists a vector field $\Psi \in \mathscr{F}^r(M)$ such that

1° $\Psi|_{\hat{\gamma}} = 0$,

2° $\Psi|_{M-U} = 0$,

3° for $s \in \boldsymbol{R}^+$ sufficiently small, γ is a hyperbolic closed orbit of the vector field $\Phi + s\Psi$.

Using this fact together with Theorem 3.4.7, we now show that if $\Phi \in \mathscr{G}^r_{3/2}(a)$, then there exists a $\Psi \in \mathscr{F}^r(M)$ such that the following conditions are fulfilled for $s \in \boldsymbol{R}^+$ small enough:

4° critical elements with periods $\leqslant a$ are the same for Φ and $\Phi + s\Psi$,

5° if γ is a critical element of Φ of period $\leqslant a$, then γ is a hyperbolic critical element of $\Phi + s\Psi$.

Indeed, Φ has finitely many critical elements with periods not exceeding a. Let $U_1, \ldots, U_k$ be pairwise disjoint neighbourhoods of these elements. For each U_i we find a vector field $\Psi_i \in \mathscr{F}^r(M)$ satisfying conditions 1°–3°, and we put

$$\Psi = \sum_{i=1}^{k} \Psi_i.$$

Then, clearly, the vector field $\Phi + s\Psi$ is in $\mathscr{G}^r_2(a)$ and fulfils conditions 4° and 5° (for $s \in \boldsymbol{R}^+$ small enough).

Our goal in this step is to prove the following statement: if $V \subset \mathscr{G}^r_1(M)$ is an open set and if $V \cap \mathscr{G}^r_{3/2}(a)$ is dense in V, then $V \cap \mathscr{G}^r_2(a)$ is also dense in V.

To this end, take any set V_1 open in V. Since $\mathscr{G}^r_{3/2}(a) \cap V$ is dense in V, there is a field $\Phi \in \mathscr{G}^r_{3/2}(a) \cap V_1$. On account of our previous considerations there exists a vector field Ψ such that $\Phi + s\Psi$ belongs to $\mathscr{G}^r_2(a)$ for small s. Since V_1 is open, we have $\Phi + s\Psi \in V_1$ for s small enough. Thus $\Phi + s\Psi \in V_1 \cap \mathscr{G}^r_2(a)$, showing that $V \cap \mathscr{G}^r_2(a)$ is dense in V, as asserted.

Step VI. Finally we are in a position to prove the theorem. Let a be an arbitrary positive number and let $\Phi \in \mathscr{G}^r_1(M)$. By the conclusion of

Step III there exist an $a_0 > 0$ and a neighbourhood V of Φ such that $\mathscr{G}_2^r(a_0) \cap V$ is dense in V. As we have shown in Step IV, $\mathscr{G}_{3/2}^r(\frac{3}{2}a_0)$ is dense in V; hence, in view of what was proved in Step V, also the set $\mathscr{G}_2^r(\frac{3}{2}a_0)$ is dense in V. It follows by induction that for every $k \in Z^+$ the set $\mathscr{G}_2^r((\frac{3}{2})^k a_0)$ is dense in V. In particular, taking k_0 such that $(\frac{3}{2})^{k_0} a_0 > a$, we get the inclusion

$$\mathscr{G}_2^r(a) \subset \mathscr{G}_2^r((\tfrac{3}{2})^{k_0} a_0),$$

which shows that $\mathscr{G}_2^r(a)$ is dense in V. Since Φ was chosen arbitrarily in $\mathscr{G}_1^r(M)$, it follows that $\mathscr{G}_2^r(a)$ is dense in $\mathscr{G}_1^r(M)$. By Theorem 3.5.1 $\mathscr{G}_1^r(M)$ is a dense subset of $\mathscr{F}^r(M)$. Consequently the set $\mathscr{G}_2^r(a)$ is dense in $\mathscr{F}^r(M)$. This set is also open in $\mathscr{F}^r(M)$ (Step I). The theorem is thus proved.

Corollary 3.6.1 *The set $\mathscr{G}_2^r(M)$ is residual in $\mathscr{F}^r(M)$. This means that Property $\mathscr{G}_2$ is C^1-generic.*

This is an immediate consequence of the last theorem and the set equality

$$\mathscr{G}_2^r(M) = \bigcap_{a \in Z^+} \mathscr{G}_2^r(a).$$

3.7. Genericity of Property $\mathscr{G}_3$

The theorem presented below and asserting the genericity of Property $\mathscr{G}_3$ has been proved independently by two authors: Kupka (1963) and Smale (1963). This theorem seems to be the most significant of all the results concerning generic properties.

Theorem 3.7.1 (Kupka–Smale) *If M is a compact manifold and if $r \geqslant 1$, then the set $\mathscr{G}_3^r(M)$ is residual in $\mathscr{F}^r(M)$, i.e., Property $\mathscr{G}_3$ is C^r-generic.*

Before passing to the proof, let us introduce some notation and recall certain facts concerning genericity.

I. Let $\Phi_0 \in \mathscr{G}_2^r(a)$ for some $a > 0$. In virtue of the Hadamard–Perron theorem, of Definition 1.9.3 and of Theorem 3.4.7, there exist:

(a) a number $\varepsilon > 0$;

(b) a neighbourhood $V(\Phi_0)$ of Φ_0 in $\mathscr{G}_2^r(a)$;

(c) open sets $U_1, \ldots, U_k \subset M$ with disjoint closures;

(d) manifolds $P_1^\sigma, \ldots, P_k^\sigma$, $\sigma = \pm$, each P_i^σ being equal to either $\boldsymbol{R}^s$ or $\boldsymbol{R}^s \times S^1$ or $\boldsymbol{R}^s \dot{\times} S^1$, where s is an integer not greater than $\dim M$;

(e) C^r-immersion $\psi_\Phi^{i,\sigma}$: $P_i^\sigma \to M$, $\Phi \in V(\Phi_0)$, such that:

(f) $V(\Phi_0) \subset \mathscr{G}_2^r(a+\varepsilon)$;

(g) every vector field $\Phi \in V(\Phi_0)$ has exactly k critical elements $\gamma_1^\Phi, \ldots, \gamma_k^\Phi$ with periods $\leqslant a+\varepsilon$, and each γ_i^Φ is contained in U_i, $i = 1, \ldots, k$;

(h) each $\psi_\Phi^{i,+}$ [resp. $\psi_\Phi^{i,-}$] is an immersion onto the stable [unstable] manifold of γ_i^Φ:

$$\psi_\Phi^{i,\sigma}(P_i^\sigma) = W^\sigma(\gamma_i^\sigma), \sigma = \pm.$$

Moreover, if $\Phi_n \to \Phi_0$ in C^1-topology, then $\psi_{\Phi_n}^{i,\sigma} \to \psi_{\Phi_0}^{i,\sigma}$ uniformly on compact subsets of P_i^σ, $\sigma = \pm$.

II. Let ψ_1 and ψ_2 be bijective immersions of manifolds P_1 and P_2 into M. We say that ψ_1 and ψ_2 are *transversal* iff the product map F: $P_1 \times P_2 \to M \times M$ given by $P(p_1, p_2) = (\psi_1(p_1), \psi_2(p_2))$ is transversal to the diagonal Δ in $M \times M$. The transversality of ψ_1 and ψ_2 is equivalent to the transversality of the submanifolds $W_1 = \psi_1(P_1)$ and $W_2 = \psi_2(P_2)$.

III. Let γ_1 and γ_2 be two critical elements of a vector field $\Phi \in \mathscr{G}_2^r(M)$. Let $W^+(\gamma_1) = \psi^+(P_1)$, $W^-(\gamma_2) = \psi^-(P_2)$, where ψ^+ and ψ^- are immersions, and P_1, $P_2 = \boldsymbol{R}^s$ or $\boldsymbol{R}^s \times S^1$ or $\boldsymbol{R}^s \dot{\times} S^1$. Let U_1 and U_2 be neighbourhoods of $\hat{\gamma}_1$ and $\hat{\gamma}_2$. Write

$$\mathscr{A}_{\Phi_0} = \Phi_0 + \mathscr{A},$$

where

$$\mathscr{A} = \{\Psi \in \mathscr{F}^r(M): \Psi(x) = 0 \text{ for } x \in U_1 \cup U_2\}.$$

$\mathscr{A}$ is a closed subset of $\mathscr{F}^r(M)$.

The construction of suitable immersions has been described in Section 1.6. We now briefly recall it.

Let $\gamma_1 = x_1$ be a critical point. Then $P_1 = \boldsymbol{R}^s$, $s = \dim W^+(x_1)$. There exists a mapping f: $U \to D$, where U is a small neighbourhood of x_1 in $W^+(x_1)$, $f(x_1) = 0$ and D is a neighbourhood of 0 in $\boldsymbol{R}^s$; the mapping f depends on the flow $\varphi_t^{\Phi_0}$ (generated by Φ_0) restricted to U. We define a map g: $D \to D$ by

$$g(p) = f \circ \varphi \circ f^{-1}(p),$$

where φ stands for $\varphi_1^{\Phi_0}$ and $p \in D$; then we extend g to the whole of $\boldsymbol{R}^s$. We have $\|dg\| < 1$, and so we may require that the extended g be a contraction of $\boldsymbol{R}^s$ to 0; denote this extension by $\hat{g}$. Further, we define

$$\psi^+_{\Phi_0}(p) = \varphi^{\Phi_0}_{-n} \circ f^{-1} \circ \hat{g}^n(p), \tag{3.7.1}$$

where n is an integer large enough to have $\hat{g}^n(p) \in D$. The mapping defined by (3.7.1) does not in fact depend on n; the value of the composition on the right of (3.7.1) is the same for any two values of n fulfilling the condition: $\hat{g}^n(p) \in D$.

Without restricting generality we may assume that $U = U_1$. The immersion $\psi^- = \psi^-_{\Phi_0}$ is constructed quite analogously.

If γ_1 is a closed orbit, the case can be reduced to the preceding one by considering suitable cross-sections.

Since the vector fields $\Phi \in \mathscr{A}_{\Phi_0}$ are identical with Φ_0 on the set $U_1 \cup U_2$, the immersions

$$\psi^+_{\Phi}(p) = \varphi^{\Phi}_{-n} \circ f^{-1} \circ \hat{g}^n(p)$$

coincide with $\psi^+_{\Phi_0}(p)$ whenever $\psi^+_{\Phi}(p) \in U_1$; similarly, $\psi^-_{\Phi}(p) = \psi^-_{\Phi_0}(p)$ whenever $\psi^-_{\Phi}(p) \in U_2$.

According to Theorem 3.3.1, the assignment $\Phi \mapsto \varphi^{\Phi}_{\pm n}$ is of class C^1 (with respect to $\Phi \in V(\Phi_0)$). Hence, a fortiori, the assignment $\Psi \mapsto \varphi^{\Phi}_{\pm n} = \varphi^{\Phi_0 + \Psi}_{\pm n}$ is of class C^1 with respect to $\Psi \in \mathscr{A}$. It follows that ψ^+_{Φ}, ψ^-_{Φ} are of class C^1, regarded as functions of $\Psi \in \mathscr{A}$, $\Psi = \Phi - \Phi_0$.

Now consider the product map

$$F_{\Phi}\colon\ P_1 \times P_2 \to M \times M$$

given by

$$F_{\Phi}(p_1, p_2) = (\psi^+_{\Phi}(p_1), \psi^-_{\Phi}(p_2)).$$

We will prove

Lemma 3.7.1 *The evaluation map*

$$\mathrm{ev}_F\colon\ \mathscr{A} \times P_1 \times P_2 \to M \times M$$

is transversal to the diagonal $\Delta \subset M \times M$.

Proof The transversality of $W^+(\gamma_1)$ and $W^-(\gamma_2)$ is invariant under the action of the flow φ^{Φ}_t, and thus it suffices to examine it at some (freely chosen) point $x \in W^+(\gamma_1) \cap W^-(\gamma_2)$. Suppose that

$$(\psi^+_{\Phi_0}(p_1), \psi^-_{\Phi_0}(p_2)) \in \Delta$$

for some p_1, p_2. Denote the common value of $\psi^+_{\Phi_0}(p_1)$ and $\psi^-_{\Phi_0}(p_2)$ by x. If $\gamma_1 \neq \gamma_2$, then (in view of $\bar{U}_1 \cap \bar{U}_2 = \emptyset$) we may assume that

$x \notin \bar{U}_1 \cup \bar{U}_2$. (Otherwise, say if $x \in \bar{U}_1$, by pushing x along the trajectory we find a moment at which the point has left $\bar{U}_1$ without entering $\bar{U}_2$; and translation along a trajectory does not impair transversality.) If $\gamma_1 = \gamma_2$, we take $U_1 = U_2$ small enough to be sure that (by hyperbolicity) there is no other orbit of Φ_0 different from γ_1 and entirely contained in U_1. Thus also in this case we may assume that $x \notin \bar{U}_1 = \bar{U}_1 \cup \bar{U}_2$.

Let us write the derivative of ev_{F} at the point $(\Psi = 0, p_1, p_2)$:

$$\begin{aligned} &\mathrm{d}(\mathrm{ev}_F)(0, p_1, p_2)(\dot{\Psi}, \dot{p}_1, \dot{p}_2) \\ &= \partial_\Psi \mathrm{ev}_F(0, p_1, p_2)(\dot{\Psi}) + \partial_{(p_1,p_2)} \mathrm{ev}_F(0, p_1, p_2)(\dot{p}_1, \dot{p}_2). \end{aligned}$$

The first summand on the right is of the form:

$$\partial_\Psi \mathrm{ev}_F(0, p_1, p_2)(\dot{\Psi}) = \big(\partial_\Psi \psi^{+}_{\Phi_0}(p_1)(\dot{\Psi}), \partial_\Psi \psi^{-}_{\Phi_0}(p_2)(\dot{\Psi})\big).$$

Now, the field $\dot{\Psi}$ can be defined arbitrarily in small neighbourhoods of $\bar{U}_1$ and $\bar{U}_2$ (provided those neighbourhoods are disjoint). Thus for every pair of vectors $\dot{x}_1, \dot{x}_2 \in T_x(M)$ there exists a vector field $\dot{\Psi} \in \mathscr{A}$ such that

$$\partial_\Psi \psi^{+}_{\Phi_0}(p_1)(\dot{\Psi}) = x_1, \quad \partial_\Psi \psi^{-}_{\Phi_0}(p_2)(\dot{\Psi}) = x_2 \tag{3.7.2}$$

(see Exercise 3.7.1 on p. 179; also Theorem 3.3.2). Hence

$$T(\Delta) + \partial_\Psi \mathrm{ev}_F(0, p_1, p_2)(\mathscr{A} \times \{0\} \times \{0\}) = T_x(M) \times T_x(M),$$

ending the proof of the lemma.

Proof of Theorem 3.7.1 Let a be any positive number. For every $\Phi_0 \in \mathscr{G}^r_2(a)$ there exists a neighbourhood $V(\Phi_0)$ satisfying conditions (a)–(e) on p. 172. By Theorem 3.6.2 $\mathscr{G}^r_2(a)$ is an open subset of the separable Banach space $\mathscr{F}^r(M)$; thus $\mathscr{G}^r_2(a)$ is a separable paracompact space. Consequently the cover $\{V(\Phi): \Phi \in \mathscr{G}^r_2(a)\}$ contains a countable subcover, which is locally finite. Let $\{V_c\}$, $c = 1, 2, \ldots$, denote such subcover. Hence, to each V_c there correspond: numbers ε_c and k_c, manifolds $P^\sigma_{i,c}$, $i = 1, \ldots, k_c$, $\sigma = \pm$, and immersions $\psi^{i,\sigma}_{\Phi,c}$ satisfying conditions (f)–(h).

Let b be any natural number. Each $P^\sigma_{i,c}$ is either $\boldsymbol{R}^s$ or $\boldsymbol{R}^s \times S^1$ or $\boldsymbol{R}^s \dot{\times} S^1$. Let $K(P^\sigma_{i,c}, b)$ be the set in $P^\sigma_{i,c}$ defined as follows:

$$K(P^\sigma_{i,c}, b) = \{e \in \boldsymbol{R}^s: \|e\| \leqslant b\} \qquad \text{if } P^\sigma_{i,c} = \boldsymbol{R}^s$$

and

$$K(P^{\sigma}_{i,c}, b) = \{(e, t) : t \in S^1,\ e \in \mathbf{R}^s,\ \|e\| \leqslant b\}$$

if $P^{\sigma}_{i,c} = \mathbf{R}^s \times S^1$ or $\mathbf{R}^s \dot{\times} S^1$.

Clearly, $K(P^{\sigma}_{i,c}, b)$ is a compact subset of $P^{\sigma}_{i,c}$.

Let $\mathscr{G}^r_3(a, b, c; i, j)$ denote the set of all vector fields $\Phi \in V_c$ such that

$$\bar{W}^+(\gamma_i^{\Phi}) \pitchfork \bar{W}^-(\gamma_j^{\Phi}) \qquad \text{on } D_1 \times D_2,$$

where

$$D_1 = \psi^{i,+}_{\Phi,c}(K(P^+_{i,c}, b)), \qquad D_2 = \psi^{j,-}_{\Phi,c}(K(P^-_{j,c}, b)).$$

On account of our previous observation (II; p. 173), the last condition expressing transversality between the stable manifold of γ_i^{Φ} and the unstable manifold of γ_j^{Φ} can be rewritten in the following alternative form:

$$\psi^{i,+}_{\Phi,c} \times \psi^{j,-}_{\Phi,c} \pitchfork_{(p_1, p_2)} \Delta$$

at every point

$$(p_1, p_2) \in K(P^+_{i,c}, b) \times K(P^-_{j,c}, b).$$

Since Δ is a closed set and $K(P^+_{i,c}, b)$, $K(P^-_{j,c}, b)$ are compact, Theorem 3.2.3 shows that the set $\mathscr{G}^r_3(a, b, c; i, j)$ is open.

Denote:

$$\mathscr{G}^r_3(a, \infty, c; i, j) := \bigcap_{b=1}^{\infty} \mathscr{G}^r_3(a, b, c; i, j).$$

We shall show that this set is dense in V_c. Let $\Phi_0 \in V_c$ and let $\mathscr{A}_{\Phi_0}$ be defined as in III. Writing $\dim M = m$, we have two possibilities:

1° $r \geqslant m+1$; since $\operatorname{codim} \Delta = m$ and

$$\dim P^+_{i,c} \times P^-_{j,c} \leqslant \dim M + \dim M = 2m,$$

we have

$$\dim P^+_{i,c} \times P^-_{j,c} - \operatorname{codim} \Delta \leqslant m < r.$$

Hence by Lemma 3.7.1 and Theorem 3.2.4 the set $\mathscr{G}^r_3(a, \infty, c; i, j)$ is residual in V_c, hence is dense in V_c.

2° $r < m+1$; then every vector field Φ_0 in V_c can be approximated by fields of class C^{r_1}, $r_1 \geqslant m+1$, and the case easily reduces to the preceding one.

Let us now set

$$\mathscr{G}_3^r(a, b) := \bigcup_{c=1}^{\infty} \bigcap_{i,j=1}^{k_c} \mathscr{G}_3^r(a, b, c; i, j).$$

It is easily seen that

$$\mathscr{G}_3^r(a, b_1) \supset \mathscr{G}_3^r(a, b_2) \qquad \text{for } b_1 < b_2.$$

We will prove the following set equality:

$$\mathscr{G}_3^r(a) = \bigcap_{b=1}^{\infty} \mathscr{G}_3^r(a, b).$$

We recall that $\mathscr{G}_3^r(a)$ is the set of all vector fields in $\mathscr{G}_2^r(a)$ for which the manifolds, stable and unstable, of critical elements with prime periods $\leqslant a$ intersect transversally.

The inclusion $\mathscr{G}_3^r(a) \subset \bigcap_b \mathscr{G}_3^r(a, b)$ is evident. To prove the opposite inclusion, assume that $\Phi \in \mathscr{G}_3^r(a, b)$, $b = 1, 2, \ldots$ This means that for every b there exists a number $c(b)$ such that

$$\Phi \in \bigcap_{i,j=1}^{k_{c(b)}} \mathscr{G}_3^r\big(a, b, c(b); i, j\big).$$

Since the cover $\{V_c\}$ is locally finite, Φ belongs to finitely many V_c, and thus some integer c_0 must enter the sequence $\{c(b)\}$ infinitely many times. Consequently,

$$\Phi \in \bigcap_{i,j=1}^{k_{c_0}} \mathscr{G}_3^r(a, b, c_0; i, j) \qquad \text{for all } b = 1, 2, \ldots$$

It follows that

$$\psi_{\Phi,c_0}^{i,+} \times \psi_{\Phi,c_0}^{j,-} \pitchfork_{(p_1,p_2)} \Delta$$

at every point (p_1, p_2), without any restriction on p_1 and p_2. Thus

$$W^+(\gamma_i^\Phi) \pitchfork W^-(\gamma_j^\Phi)$$

for any two critical elements γ_i^Φ and γ_j^Φ with prime periods not greater than a.

Each of the sets $\mathscr{G}_3^r(a, b, c; i, j)$ is open; hence so is the set $\mathscr{G}_3^r(a, b)$, for any b. Since $\mathscr{G}_3^r(a, \infty, c; i, j)$ is dense in V_c, automatically $\mathscr{G}_3^r(a, b, c; i, j)$ is also dense in V_c. This shows that $\mathscr{G}_3^r(a, b)$ arises from open dense sets by taking finite intersections and a countable union. It follows that $\mathscr{G}_3^r(a, b)$ is dense in V_c.

The family V_c is an open cover of $\mathscr{G}_2^r(a)$, and $\mathscr{G}_2^r(a)$ is dense in $\mathscr{F}^r(M)$ (Theorem 3.6.1). Thus each $\mathscr{G}_3^r(a, b)$, $b = 1, 2, \ldots$, is dense in $\mathscr{F}^r(M)$, and so $\mathscr{G}_3^r(a)$ is residual in $\mathscr{F}^r(M)$.

Now, a has been chosen arbitrarily. We have

$$\mathscr{G}_3^r(a_1) \supset \mathscr{G}_3^r(a_2) \qquad \text{for } a_1 < a_2.$$

Therefore

$$\mathscr{G}_3^r(M) = \bigcap_{a=1}^{\infty} \mathscr{G}_3^r(a).$$

Each of the sets $\mathscr{G}_3^r(a)$ is residual in $\mathscr{F}^r(M)$; thus $\mathscr{G}_3^r(M)$ is also residual in $\mathscr{F}^r(M)$, and this is precisely the assertion of our theorem. The proof is complete.

The Kupka–Smale theorem admits certain modifications. For instance, let M be a compact Riemannian manifold. We denote by $\text{grad}^r(M)$ the space of all gradient vector fields of class C^r on M. Clearly, $\text{grad}^r(M)$ is a closed linear subspace of $\mathscr{F}^r(M)$; in fact, $\text{grad}^r(M)$ is isomorphic to the space $C^r(M)$ of all real-valued C^r-functions on M. The proof of Theorem 3.7.1 works without any essential changes in the case of the space $\text{grad}^r(M)$. And thus, we have:

Theorem 3.7.2 If M is compact, then Property $\mathscr{G}_3$ is generic in the space $\text{grad}^r(M)$, $r \geqslant 1$.

Let us now turn our attention to the space $\text{Diff}^r(M)$; this is a complete metric space. The concept of genericity is introduced in the space of diffeomorphisms exactly in the same way as in the space of vector fields. The Kupka–Smale theorem for diffeomorphisms reads as follows:

Theorem 3.7.3 If M is a compact manifold, then the following properties are generic in the space $\text{Diff}^r(M)$, $r \geqslant 1$:

(1) *every periodic point (of the diffeomorphism under consideration) is hyperbolic;*

(2) *for every pair of periodic points $p, q \in M$, the manifolds $W^+(p)$ and $W^-(p)$ intersect transversally.*

Theorem 3.7.3. is easily derived from Theorem 3.7.1 by passing to suspensions of diffeomorphisms $\varphi \in \text{Diff}^r(M)$.

Exercise

3.7.1.A. Prove the statement including formulas (3.7.2) in the proof of Lemma 3.7.1.

3.8. Non-Wandering Points. Closing Lemma. Genericity of Properties $\mathscr{G}_4$ and $\mathscr{G}_5$

We now introduce the important concept of non-wandering points. Let $\Phi \in \mathscr{F}^r(M)$ and let φ_t denote as usual the flow on M generated by Φ.

Definition 3.8.1 A point $x \in M$ is called *wandering* iff there exist a neighbourhood U of x and a number $t_0 > 0$ such that

$$\bigcup_{|t| > t_0} \varphi_t(U) \cap U = \emptyset.$$

It immediately follows from this definition that the set of wandering points is open and φ_t-invariant.

Definition 3.8.2 A point $x \in M$ which is not wandering is called *non-wandering*. The set of all non-wandering points with respect to a vector field Φ will be denoted by $\Omega(\Phi)$.

In other words: a point $x \in M$ is non-wandering iff for every neighbourhood U of x there is a sequence (t_n), $|t_n| \to +\infty$, such that each $\varphi_{t_n}(U)$ intersects U (see Fig. 3.8.1).

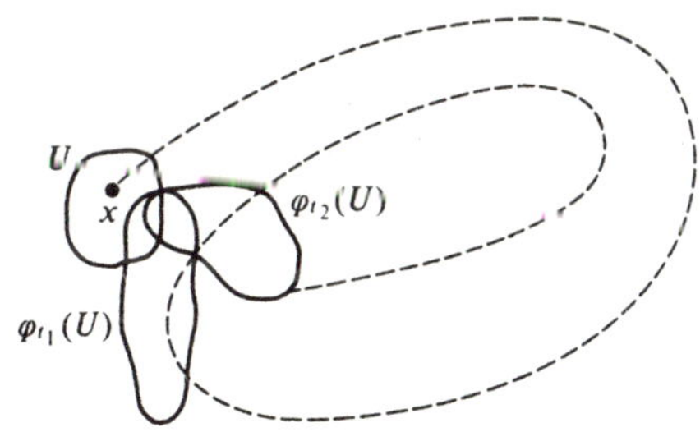

FIG. 3.8.1

Definition 3.8.3 A point $x \in M$ is called *recurrent* iff $x \in \omega(\gamma_x) \cup \alpha(\gamma_x)$, where γ_x is the trajectory through x and $\omega(\gamma_x)$, $\alpha(\gamma_x)$ denote the limit sets defined in Section 1.11.

If $x \in \omega(\gamma_x)$, we say that x is *ω-recurrent*; if $x \in \alpha(\gamma_x)$, we say that x is *α-recurrent*.

It is easy to note that every recurrent point is non-wandering. The converse is not necessarily true. E.g., consider a flow (on a two-dimensional manifold M) behaving on a fragment of M as shown in Fig. 3.8.2 (x_1, x_2, x_3, x_4 are critical points). Then all points which lie on the trajectories $\gamma_1, \gamma_2, \gamma_3, \gamma_4$ are non-wandering, yet not α-recurrent.

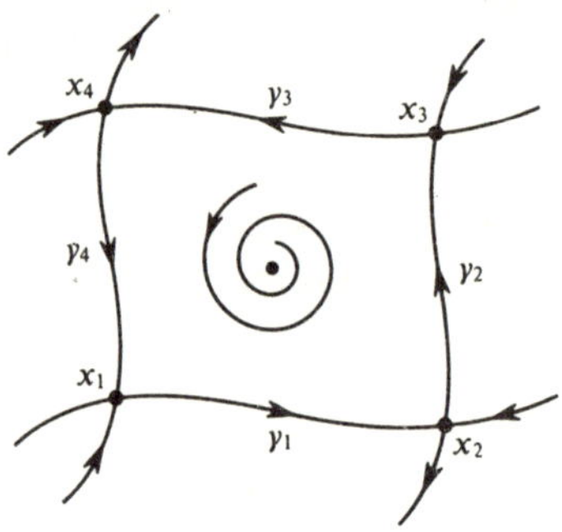

FIG. 3.8.2

Exercises

3.8.1.A. Show that Definition 3.8.2 is equivalent to the following: a point $x \in M$ is non-wandering iff for every neighbourhood U of x there is a sequence (t_n), $|t_n| \to +\infty$, such that $\varphi_{t_n}(U) \cap U \neq \emptyset$, $n = 1, 2, \ldots$

3.8.2.B. Find an example of a point which is ω-recurrent but not α-recurrent.

Hint. Take a constant vector field on the torus T^2 without closed orbits (then every orbit is dense in T^2); multiply that field by a function vanishing at exactly one point.

Non-wandering points are defined quite similarly for the case of a single diffeomorphism $\varphi \in \mathrm{Diff}^r(M)$.

Definition 3.8.4 A point $x \in M$ is called *wandering* iff there exists a neighbourhood U of x such that

$$\bigcup_{|n|>0} \varphi^n(U) \cap U = \emptyset.$$

Definition 3.8.5 A point $x \in M$ which is not wandering is called *non-wandering*. The set of all non-wandering points with respect to a diffeomorphism φ will be denoted by $\Omega(\varphi)$.

As before, $\Omega(\varphi)$ is a closed set, invariant under φ.

We now state a theorem on non-wandering points, the famous "Closing Lemma" proved by Pugh (1967a, 1967b). The formulation is astonishingly simple, unlike the proof, which is long and extremely difficult; the two papers have jointly some 60 pages.

Theorem 3.8.1 (Closing Lemma) Let $\Phi_0 \in \mathscr{F}^1(M)$, M compact. Suppose that $x \in M$ is a non-wandering point for Φ_0. Then every neighbourhood V of Φ_0 in $\mathscr{F}^1(M)$ contains a vector field Φ such that x lies on a certain closed orbit of Φ.

We do not present a proof.

It should be emphasized that the theorem is concerned exclusively with vector fields of class C^1. An analogous statement for fields of class C^0 (i.e. continuous) is very easy to prove (Exercise 3.8.3); as regards fields of a higher class of regularity, the question is still open.

Problem Is the Closing Lemma true for $\Phi_0 \in \mathscr{F}^r(M)$, where $r \geqslant 2$? More precisely: let $x \in M$ be a non-wandering point for a vector field $\Phi_0 \in \mathscr{F}^r(M)$, $r \geqslant 2$. Does every neighbourhood V of Φ_0 in $\mathscr{F}^r(M)$ necessarily contain a field Φ such that x lies on a closed orbit of Φ?

Exercise

3.8.3.B. Let $\Phi_0 \in \mathscr{F}^0(M)$ be a vector field generating a flow φ_t. Suppose that $x \in M$ is a non-wandering point. Show that in every neighbourhood V of Φ_0 in $\mathscr{F}^0(M)$ there is a field $\Phi \in V$ such that x lies on some closed orbit of Φ.

Hint. Perturbate Φ_0 locally until the orbit through x closes.

We recall that the symbol Γ_Φ denotes the set of all points of M belonging to critical elements of Φ.

Definition 3.8.6 We say that the field $\Phi \in \mathscr{F}^r(M)$ has *Property* $\mathscr{G}_4$ iff $\Omega(\Phi) = \overline{\Gamma_\Phi}$.

We are going to show that Property $\mathscr{G}_4$ is generic. The proof requires some facts of general topology, which we now briefly discuss.

Let Y be a compact metric space. Denote by FY the class of all closed subsets of Y. Then FY is a metric space with metric

$$d(A, B) = \max\left(\sup_{a\in A} \varrho(a, B),\ \sup_{b\in B} \varrho(A, b)\right), \tag{3.8.1}$$

ϱ denoting the metric in Y. We call d the *Hausdorff metric* in FY.

Exercise

3.8.4.A. Show that formula (3.8.1) defines a metric in FY.

For a sequence of sets $A_n \in FY$ define

$$\liminf_{n\to\infty} A_n := \{y \in Y\colon\ y = \lim y_n,\ \text{where } y_n \in A_n\}.$$

This set is an element of FY.

Let $g\colon X \to FY$ be a set-valued function; X is assumed to be a metric space.

Definition 3.8.7 The function g is said to be *lower semi-continuous* (abbr.: l.s.c.) iff for every $x \in X$ and every sequence (x_n) converging to x we have

$$\liminf_{n\to\infty} g(x_n) \supset g(x).$$

Lemma 3.8.1 For Y compact, every monotonically increasing sequence of l.s.c. functions converges to an l.s.c. function. This means that if $g_n\colon X \to FY$ *are l.s.c. and* $g_n(x) \subset g_{n+1}(x)$, $n = 1, 2, \ldots$, *then the limit g given by*

$$g(x) = \lim_{n\to\infty} g_n(x) := \overline{\bigcup_n g_n(x)}$$

is l.s.c.

Lemma 3.8.2 For Y compact, the set of points of continuity of any l.s.c. function is residual in X.

We omit the proofs of the two lemmas; they can be found in Kuratowski (1961), pp. 38–40.

In the application that follows, $\mathscr{F}^1(M)$ will be taken for X and M for Y (by assumption, M is compact). Let $\Gamma_n(\Phi)$ denote the set of all periodic points of Φ with periods $\leqslant n$ (including critical points). As before, let $\mathscr{G}_3^1(M)$ denote the set of all vector fields $\Phi \in \mathscr{F}^1(M)$ sharing Property $\mathscr{G}_3$. By Theorem 3.8.1 $\mathscr{G}_3^1(M)$ is a residual subset of $\mathscr{F}^1(M)$.

Lemma 3.8.3 *The assignment $\Phi \mapsto \Gamma_n(\Phi)$ is an l.s.c. map from $\mathscr{G}_3^1(M)$ into FM.*

Proof Let $\Phi_0 \in \mathscr{G}_3^1(M)$; hence, a fortiori, $\Phi_0 \in \mathscr{G}_\Delta^1(M)$. In view of Theorem 3.4.6, Φ_0 has finitely many critical points and closed orbits with periods $\leqslant n$. Moreover, if Φ is sufficiently close to Φ_0, then Φ has precisely as many critical elements with periods $\leqslant n$ as Φ_0. Consequently, if $\Phi_k \to \Phi_0$, then for any critical element of Φ_0 (denote it by γ_0) with period $\leqslant n$ and for any sufficiently small neighbourhood U of $\hat{\gamma}_0$ there exists exactly one critical element of Φ_k (say γ_k) with period $\leqslant n$, contained in U and thus close to γ_0 in the metric d. Hence

$$\liminf_{k\to\infty} \Gamma_n(\Phi_k) \supset \Gamma_n(\Phi_0),$$

which ends the proof of the lemma.

Now we define a mapping

$$\bar{\Gamma}\colon\ \mathscr{G}_3^1(M) \to FM$$

by:

$$\bar{\Gamma}(\Phi) = \overline{\bigcup_{n=1}^{\infty} \Gamma_n(\Phi)}\,.$$

Lemma 3.8.4 *The set of continuity points of $\bar{\Gamma}$ is residual in $\mathscr{G}_3^1(M)$.*

Proof $\bar{\Gamma}$ is the limit of the increasing sequence (Γ_n). By Lemma 3.8.3 each Γ_n is l.s.c., and so is Γ, by Lemma 3.8.1.

Theorem 3.8.2 (*Pugh*) *Let G be the set of continuity points of the mapping $\bar{\Gamma}\colon \mathscr{G}_3^1(M) \to FM$. If $\Phi \in G$ then $\Omega(\Phi) = \bar{\Gamma}(\Phi)$.*

Proof It is easy to see that $\Omega(\Phi) \supset \bar{\Gamma}(\Phi)$ for every $\Phi \in \mathscr{F}^1(M)$. Assume that the inclusion is strict for some Φ_0. Thus let $x_0 \in \Omega(\Phi_0) - \bar{\Gamma}(\Phi_0)$. By the Closing Lemma there exists a vector field $\Phi \in \mathscr{F}^1(M)$ arbitrarily close to Φ_0 and such that the trajectory of Φ passing through x_0 is closed. Moreover, it may be assumed that $\Phi \in \mathscr{G}_3^1(M)$; this is a consequence of the Kupka–Smale theorem (we leave the details to the reader). In this way we find a sequence of vector fields $\Phi_n \in \mathscr{G}_3^1(M)$ such that x_0 belongs to a closed orbit of each of them and $\Phi_n \to \Phi_0$ in $\mathscr{F}^1(M)$. Then $x_0 \in \bar{\Gamma}(\Phi_n)$. On the other hand, we have assumed that $x_0 \notin \bar{\Gamma}(\Phi_0)$. This shows that the mapping $\bar{\Gamma}$ is discontinuous at Φ_0; the theorem is thus proved.

Corollary 3.8.1 (*Pugh*) *Property $\mathscr{G}_4$ is C^1-generic.*

Proof In virtue of the Kupka–Smale theorem the set $\mathscr{G}_3^1(M)$ is residual in $\mathscr{F}^1(M)$. By Lemmas 3.8.2 and 3.8.4 the set G of continuity points of $\bar{\Gamma}$ is residual in $\mathscr{G}_3^1(M)$. A residual subset of a residual subset is itself residual. Hence G is residual in $\mathscr{F}^1(M)$. On account of Theorem 3.8.2, this implies that Property $\mathscr{G}_4$ is generic.

Definition 3.8.8 By a *regular first integral* of a vector field $\Phi \in \mathscr{F}^r(M)$ we understand any real-valued function f of class C^m on M, $m = \dim M$, which is non-constant on any open set in M and which is constant along the trajectories of Φ.

Definition 3.8.9 We say that a field $\Phi \in \mathscr{F}^r(M)$ has *Property* $\mathscr{G}_5$ iff Φ has no regular first integral.

Theorem 3.8.3 (*Thom*) *If M is compact, then Property $\mathscr{G}_5$ is C^1-generic.*

Proof It suffices to show that any field $\Phi \in \mathscr{G}_3^1(M) \cap \mathscr{G}_4^1(M)$ has no regular first integral (as usual, $\mathscr{G}_4^1(M) := \{\Phi \in \mathscr{F}^1(M)\colon \Phi$ has Property $\mathscr{G}_4\}$).

Assume, on the contrary, that $\Phi_0 \in \mathscr{G}_3^1(M) \cap \mathscr{G}_4^1(M)$ and that Φ_0 has a regular first integral f. Write $f(M) = \langle a, b \rangle$. By virtue of Sard's theorem, there is an interval $(\alpha, \beta) \subset \langle a, b \rangle$ such that all numbers in (α, β) are regular values of f. Let $y_0 \in (\alpha, \beta)$. Then $f^{-1}(y_0)$ is a submanifold of M of codimension 1.

Observe that $f^{-1}(y_0)$ is disjoint with Γ_{Φ_0}. Indeed, supposing that some critical element γ has a point in common with $f^{-1}(y_0)$ and taking into account that f is constant along γ, we see that $\hat{\gamma} \subset f^{-1}(y_0)$ and $W^+(\gamma)$, $W^-(\gamma) \subset f^{-1}(y_0)$, whence

$$\dim\big(T_x(W^+(\gamma)) + T_x(W^-(\gamma))\big) \leqslant \dim f^{-1}(y_0) = m-1 \tag{3.8.2}$$

for any $x \in \hat{\gamma}$. On the other hand, Φ_0 has Property $\mathscr{G}_3$ (by assumption), and so $W^+(\gamma) \pitchfork W^-(\gamma)$, which yields

$$\dim\big(T_x(W^+(\gamma)) + T_x(W^-(\gamma))\big) = m. \tag{3.8.3}$$

The contradiction between (3.8.2) and (3.8.3) shows that $f^{-1}(y_0) \cap \Gamma_{\Phi_0} = \emptyset$, as claimed.

Since y_0 was chosen arbitrarily in (α, β), we obtain

$$f^{-1}((\alpha, \beta)) \cap \Gamma_{\Phi_0} = \emptyset.$$

Now, let $x_0 \in f^{-1}(y_0)$, $y_0 \in (\alpha, \beta)$, and let $\bar{x} \in \omega(\gamma_{x_0})$ where γ_{x_0} denotes the trajectory of Φ_0 passing through x_0. Since $f|\hat{\gamma}_{x_0} = \text{const}$, we have $f(\bar{x}) = f(x_0)$, i.e., $\bar{x} \in f^{-1}(y_0)$. The point $\bar{x}$ is non-wandering; consequently, the assumption that $\Phi_0 \in \mathscr{G}_4^1(M)$ implies that x is a cluster point of the set Γ_{Φ_0}. However, this is impossible because $\bar{x}$ belongs to the set $f^{-1}((\alpha, \beta))$, which is open in M and disjoint with Γ_{Φ_0}. The proof is thus complete.

4

Structural Stability. Homoclinic Points

4.1. The Idea of Stability. Structural Stability

One of the most essential problems of mechanics is to what extent a small change of initial data or of external forces influences the orbits of motion. In the case where a small change of the above parameters involves a small change of the trajectories, the motion is said to be stable. Of course, such a formulation lacks precision; this is just putting the matters in terms of "common sense language". Exact definitions are given below.

The dynamical systems which are actually encountered in the natural world are stable, as a rule. Otherwise they alter under the action of casual perturbations until they become stable. All the same, certain unstable motions are in fact observed. E.g. in celestial mechanics the comets are regarded as highly unstable bodies, contrary to the planets, whose motion about the sun may be justifiably called stable; it has lasted, without serious changes, for some 4 milliard years and nothing indicates, in spite of certain minor perturbations of random nature, that the situation might change essentially. By "minor perturbations" we mean here events like collisions between planetoids or the bombardment of the earth by meteorites, e.g. the Arizonian or the Tungusian (the former was in fact a small planetoid).[41]

Of course, empirically, one can hardly require a better proof of the stability of the solar system than the 4 milliard years of its life. But this does not answer the question whether, from the theoretical viewpoint, the system is indeed stable, or whether it is just a matter of luck that a serious catastrophe has been so far avoided. This natural question was posed by astronomers, mathematicians and philosophers of the 18-th century. Many scientists have attempted to solve this problem, Laplace, Lagrange, Poisson, Dirichlet among them. All of them have claimed to know the solution, but in reality, Dirichlet was the only one who, maybe, found some proof. Yet he died without having written it down.

In the second half of the 18-th century the king of Sweden founded a reward for a solution to this problem. Eventually, the solution given by Poincaré in 1889 was accepted as convincing. From the viewpoint of present-day mathematics, Poincaré's reasoning contains several omissions and inaccuracies; we would hardly call it a proof today. Nevertheless, Poincaré's work contains various ingenious intuitions; it has given an inspiring impulse to the research work of many mathematicians of today.

Further significant results in the study of the stability of the solar system have been obtained by Arnold, Kolmogorov and Moser. More information covering the subject can be found in Moser (1973).

There exist several definitions of stability, referring to various types of objects. And thus, one can speak of the stability of particular orbits (e.g. stability in the sense of Lyapunov), as well as of the stability of the whole system of orbits of a flow on a manifold, i.e., stability of the vector field generating the flow. It is the last concept that will be our concern in this book.

Definition 4.1.1 (1) Let $\Phi_0 \in \mathscr{F}^r(M)$ be a given vector field on a compact manifold M. The field Φ_0 is called *C^r-structurally stable*, $r \geqslant 1$, iff there exists a neighbourhood V of Φ_0 in $\mathscr{F}^r(M)$ such that for every $\Phi \in V$ the vector fields Φ_0 and Φ have equivalent phase portraits.

(2) Let $\varphi_0 \in \mathrm{Diff}^r(M)$. The diffeomorphism φ_0 is called *C^r-structurally stable* iff there exists a neighbourhood V of φ_0 in $\mathrm{Diff}^r(M)$ such that for every $\varphi \in V$ the diffeomorphisms φ_0 and φ are topologically conjugate.

Given a vector field Φ_0 on M which is C^r-stable in the sense of Definition 4.1.1, one can additionally demand that the following condition be satisfied:

(ε) For every neighbourhood W of the identity in the space of all homeomorphisms of M onto M there exists a neighbourhood $V = V(W)$ of Φ_0 in $\mathscr{F}^r(M)$ such that for any $\Phi \in V$ there is a homeomorphism $h \in W$ realizing the equivalence between the phase portraits of Φ_0 and Φ.

In the case of M being a Riemannian manifold, condition (ε) can be alternatively formulated as follows:

For every $\varepsilon > 0$ there exists a neighbourhood $V = V(\varepsilon)$ of Φ_0 in $\mathscr{F}^r(M)$ such that for any $\Phi \in V$ there is a homeomorphism h realizing the equivalence between the phase portraits of Φ_0 and Φ and fulfilling the condition:

$$\varrho(x, h(x)) < \varepsilon \qquad \text{for all } x \in M.$$

Problem Is the definition of structural stability with condition (ε) essentially more restrictive than the one in which that condition is not imposed?

The set of all C^r-structurally stable vector fields on a manifold M will be denoted by the symbol $\Sigma^r(M)$. The set of all C^r-structurally stable diffeomorphisms of M will be denoted by $\sigma^r(M)$.

Examples

4.1.1. Let $M = S^1 = \langle 0, 1\rangle$ (identify 0 with 1) and let φ be defined as in Fig. 4.1.1. If ψ is C^1-close to φ, then ψ has exactly two fixed points: the one close to 0, the other close to x_0. Denote these fixed points by a and b, respectively. According to the conclusion of Example 1.2.1 on p. 17, the systems $(\langle 0, x_0\rangle, \varphi)$ and $(\langle a, b\rangle, \psi)$ are topologically conjugate, as are also the systems $(\langle x_0, 1\rangle, \varphi)$ and $(\langle b, a\rangle, \psi)$. It is not difficult to verify that the homeomorphisms which establish topological conjugacy on the two intervals can be pieced together to a homeomorphism of the entire circle. It follows that the systems (S^1, φ) and (S^1, ψ) are topologically conjugate, and thus the system (S^1, φ) is structurally stable.

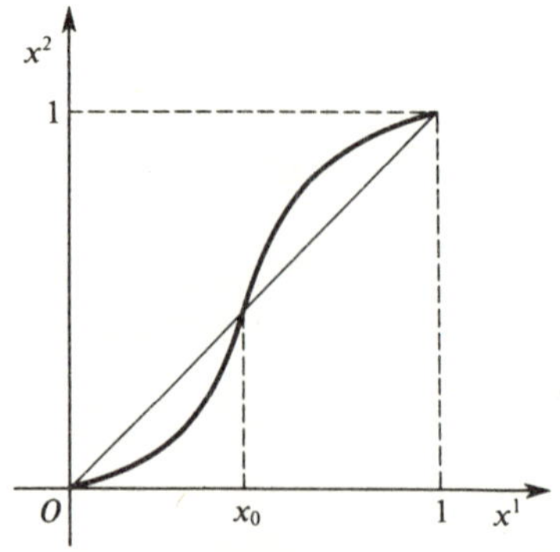

FIG. 4.1.1

4.1.2. Let $M = S^1$ and let r be a rotation; r is not structurally stable. Indeed, in any neighbourhood of r there are rational rotations and irrational rotations. The two types of rotation are topologically quite different (the orbits are periodic in the one case and dense in the other); hence r is not structurally stable.

4.1.3. Recall examples (a), (b) and (c) considered in Exercise 1.4.4 on p. 36 (the torus in 3 positions). The flows (a) and (b) are not structurally stable: in the case of the (b)-flow, the effect of a small change of the angle α is that the resulting flow (position (c)) has finitely many critical points, whereas the original flow has infinitely many such points. As regards case (a), observe that the critical points are connected by trajectories as follows: x_4 with x_3, x_3 with x_2, x_2 with x_1. Contrary to this, none of the positions (c) admits a connection of x_3 and x_2 by a trajectory (why?), and so there is no four-link chain of connections between critical points. Clearly, such a chain is an invariant of phase portrait equivalence. Hence the fields Φ_a and Φ_c do not have equivalent phase portraits. Since any neighbourhood $V \subset \mathscr{F}^r(T^2)$ of Φ_a contains vector fields of type Φ_c, the field Φ_a is not structurally stable.

A vector field of type Φ_c is structurally stable; this follows from certain general theorems. We shall return to this example later on (in Section 4.6, Example 4.6.1).

Exercise

4.1.1.B. Show that the mapping φ considered in Exercise 1.3.6 on p. 30 is C^1-structurally stable (φ: $S^2 \to S^2$ is of the following form in the stereographic coordinates: $\varphi(x) = Ax$, $A = \begin{bmatrix} \lambda & 0 \\ 0 & \lambda \end{bmatrix}$, $\lambda > 1$).

Hint. φ has just two fixed points; both are hyperbolic. Thus if φ_1 is sufficiently C^1-close to φ, then also φ_1 has exactly two fixed points, which are hyperbolic. Now, let D be a ball with centre at N. We have $\varphi(D) \subset D$, whence $\varphi_1(D) \subset D$. Construct a homeomorphism h in a way similar to that in Example 1.2.1.

4.2. Structural Stability of Diffeomorphisms of the Circle

Throughout this section we assume that φ: $S^1 \to S^1$ is a C^r-diffeomorphism, $r \geqslant 1$, which preserves orientation (i.e. $\varphi'(x) > 0$). We employ additive notation.

It is easy to note that if φ has at least one periodic point, then any other periodic point has the same prime period.

Definition 4.2.1 A periodic point $x_0 \in S^1$ of prime period p is called:
(1) a *sink* iff it has a neighbourhood I such that

$$\lim_{n\to\infty} \varphi^{np}(x) = x_0 \qquad \text{for all } x \in I;$$

(2) a *source* iff x_0 is a sink for φ^{-1}.

Clearly, these terms make sense with regard to diffeomorphisms and flows on arbitrary manifolds.

If $x_0 \in S^1$ is a periodic point, $\varphi^p(x_0) = x_0$, at which

$$\frac{d\varphi^p}{dx}(x_0) < 1,$$

then x_0 is a sink:

$$|\varphi^p(x) - x_0| = |\varphi^p(x) - \varphi^p(x_0)| \leqslant \frac{d\varphi^p}{dx}(\bar{x})|x - x_0|, \qquad \bar{x} \in (x_0, x),$$

whence it follows that for x close to x_0 we have

$$|\varphi^p(x) - x_0| \leqslant \alpha |x - x_0|,$$

where α is a constant, $0 < \alpha < 1$.

Analogously, if $x_0 \in S^1$ is periodic: $\varphi^p(x_0) = x_0$, and such that

$$\frac{d\varphi^p}{dx}(x_0) > 1,$$

then x_0 is a source.

Of course, these two cases do not exhaust all the possible situations. E.g. the condition

$$\frac{d\varphi^0}{dx}(x_0) = 1$$

may be satisfied at a sink as well as at a source; also it can happen that a periodic point is a left-sided sink and a right-sided source or conversely.

Suppose that $I \subset S^1$ is an open interval such that $\varphi^p(I) = I$ and $\varphi^k(I) \cap I = \emptyset$ for $k = 1, \ldots, p-1$. The family of intervals $\{I, \varphi(I), \ldots, \varphi^{p-1}(I)\}$ is called a *cycle.* If I contains no subinterval with the same property, the cycle is called *prime*. This happens if there are no periodic points inside I. If $I \neq S^1$ is maximal, the cycle is called *maximal*. This happens if $\overline{\bigcup_{k=0}^{p-1} \varphi^k(I)} = S^1$.

Assume that φ has finitely many, say m, periodic points and denote by p their prime period. Then S^1 is the union of intervals $\bar{K}_1, \ldots, \bar{K}_m$ without interior points in common and with endpoints at the periodic points. Let r be the number of prime cycles for the system (S^1, φ). Thus $m = rp$. Assume that the numbering of the K_i's agrees with their arrangement on the circle, in positive sense (anti-clockwise). Let $I_1, \ldots, I_p$ be the intervals entering the maximal cycle, numbered according to their arrangement on the circle, the initial point of I_1 coinciding with that of K_1. It is not difficult to see that $I_1 = \bigcup_{i=1}^{r} K_i$.

Let s be the index defined by:

$$\varphi(I_1) = I_s. \tag{4.2.1}$$

Thus s indicates how far I_1 has "jumped" under the action of φ. Evidently, each I_k then "jumps" just as far: $\varphi(I_k) = I_{k+s-1}$.

Theorem 4.2.1 Let φ be as above and let $\psi\colon S^1 \to S^1$ be another diffeomorphism with m periodic points of prime period p. Suppose that the index s defined by (4.2.1) *is the same for φ and for ψ, and that each periodic point is either a source or a sink* (*absence of points of other types*). *Then the systems (S^1, φ) and (S^1, ψ) are topologically conjugate.*

Proof Denote the intervals which constitute the maximal cycles for φ and ψ by $I_1, \ldots, I_p$ and $J_1, \ldots, J_p$ respectively. Let their numbering be such that the corresponding endpoints of I_k and J_k, $k = 1, \ldots, p$, be periodic points of the same type (sink or source); this is possible, because sinks and sources lie alternately on the circle. Since ψ has just as many periodic points in J_1 as φ has in I_1 and since the corresponding endpoints of I_1 and J_1 are periodic points of the same character, the systems (I_1, φ^p) and (J_1, ψ^p) are topologically conjugate, in view of Example 1.2.1 on p. 17, and the conjugating homeomorphism h_0 is orientation preserving. Put

$$h(x) = \psi^i \circ h_0 \circ \varphi^{-i}(x) \qquad \text{for } x \in \varphi^i(I_1).$$

Clearly, h is continuous in the interior of each I_k. Since s is the same for φ and for ψ, we have $h(I_k) = J_k$, and so h is also continuous at the endpoints of the I_k's. Thus h is a homeomorphism of S^1 onto itself (bijectivity is obvious). By the definition of h we have the equality

$$h \circ \varphi(x) = \psi \circ h(x) \qquad \text{for } x \in S^1$$

and thus the two systems are topologically conjugate.

Exercise

4.2.1.B. Show that the ratio s/p is equal to the rotation number of φ (the meaning of s and p is as above).

Hint. p and s are relatively prime; hence the minimal integer n for which $sk = pn$ is $n = s$ (then $k = p$). This means that in p iterates the point makes s rounds, and so $\varphi = s/p$.

Corollary 4.2.1 Suppose that $\varphi \in \mathrm{Diff}^r(S^1)$ satisfies the conditions:

(1) *the set of periodic points of φ is non-empty and finite,*

(2) *at each periodic point x_0 we have*

$$\frac{d\varphi^p}{dx}(x_0) \neq 1,$$

p denoting the prime period.

Then the system (S^1, φ) is C^r-structurally stable.

This is an immediate consequence of Theorem 4.2.1.

Theorem 4.2.2 The set of all diffeomorphisms $\varphi \in \mathrm{Diff}^r(S^1)$ satisfying conditions (1) *and* (2) *of Corollary 4.2.1 is dense in* $\mathrm{Diff}^r(S^1)$.

Before proving the theorem we have to prepare two lemmas. The first of them is in fact a version of the Closing Lemma for $\mathrm{Diff}^r(S^1)$, $r \geq 2$.

Lemma 4.2.1 Let $\varphi_0 \in \mathrm{Diff}^r(S^1)$, $x_0 \in S^1$, $\varepsilon > 0$. Suppose that for some $p \in Z^+$ we have $\mathrm{dist}(x_0, \varphi_0^p(x_0)) < \varepsilon$. Then there exists a diffeomorphism $\varphi \in \mathrm{Diff}^r(S^1)$ such that $d_r(\varphi_0, \varphi) < \varepsilon$ and $\varphi^p(x_0) = x_0$. Here d_r denotes the distance in $\mathrm{Diff}^r(S^1)$, *i.e.,*

$$d_r(\varphi_0, \varphi) = \sup_{x \in S^1} \max_{0 \leq i \leq r} \left| \frac{d^i\varphi_0}{dx^i}(x) - \frac{d^i\varphi}{dx^i}(x) \right|.$$

Proof The reasoning is carried out in the covering space $\boldsymbol{R}$. Let f_0 denote a lift of φ_0. Then

$$f_0(y+1) = f_0(y)+1 \qquad \text{for } y \in \boldsymbol{R};$$

f_0 is increasing, since φ_0 is orientation preserving. Write

$$f_t(y) = f_0(y)+t, \qquad |t| < \varepsilon.$$

We claim that the intervals $(f_0^n(y), f_t^n(y))$ and $(f_0^n(y), f_{-t}^n(y))$, $n \in Z^+$, are of length not less than $|t|$. Indeed, for $n = 1$ this is clear. For $n = 2$ and for $t > 0$ we have

$$f_t^2(y) - f_0^2(y) = f_0(f_0(y)+t)+t-f_0(f_0(y)) \geqslant t,$$

since f_0 is increasing. The argument is similar for $t < 0$. Proceeding analogously we prove the claim by induction. In particular, taking $n = p$ we see that the intervals $(f_0^p(y), f_\varepsilon^p(y))$ and $(f_0^p(y), f_{-\varepsilon}^p(y))$ are not shorter than ε.

The point $x_0 \in S^1$ at which we have $\operatorname{dist}(x_0, \varphi_0^p(x_0)) < \varepsilon$ is the projection of a $y_0 \in \boldsymbol{R}$; the lift f_0 then satisfies at y_0 the inequality

$$|f_0^p(y_0) - y_0| < \varepsilon. \tag{4.2.2}$$

Examine the set $\{f_t^p(y_0): |t| < \varepsilon\}$. This is an interval and, by the foregoing considerations, it contains the interval $(f_0^p(y_0)-\varepsilon, f_0^p(y_0)+\varepsilon)$. Hence, in view of (4.2.2), there exists a t_0, $|t_0| < \varepsilon$, such that $f_{t_0}^p(y_0) = y_0$. Then $d_r(f_{t_0}, f_0) = |t_0| < \varepsilon$. Projecting f_{t_0} onto S^1 we define a diffeomorphism $\varphi \in \operatorname{Diff}^r(S^1)$ with the required properties.

Corollary 4.2.2 *If f_0 is of the form: $f_0(y) = y + T(y)$, where T is a trigonometric polynomial with period* 1, *then f_{t_0}, the function occurring in the proof of Lemma* 4.2.1, *is of the form*

$$f_{t_0}(y) = y + T(y) + t = y + T_1(y),$$

where T_1 is also a trigonometric polynomial.

Lemma 4.2.2 Let $\varphi_0 \in \operatorname{Diff}^r(S^1)$ and let $x_0 \in S^1$ be an isolated periodic point, $\varphi_0^p(x_0) = x_0$. Suppose that

$$\frac{d\varphi_0^p}{dx}(x_0) = 1.$$

Then for every $\varepsilon > 0$ *there are a neighbourhood* U *of the trajectory* $\hat{\gamma}(x_0)$ *and a diffeomorphism* $\varphi \in \mathrm{Diff}^r(S^1)$ *such that*

(1) $d_r(\varphi_0, \varphi) < \varepsilon$,

(2) $\varphi_0(x) = \varphi(x)$ *for* $x \in S^1 - U$,

(3) $\mathrm{Per}\,\varphi \subset \mathrm{Per}\,\varphi_0$,

and moreover:

1° *if* x_0 *is of the source-sink or sink-source type for* φ_0^p *then* φ *has no periodic point within* U;

2° *if* x_0 *is a sink [resp. source] for* φ_0^p, *then* $\varphi^p(x_0) = x_0$ *and*

$$\frac{\mathrm{d}\varphi^p}{\mathrm{d}x}(x_0) < 1 \quad [\textit{resp.} > 1].$$

The proof is left to the reader as an exercise.

Proof of Theorem 4.2.2 Let $\varphi_0 \in \mathrm{Diff}^r(S^1)$ be any C^r-diffeomorphism. For any $x_0 \in \Delta = \Omega(\varphi_0)$ the set $\omega(x_0)$ is non-empty; if x_0 is not periodic, then for every $\varepsilon > 0$ there exists $p \in Z^+$ such that $\mathrm{dist}(x_0, \varphi_0^p(x_0)) < \varepsilon$. In view of Lemma 4.2.1 the ball $V = \{\varphi \in \mathrm{Diff}^r(S^1)\colon d_r(\varphi, \varphi_0) < \varepsilon\}$ contains a diffeomorphism φ such that $\varphi^p(x_0) = x_0$. According to Corollary 4.2.2, on account of the fact that trigonometric polynomials are dense in $C^r(S^1)$, we may assume that the lift of φ to $\boldsymbol{R}$ is a function of the form:

$$f(y) = y + T(y),$$

T denoting a trigonometric polynomial with period 1.

The periodic points of φ are precisely those points whose lifts $y \in \langle 0, 1)$ satisfy the equation

$$f^p(y) = y + m, \tag{4.2.3}$$

m being a positive integer. Since f^p is increasing and since

$$f^p(y+1) = f^p(y) + 1,$$

m can take on at most 2 values. The function f is analytic; hence also f^p is analytic. Consequently, equation (4.2.3) can have only finitely many solutions in $\langle 0, 1)$. It follows that the set of periodic points of φ is non-empty and finite.

Applying Lemma 4.2.2 to the periodic points of φ, we can find for every $\varepsilon > 0$ a diffeomorphism $\psi \in \mathrm{Diff}^r(S^1)$ such that $d_r(\varphi, \psi) < \varepsilon$,

$\mathrm{Per}\,\psi \subset \mathrm{Per}\,\varphi$ and $d\psi^p/dx \neq 1$ at each point of $\mathrm{Per}\,\psi$. This shows that the set of all C^r-diffeomorphisms satisfying conditions (1) and (2) of Corollary 4.2.1 is dense in $\mathrm{Diff}^r(S^1)$, as asserted.

Theorem 4.2.3 (*Arnold, 1961*) *A diffeomorphism* $\varphi \in \mathrm{Diff}^r(S^1)$ *is structurally stable if and only if it satisfies conditions* (1) *and* (2) *of Corollary 4.2.1.*

Proof The sufficiency of those conditions is stated in Corollary 4.2.1. We prove the necessity.

Let φ be C^r-structurally stable. By Theorem 4.2.2 every neighbourhood of φ contains diffeomorphisms with finitely many periodic points. Hence $\mathrm{card}\,\mathrm{Per}\,\varphi < +\infty$.

Suppose that at some periodic point $x_0 \in S^1$ with prime period p we have

$$\frac{d\varphi^p}{dx}(x_0) = 1\,;$$

x_0 is a fixed point for φ^p. If it is a fixed point of the source-sink or sink-source type (see Fig. 4.2.1), then by Lemma 4.2.2 every neighbourhood of φ contains a diffeomorphism ψ which has p periodic

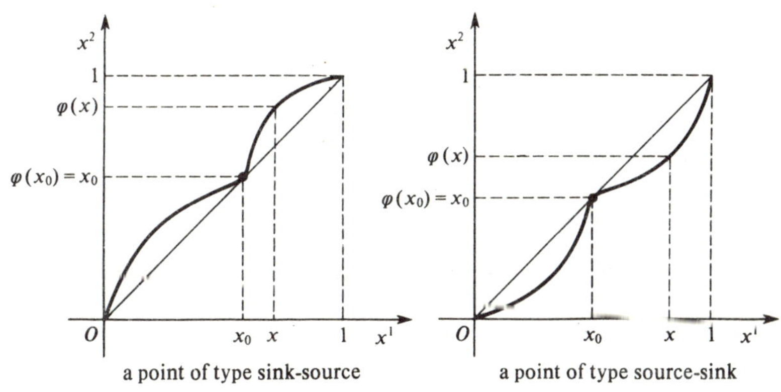

FIG. 4.2.1

points less than φ; such a ψ cannot be topologically conjugate to φ. If x_0 is a sink, then proceeding as in Lemma 4.2.2 we find, for any $\varepsilon > 0$, a neighbourhood U of $\hat{\gamma}$ (the trajectory through x_0) and a diffeomorphism $\psi \in \mathrm{Diff}^r(S^1)$ such that:

1° U is disjoint with the set of the remaining periodic points,

2° $d_r(\varphi, \psi) < \varepsilon$ and $\psi(x) = \varphi(x)$ for $x \in S^1 - U$,

3° $\psi^p(x_0) = x_0$ and $\dfrac{d\psi^p}{dx}(x_0) > 1$.

By assumption, φ and ψ are topologically conjugate, provided ε is small enough. Let x_1 and x_2 be fixed points of φ^p such that the arc (x_1, x_2) contains no fixed point of φ^p other than x_0. Evidently, x_1 and x_2 are sources. By the construction of ψ, x_1 and x_2 are also sources for ψ^p. Consequently, each of the arcs (x_1, x_0) and (x_0, x_2) necessarily contains a fixed point of ψ^p. Thus ψ^p has actually more fixed points than φ^p, which contradicts the topological conjugacy of φ and ψ.

Corollary 4.2.3 The set $\sigma^r(S^1)$, i.e., the set of C^r-structurally stable diffeomorphisms of S^1, is dense in $\mathrm{Diff}^r(S^1)$.

Theorem 4.2.1 and Exercise 4.2.1 imply:

Corollary 4.2.4 The following three quantities constitute a complete system of topological invariants for $\varphi \in \sigma^r(S^1)$:

(1) *the number of periodic points m,*

(2) *their prime period p,*

(3) *the rotation number ϱ.*

Exercise

4.2.2.B. Denote by $H(S^1)$ the space of all homeomorphisms of the circle. Consider the function f: $H(S^1) \to \boldsymbol{R}$ given by $f(\varphi) = \varrho$, the rotation number of φ. Prove that f is continuous (Arnold, 1961).

4.3. Structural Stability. Necessary Conditions

To begin with, let us examine the consequences of certain apparently slight modifications introduced in the definition of structural stability.

Firstly, observe that, by definition, the set $\Sigma^r(M)$ [resp. $\sigma^r(M)$] is open in $\mathscr{F}^r(M)$ [resp. $\mathrm{Diff}^r(M)$]. Hence, if Φ [resp. φ] is structurally stable, then Φ [resp. φ] necessarily enjoys all the generic properties which are invariant under phase portrait equivalence [resp. topological conjugacy].

E.g., the property of possessing critical elements is generic (the

Kupka–Smale theorem) and is invariant under phase portrait equivalence. Hence follows:

Corollary 4.3.1 Every dynamical system which is C^1-structurally stable has a non-empty set of critical elements.

Observe that if the condition of topological conjugacy in the definition of structural stability for cascades were replaced by the requirement of diffeomorphic conjugacy, then it might happen that no diffeomorphism would be structurally stable on the manifold considered. Take for example $M = S^1$ and $\varphi \in \mathrm{Diff}^r(S^1)$. The values of the derivatives $d\varphi^p(x)/dx$ at a periodic point are invariant under diffeomorphic conjugacy; on the other hand, every diffeomorphism φ van be C^1-approached (even C^r-approached, $r \geqslant 1$) by diffeomorphisms for which those values are different from those assumed by φ.

Also, if we considered $\mathrm{Diff}^r(M)$ with C^0-topology, no diffeomorphism would be structurally stable (in the modified meaning); diffeomorphisms which are close in C^0-topology may have different numbers (or sets) of periodic points.

Now we can readily see that if the definition of structural stability for flows imposed topological conjugacy in place of phase portrait equivalence, we would arrive at a concept of little use, if any. The periods of closed orbits would be invariants. Thus e.g. every vector field Φ having, say, finitely many closed orbits would be unstable (in that sense): the field $(1+\varepsilon)\Phi$ is arbitrarily close to Φ, but the periods of closed orbits of Φ and $(1+\varepsilon)\Phi$ are different.

Exercise

4.3.1.B. Let Φ be a constant vector field on the torus T^2. Show that $\Phi \notin \Sigma^r(T^2)$.

Hint. Applying an arbitrarily small perturbation we can obtain either a field whose orbits are all closed or a field whose orbits are all dense, according as the trajectories of the perturbed field intersect the coordinate axes at an angle α with $\tan\alpha$ rational or irrational.

After these preliminary remarks we discuss certain conditions necessary for a system to be structurally stable.

M denotes a compact manifold.

Definition 4.3.1 A closed orbit γ of a vector field $\Phi \in \mathscr{F}^r(M)$ is said to be *tubularly isolated* iff there exists a tubular neighbourhood U of $\hat{\gamma}$ such that no other closed orbit is entirely contained within U (we admit, however, a non-empty intersection with U).

Lemma 4.3.1 If $\Phi \in \Sigma^r(M)$, then every closed orbit of Φ is tubularly isolated.

Proof Each hyperbolic closed orbit is tubularly isolated. Thus the lemma is true for fields with Property $\mathscr{G}_2$. Since this property is generic, $\mathscr{G}_2(M)$ is dense in $\mathscr{F}^r(M)$. Consequently, for every $\Phi \in \Sigma^r(M)$ one can find $\Phi_1 \in \mathscr{G}_2(M)$ such that Φ and Φ_1 have equivalent phase portraits. It is easy to see that tubular isolation is invariant under phase portrait equivalence. Hence the assertion.

Theorem 4.3.1 (Markus, 1961) If $\Phi_0 \in \mathscr{F}^r(M)$ is structurally stable and satisfies condition (ε), *then Φ_0 has Property $\mathscr{G}_2$.*

Proof Step I. We first prove that all critical points of Φ_0 are hyperbolic. Since $\mathscr{G}_1^r(M)$ is a dense subset of $\mathscr{F}^r(M)$ and since every field with Property $\mathscr{G}_1$ has finitely many critical points, the field Φ_0 also has finitely many critical points. Denote these points by $x_1, \ldots, x_s$.

For each x_i, $i = 1, \ldots, s$, consider the Hessian $\dot{\Phi}_0(x_i)$ and examine its eigenvalues. Suppose that $\dot{\Phi}_0(x_i)$ has k_i eigenvalues λ with $\operatorname{Re}\lambda < 0$, l_i eigenvalues with $\operatorname{Re}\lambda > 0$ and j_i eigenvalues with $\operatorname{Re}\lambda = 0$.

Fix an x_i and carry our consideration locally in a chart. Thus assume that $x_i \in \boldsymbol{R}^m$ and that Φ_0 is defined in a neighbourhood U of x_i in $\boldsymbol{R}^m$. Let U_2 be a neighbourhood of x_i containing no critical points of Φ_0 other than x_i. Let $g(x)$ be a C^r-function which is equal to 0 outside U_2 and to 1 in some smaller neighbourhood of x_i. For any $\eta \in \boldsymbol{R}$ write

$$\tilde{Q}_i^\eta(x) = \eta(x - x_i) \qquad \text{for } x \in U_2,$$

$$Q_i^\eta(x) = g(x)\tilde{Q}_i^\eta(x);$$

Q_i^η is a C^r-field defined for all $x \in U$. Define;

$$\Psi_i^\eta(x) = \Phi_0(x) + Q_i^\eta(x).$$

Clearly $\Psi_i^\eta(x_i) = 0$. Further, we have

$$\lambda I - \dot{\Psi}_i^\eta(x_i) = \lambda I - \dot{\Phi}_0(x_i) - \eta I = (\lambda - \eta)I - \dot{\Phi}_0(x_i).$$

Thus

$$\operatorname{sp}\dot{\Psi}_i^\eta(x_i) = \operatorname{sp}\dot{\Phi}_0(x_i)+\eta.$$

Choose η so as to have

$$\operatorname{Re}\lambda \neq 0 \qquad \text{for } \lambda \in \operatorname{sp}\dot{\Psi}_i^\eta(x_i)$$

(then x_i is a hyperbolic critical point of Ψ_i^η). If η is close to 0 and negative, $\dot{\Psi}_i^\eta(x_i)$ has k_i+j_i eigenvalues λ with $\operatorname{Re}\lambda < 0$; if η is close to 0 and positive, $\dot{\Psi}_i^\eta(x_i)$ has k_i eigenvalues λ with $\operatorname{Re}\lambda < 0$.

Now, returning to the manifold, we may regard Q_i^η as a C^r-vector field on M, equal to Φ_0 outside a neighbourhood of x_i. Put

$$\Phi^\eta = \Phi_0 + \sum_{i=1}^{s} Q_i^\eta.$$

For $|\eta|$ small enough the field Φ^η is arbitrarily close to Φ_0; any two fields Φ^η have equivalent phase portraits, provided that $|\eta|$ is sufficiently small. Further, for small $|\eta|$, the points $x_1, \ldots, x_s$ are the only critical points of Φ^η. Write

$$d(\Phi^\eta) = \sum_{i=1}^{s} \dim W^+(x_i, \Phi^\eta).$$

Obviously, this number is invariant under phase portrait equivalence of the Φ^η. Supposing that $|\eta|$ is small enough, we have:

$$d(\Phi^\eta) = \begin{cases} \sum_{i=1}^{s} (k_i+j_i) & \text{if } \eta > 0, \\ \sum_{i=1}^{s} k_i & \text{if } \eta < 0. \end{cases}$$

Hence $j_i = 0$ for $i = 1, \ldots, s$, and this means that each critical point x_i is hyperbolic.

We emphasize the fact that condition (ε) has not been used in this part of the proof.

Step II. Now let γ be a closed orbit of Φ_0 with prime period τ. Let U be a tubular neighbourhood of $\hat{\gamma}$. By Lemma 4.3.1 we may assume that U is a neighbourhood realizing tubular isolation.

The set $\tilde{U}$ defined below is a covering space for U:

$$\tilde{U} = \{x \in \boldsymbol{R}^m\colon |x^i| < 1,\ i = 1, \ldots, m-1\}.$$

Projection $\pi\colon \tilde{U} \to U$ is defined as follows:

$$x^1 = \begin{cases} \tilde{x}^1 & \text{if } M \text{ is orientable,} \\ -\tilde{x}^1 & \text{if } M \text{ is non-orientable,} \end{cases}$$

$$x^i = \tilde{x}^i \qquad \text{for } i = 2, \dots, m-1,$$

$$x^m = \tilde{x}^m (\operatorname{mod} \tau).$$

The lift $\tilde{\gamma}$ of γ is identified with the set $\{x^1 = \dots = x^{m-1} = 0\}$.

The flow φ_t can be investigated in a neighbourhood $\tilde{U}_1$ of $\tilde{\gamma}$ in $\tilde{U}$. The lift of φ_t (for simplicity, we will denote it also by φ_t) is given by a system of differential equations:

$$\frac{\mathrm{d}\varphi_t(x)}{\mathrm{d}t} = f(\varphi_t(x)), \tag{4.3.1}$$

where

$$f(x) = (f_1(x), \dots, f_{m-1}(x), f_m(t)),$$
$$x = (x^1, \dots, x^{m-1}, t).$$

If x is sufficiently close to 0, the piece of trajectory $\{\varphi_t(x)\colon 0 \leqslant t \leqslant 2\tau\}$ is entirely contained in $\tilde{U}_1$. Differentiating (4.3.1) with respect to x we obtain

$$\frac{\mathrm{d}}{\mathrm{d}t}\mathrm{d}\varphi_t(x) = \mathrm{d}f(\varphi_t(x))\mathrm{d}\varphi_t(x).$$

Setting $x = 0$, we get

$$\frac{\mathrm{d}}{\mathrm{d}t}\mathrm{d}\varphi_t(0) = \mathrm{d}f(0, \dots, 0, t)\mathrm{d}\varphi_t(0). \tag{4.3.2}$$

Let $\tilde{U}_2$ be a small neighbourhood of $\tilde{\gamma}$, $\bar{\tilde{U}}_2 \subset \tilde{U}_1$. Let $g(x)$ be a C^r-function, $g(x) = 1$ for $x \in \tilde{U}_2$ and $g(x) = 0$ for $x \notin \tilde{U}_1$. Consider the vector field

$$\Psi^\eta(x) = \Phi_0(x) + \eta g(x)x,$$

where η is some real number. The flow ψ_t generated by Ψ^η satisfies the equation

$$\frac{\mathrm{d}\psi_t(x)}{\mathrm{d}t} = f(\psi_t(x)) + \eta g(\psi_t(x))\psi_t(x). \tag{4.3.3}$$

Differentiating (4.3.3) with respect to x and setting $x = 0$, we obtain

$$\frac{\mathrm{d}}{\mathrm{d}t}\mathrm{d}\psi_t(0) = \big(\mathrm{d}f(0, \ldots, 0, t)+\eta\big)\mathrm{d}\psi_t(0). \tag{4.3.4}$$

A direct verification shows that the following is a solution of (4.3.4):

$$\mathrm{d}\psi_t(0) = \mathrm{e}^{\eta t}\mathrm{d}\varphi_t(0).$$

Let k, l, j denote the numbers of eigenvalues λ of the operator $\mathrm{d}\varphi_\tau(0)$ such that $|\lambda| < 1$, $|\lambda| > 1$, $|\lambda| = 1$, respectively. Fix $\eta > 0$ and denote the corresponding flow ψ_t by ψ_t^+; then fix $\eta < 0$ and denote the corresponding flow ψ_t by ψ_t^-. If these values of η are sufficiently close to zero, then γ is a hyperbolic closed orbit of the flows $\psi_t^\pm$ such that the dimensions of the stable and unstable manifold of γ are, respectively:

k and $l+j$, for γ regarded relative to ψ_t^+,

$k+j$ and l, for γ regarded relative to ψ_t^-.

Denote by Ψ^+, Ψ^- the vector fields associated with the flows ψ_t^+, ψ_t^-. Let ε be any number in the interval $0 < \varepsilon < \varrho(\hat{\gamma}, \mathrm{Fr}\, U)$ and let $V = V(\varepsilon)$ be the neighbourhood of Φ_0 in $\mathscr{F}^r(M)$ chosen to ε according to condition (ε) of structural stability (see p. 187). The fields Ψ^+ and Ψ^- belong to V, provided that the values of $\eta > 0$ and $\eta < 0$ defining Ψ^+ and Ψ^- have been chosen sufficiently close to 0. Then there exist homeomorphisms h^+ and h^- realizing the equivalence between the phase portraits of Φ_0 and $\Psi^\pm$ and such that $\varrho\big(x, h^\pm(x)\big) < \varepsilon$ for all x. It follows that $h^\pm(\hat{\gamma}) \subset U$.

According to our assumption, U contains only one closed orbit of Φ_0, namely, γ. Hence $h^\pm(\hat{\gamma}) = \hat{\gamma}$. The homeomorphism $h_1 = (h^-)^{-1} \circ h^+$ establishes equivalence between the phase portraits of Ψ^+ and Ψ^-. Clearly, h_1 carries γ into itself. Now, the dimensions of the stable and unstable manifold of γ viewed first as an orbit of Ψ^+ and then as an orbit of Ψ^-, are equal, respectively, to k and $l+j$ (in the case of Ψ^+) and to $k+j$ and l (in the case of Ψ^-). Evidently, these dimensions are invariant under phase portrait equivalence. Hence $j = 0$, and this ends the proof.

Remark 4.3.1 In the case of $\dim M = 2$ condition (ε) is redundant in Markus' theorem; closed orbits are then isolated and one can just use the argument applied in Step I of the above proof.

Quite similarly the following theorem is proved.

Theorem 4.3.2 *If $\varphi \in \sigma^r(M)$, then every closed orbit is hyperbolic.*

Theorem 4.3.3 *If $\Phi \in \Sigma^1(M)$, then Φ has Property $\mathscr{G}_4$; in other words, the set of non-wandering points then coincides with the closure of the set of periodic points. The same is true for diffeomorphisms $\varphi \in \sigma^1(M)$.*

Proof Property $\mathscr{G}_4$ is C^1-generic (see Corollary 3.8.1) and is invariant under phase portrait equivalence; hence $\Sigma^1(M) \subset \mathscr{G}_4^1(M)$.

Also the following two theorems are true.

Theorem 4.3.4 (*Smale, 1970*) *If a smooth dynamical system is C^1-structurally stable, then the stable and unstable manifolds of critical elements intersect transversally.*

Theorem 4.3.5 *If $\Phi \in \Sigma^1(M)$, then Φ has Property $\mathscr{G}_5$, that is to say, Φ has no regular first integral.*

Proof Theorems 4.3.1, 4.3.3 and 4.3.4 jointly lead to the conclusion that Properties $\mathscr{G}_3$ and $\mathscr{G}_4$ are necessary for structural stability. As follows from the proof of Theorem 3.8.3 on p. 184, vector fields $\Phi \in \mathscr{G}_3(M) \cap \mathscr{G}_4(M)$ do not have regular first integrals.

4.4. Morse–Smale Systems

We shall now devote more attention to a certain specific class of dynamical systems.

Let M be a compact manifold and let $S^r(M)$ denote the set of all vector fields $\Phi \in \mathscr{F}^r(M)$ which fulfil the following conditions:

(1) the set $\Omega(\Phi)$ is the union of a finite set of critical points and a finite number of closed orbits,

(2) all critical elements are hyperbolic,

(3) the stable and unstable manifolds of critical elements intersect transversally.

Definition 4.4.1 A vector field fulfilling these conditions is termed a *Morse–Smale field*. Conditions (1)–(3) are called the *Morse–Smale conditions.*

This definition has a counterpart referring to diffeomorphisms.

Definition 4.4.2 A diffeomorphism $\varphi \in \mathrm{Diff}^r(M)$ is called a *Morse–Smale diffeomorphism* iff:

(1) the set $\Omega(\varphi)$ is finite,

(2) all periodic points are hyperbolic,

(3) the stable and unstable manifolds of periodic points intersect transversally.

Condition (1) of Definition 4.4.2 implies that then $\Omega(\varphi)$ is just the set of periodic points.

The set of all Morse–Smale C^r-diffeomorphisms will be denoted by the symbol $s^r(M)$.

Examples

4.4.1. Every diffeomorphism of the circle which is structurally stable is Morse–Smale (see Theorem 4.2.3).

4.4.2. Let $M = S^2$ and let $\Phi \in \mathscr{F}^r(S^2)$ be the vector field whose integral curves are as shown in Fig. 4.4.1; Φ is a Morse–Smale field.

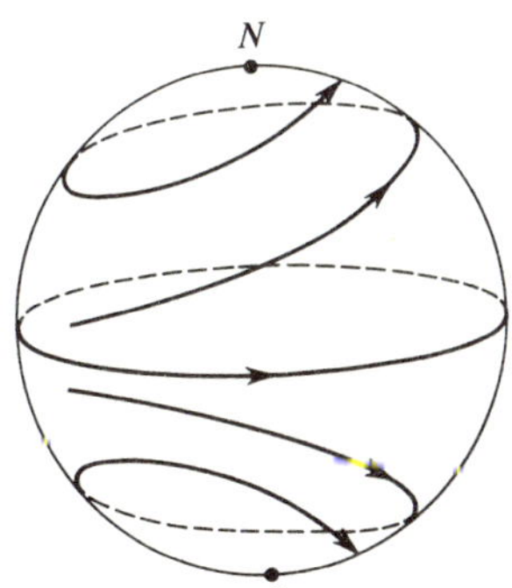

FIG. 4.4.1

4.4.3. Let $M = S^2$ and let $\Phi \in \mathscr{F}^r(S^2)$ be such that the equator is a hyperbolic closed orbit and the integral curves on the north and south hemispheres are as shown in Fig. 4.4.2, with foci a_N, b_N, a_S, b_S and saddles N and S; Φ is a Morse–Smale field.

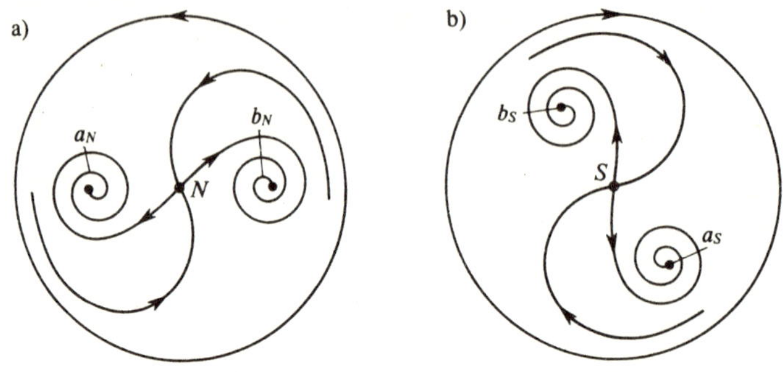

FIG. 4.4.2

We shall now study the properties of Morse–Smale systems (see Palis, 1969).

Let W_1 and W_2 be immersed submanifolds of M; that is to say, let $W_1 = i_1(P_1)$, $W_2 = i_2(P_2)$, where P_1, P_2 are manifolds and i_1, i_2 are bijective immersions.

Definition 4.4.3 We say that W_1 is *ε-C^1-close* to W_2 iff there exists a smooth mapping $h\colon P_1 \to P_2$ such that

$$d_1(i_1, i_2 \circ h) \leqslant \varepsilon,$$

d_1 denoting the distance in C^1-topology.

E.g., if $M = \boldsymbol{R}^2$ and W_1, W_2 are graphs of some smooth functions f and g, then W_1 is ε-C^1-close to W_2 if

$$|f(x)-g(x)|+|f'(x)-g'(x)| \leqslant \varepsilon \qquad \text{for all } x \in \boldsymbol{R};$$

in this case $P_1 = P_2 = \boldsymbol{R}$ and h is the identity.

The relation just introduced is not symmetric. E.g. a straight line can be ε-C^1-close to a plane, but not conversely.

Definition 4.4.4 Let $M_1 \subset M$ be an immersed submanifold of M. A set $B \subset M_1$ is called a *cell* iff B regarded as a subset of M_1 is a simply connected set with non-empty interior and smooth boundary. In other words: if $i\colon P \to M$ is a one-to-one immersion, P denoting some manifold, $i(P) = M_1$, then a set $B \subset M_1$ is a cell iff the set $i^{-1}(B) \subset P$ has non-empty interior and smooth boundary and is simply connected;

in the case of $P = \boldsymbol{R}^s$ this means that $i^{-1}(B)$ is diffeomorphic to a ball in $\boldsymbol{R}^s$.

The theorem which we now present is known as the λ-Lemma (see Palis, 1969); it is one of the crucial theorems of the theory of smooth dynamical systems and will be often resorted to in the sequel.

Let $\varphi \in \mathrm{Diff}^r(M)$ and let $p \in M$ be a hyperbolic fixed point of φ. As usual, $W^+(p)$ and $W^-(p)$ denote the stable and the unstable manifolds of p.

Theorem 4.4.1 (λ-*Lemma*) *Suppose that* $\dim W^-(p) = s$, $0 < s < m = \dim M$, *and let* S *be an immersed submanifold of* M *which is* φ-*invariant. Suppose that* S *has a point of transversal intersection with* $W^+(p)$. *Then for every cell neighbourhood* B^s *of* p *in* $W^-(p)$ *and for every* $\varepsilon > 0$ *there is a cell* $D^s \subset S$ *such that* B^s *is* ε-C^1-*close to a suitable component of* $\varphi^n(D^s)$, *for* $n = 0, 1, 2, \ldots$

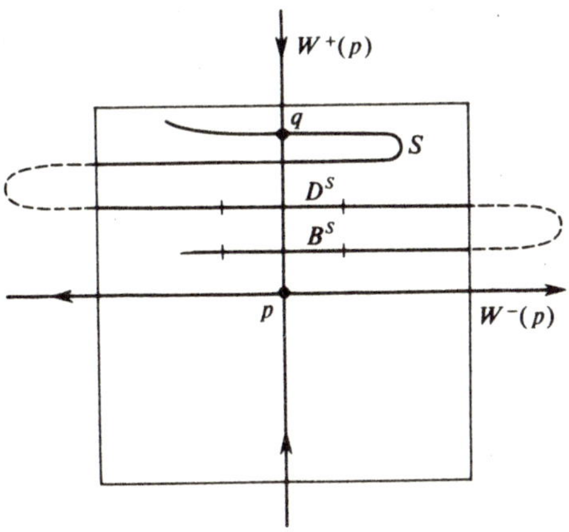

FIG. 4.4.3

Proof Let q be the point of transversal intersection of S and $W^+(p)$ Then each $q_n = \varphi^n(q)$ is such a point, as well, and we have $q_n \to p$ as $n \to +\infty$. Consequently we may assume that p and q lie within a common chart, which allows us to carry over our considerations to the space $\boldsymbol{R}^m$.

Thus assume that $p = 0 \in \boldsymbol{R}^m$ and choose the coordinate system so that $W^+(p)$ and $W^-(p)$ will be spanned by some coordinate axes. In a neighbourhood U of 0 the diffeomorphism φ is of the form

$$\varphi(x^1, x^2) = \left(Bx^1 + f^1(x^1, x^2),\ Ax^2 + f^2(x^1, x^2)\right),$$

where

$$x^1 \in W^-(0), \qquad x^2 \in W^+(0),$$

$$\mathrm{d}\varphi(0, 0) = \begin{bmatrix} B & 0 \\ 0 & A \end{bmatrix},$$

$$f^i(0, 0) = 0, \quad \mathrm{d}f^i(0, 0) = 0 \qquad \text{for } i = 1, 2,$$

and

$$\frac{\partial f^1}{\partial x^2}(0, x^2) = \frac{\partial f^2}{\partial x^1}(x^1, 0) = 0. \tag{4.4.1}$$

By an appropriate choice of the scalar product in $\boldsymbol{R}^m$ we may assume that

$$\|A\| = a < 1, \qquad \|B^{-1}\| = 1/b < 1. \tag{4.4.2}$$

There is a neighbourhood of zero $U_1 \subset U$ such that the number

$$k := \max_{i,j=1,2} \sup_{U_1} \left\| \frac{\partial f^i}{\partial x^j} \right\|$$

satisfies the inequalities:

$$\begin{aligned} &0 < k < 1, \\ &b_1 = b - k > 1, \qquad a_1 = a + k < 1, \\ &k < \tfrac{1}{4}(b_1 - 1)^2. \end{aligned} \tag{4.4.3}$$

Now let B^s be a cell neighbourhood of 0 in $W^-(0)$. Assume that $B^s \subset U_1$; this does not affect generality, since such an inclusion is certainly satisfied for some $\varphi^{-n}(B^s)$ in place of B^s, and we might just take that set for B^s.

The metric in U_1 is inherited from $\boldsymbol{R}^m$. Let v_0 be a vector of length 1 in $T_x(S)$, where x denotes some point in $S \cap U_1$. Write v_0 in the form: $v_0 = v_0^+ + v_0^-$, the summands on the right denoting the components of v_0 relative to $W^+(0)$ and $W^-(0)$. Define:

$$\lambda(v_0) := \frac{\|v_0^+\|}{\|v_0^-\|} \qquad \text{whenever} \qquad v_0^- \neq 0;$$

we call this number the *tilt* of v_0.

Assume the notation:

$$x_n = \varphi^n(x), \qquad v_n = \mathrm{d}\varphi(x_{n-1})(v_{n-1}),$$
$$q_n = (0, q_n^2) = \varphi^n(q),$$
$$F^{i,j}(x) = \frac{\partial f^i}{\partial x^j}(x), \qquad i, j = 1, 2.$$

Write the derivative $\mathrm{d}\varphi(x)$ in the matrix form:

$$\mathrm{d}\varphi(x) = \begin{bmatrix} \dfrac{\partial \varphi^1}{\partial x^1}(x) & \dfrac{\partial \varphi^1}{\partial x^2}(x) \\ \dfrac{\partial \varphi^2}{\partial x^1}(x) & \dfrac{\partial \varphi^2}{\partial x^2}(x) \end{bmatrix} = \begin{bmatrix} B + F^{1,1}(x) & F^{1,2}(x) \\ F^{2,1}(x) & A + F^{2,2}(x) \end{bmatrix}. \tag{4.4.4}$$

Condition (4.4.1) implies that $F^{1,2}(x) = 0$ for $x = (0, x^2)$. Thus, putting $x = q_n$, we obtain

$$\mathrm{d}\varphi(q_n) = \begin{bmatrix} B + F^{1,1}(q_n) & 0 \\ F^{2,1}(q_n) & A + F^{2,2}(q_n) \end{bmatrix}. \tag{4.4.5}$$

Let $\varepsilon > 0$ be arbitrary. Using (4.4.1), we can find a neighbourhood of 0 of the form:

$$U_2 = B^s \times \delta I \subset U,$$

where δI denotes the ball of radius δ in $W^+(0)$, δ being sufficiently small for the following estimate to be valid:

$$k_1 := \sup_{U_2} \left\| \frac{\partial f^2}{\partial x^1} \right\| < \min(\varepsilon, k). \tag{4.4.6}$$

Applying the representation of $\mathrm{d}\varphi(x)$ in the form (4.4.4) together with inequalities (4.4.2) and (4.4.3), we obtain the following estimate, holding for any $v_0 \in T_x(S)$, $x \in S$ such that $\varphi^i(x) \in S \cap U_2$, $i = 0, \ldots, n$:

$$\begin{aligned} \lambda(v_n) &= \frac{\|(\mathrm{d}\varphi(x_{n-1})(v_{n-1}))^+\|}{\|(\mathrm{d}\varphi(x_{n-1})(v_{n-1}))^-\|} \\ &= \frac{\|(A + F^{1,1}(x_{n-1}))(v_{n-1}^+) + F^{2,1}(x_{n-1})(v_{n-1}^-)\|}{\|F^{1,2}(x_{n-1})(v_{n-1}^+) + (B + F^{2,2}(x_{n-1}))(v_{n-1}^-)\|} \\ &\leqslant \frac{\|v_{n-1}^+\| + k_1\|v_{n-1}^-\|}{b_1\|v_{n-1}^-\| - k_1\|v_{n-1}^+\|} \leqslant \frac{\lambda(v_{n-1}) + k_1}{b_1 - k\lambda(v_{n-1})}. \end{aligned} \tag{4.4.7}$$

Setting $x = q$ we get, in view of (4.4.5),

$$\lambda(v_n) = \frac{\|(A + F^{1,1}(q_{n-1}))(v_{n-1}^+) + F^{2,1}(q_{n-1})(v_{n-1}^-)\|}{\|(B + F^{2,2}(q_{n-1}))(v_{n-1}^-)\|}$$

$$\leqslant \frac{\|v_{n-1}^+\| + k_1\|v_{n-1}^-\|}{b_1\|v_{n-1}^-\|} \leqslant \frac{\lambda(v_{n-1}) + k}{b_1}. \tag{4.4.8}$$

Hence by induction

$$\lambda(v_n) \leqslant \frac{\lambda(v_0)}{b_1^n} + k_1 \sum_{i=1}^{n} \frac{1}{b_1^i} \leqslant \frac{\lambda(v_0)}{b_1^n} + \frac{k_1}{b_1 - 1}.$$

Since $k_1 < k < \frac{1}{4}(b_1 - 1)^2$, we get for n sufficiently large, say, $n \geqslant n_0$, the estimate:

$$\lambda(v_n) \leqslant \frac{\lambda(v_0)}{b_1^n} + \frac{b_1 - 1}{4} < \frac{b_1 - 1}{3}. \tag{4.4.9}$$

The value of $\lambda(v_0)$ depends continuously on x and v_0, $v_0 \in T_x(S)$. Thus, on account of (4.4.9), there exists a cell neighbourhood $D_0 \subset U_2$ of q_{n_0} in S such that for any $x \in D_0$ and any $v_0 \in T_x(S)$ of length 1 we have

$$\lambda(v_0) \leqslant \tfrac{1}{2}(b_1 - 1).$$

We claim that if $\varphi^i(x) \in U_2$ for $i = 0, \ldots, n$, then

$$\lambda(v_n) \leqslant \tfrac{1}{2}(b_1 - 1). \tag{4.4.10}$$

Assume this for some n. Then in view of (4.4.3), (4.4.6) and the joint inequality $k_1 < k < 1$ we get

$$\lambda(v_{n+1}) \leqslant \frac{\lambda(v_n) + k_1}{b_1 - k\lambda(v_n)} \leqslant \frac{\lambda(v_n) + k_1}{b_1 - \lambda(v_n)}$$

$$\leqslant \frac{\frac{1}{2}(b_1 - 1) + \frac{1}{4}(b_1 - 1)^2}{b_1 - \frac{1}{2}(b_1 - 1)} = \tfrac{1}{2}(b_1 - 1),$$

proving the claim.

We now estimate $\lambda(v_n)$ uniformly with respect to $x \in D_0$ and $v_0 \in T_x(S)$. Applying again (4.4.6) and (4.4.10), we obtain

$$\lambda(v_n) \leqslant \frac{\lambda(v_{n-1}) + k_1}{b_1 - k\lambda(v_{n-1})} \leqslant \frac{\lambda(v_{n-1}) + k_1}{b_1 - \frac{1}{2}(b_1 - 1)}$$

$$= \frac{\lambda(v_{n-1}) + k_1}{\frac{1}{2}(1 + b_1)} = \frac{\lambda(v_{n-1}) + k_1}{b_2},$$

where $b_2 = \frac{1}{2}(1+b_1) > 1$. Hence by induction

$$\lambda(v_n) \leqslant \frac{\lambda(v_0)}{b_2^n} + \sum_{i=1}^{n-1} \frac{k_1}{b_2^i} \leqslant \frac{\lambda(v_0)}{b_2^n} + \frac{k_1}{b_1 - 1}.$$

By (4.4.6), $k_1 \leqslant \varepsilon$; and so we have

$$\lambda(v_n) \leqslant \frac{\lambda(v_0)}{b_2^n} + \frac{\varepsilon}{b_2 - 1} \leqslant \varepsilon\left(1 + \frac{1}{b_2 - 1}\right) \tag{4.4.11}$$

for n large enough, say, $n \geqslant n_1$, for any $x \in D_0$ such that $\varphi^i(x) \in U_2$, $i = 0, \ldots, n$, and for any $v_0 \in T_x(S)$ with $\|v_0\| = 1$.

Let D_{n_1} be the connected component of the set $\varphi^{n_1}(D_0) \cap U_2$ which contains the point q_{n_1}.

Denote by π the projection onto $W^-(0)$:

$$\pi((x^1, x^2)) := x^1.$$

We may assume that $\pi(D_{n_1}) \supset B^s$; otherwise, the map $\varphi|_{W^-(0)}$ being expanding, we could find $n_2 > n_1$ such that $\varphi^{n_2}(\pi(D_{n_1}))$ does contain B^s.

Now, if $\dim S = s$, then for every $(x^1, 0) \in B^s$ there is exactly one point $(x^1, x^2) \in D_{n_1}$; for if this were not the case, then by the connectedness of D_{n_1} there would exist $x \in D_{n_1}$ such that $v_0^- = 0$ for some $v_0 \in T_x(S)$, and this contradicts (4.4.11).

For any $(x^1, 0) \in B^s$ denote by $h(x^1)$ the unique point $(x^1, x^2) \in D_{n_1}$; this definition is correct by the preceding remark. Now it is not difficult to verify, using the estimate (4.4.11), that the mapping h defined in this way satisfies the condition which establishes that B^s is $\varepsilon(1+(b_2-1)^{-1})$-close to D_{n_1}. Furthermore, for every $n \geqslant n_1$ the set D_{n_1} contains a cell D'_{n_1} such that B^s is $\varepsilon(1+(b_2-1)^{-1})$-close to D'_{n_1}.

If $\dim S > s$, we draw a hyperplane H through $W^-(0)$ and the point q_{n_1}; considering the submanifold $S_1 = H \cap S$ and appealing to the conclusion of the preceding case we can find a cell $D'_{n_1} = D_{n_1} \cap H$ such that B^s is $\varepsilon(1+(b_2-1)^{-1})$-close to D'_{n_1}. The proof of the λ-Lemma is complete.

Corollary 4.4.1 Let p_1, p_2, p_3 be three hyperbolic fixed points of φ. Suppose that $W^+(p_1)$ and $W^-(p_2)$ have a point of transversal intersection and so do $W^+(p_2)$ and $W^-(p_3)$. Then also $W^+(p_1)$ and $W^-(p_3)$ have a point of transversal intersection.

Proof Let $a \in M$ be a point of transversal intersection of $W^+(p_1)$ and $W^-(p_2)$. Let $A, B \subset W^-(p_2)$ be cell neighbourhoods of a and p_2, respectively. Take $n_0 \in Z^+$ such that $\varphi^{-n_0}(A) \subset B$; such an n_0 certainly exists because $\varphi^{-1}|_{W^-(p_2)}$ is a contraction (to p_2). In virtue of the λ-Lemma, for any $\delta > 0$ there exists a cell neighbourhood $D \subset W^-(p_3)$ such that B is δ-C^1-close to D. Thus given $\varepsilon > 0$, we may assert that A is ε-C^1-close to $\varphi^{n_0}(D)$, just by choosing δ appropriately to ε. Since ε may be arbitrary, we conclude that $\varphi^{n_0}(D)$ intersects $W^+(p_1)$ transversally at some point $a_1 \in \varphi^{n_0}(D)$ lying close to a (see Fig. 4.4.4).

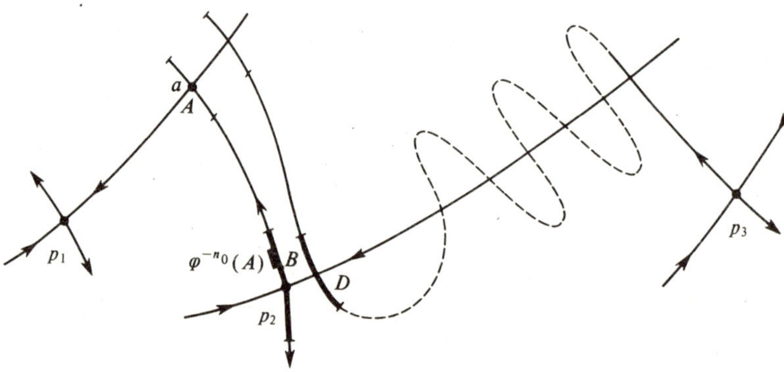

FIG. 4.4.4

Corollary 4.4.2 Let $p \in M$ be a hyperbolic fixed point of a diffeomorphism $\varphi \in \mathrm{Diff}^r(M)$, $r \geqslant 1$. Suppose that for some $q \in M$, $q \neq p$, we have $W^+(p) \pitchfork_q W^-(p)$. Then the set $\Omega(\varphi)$ is infinite.

Proof Let $B \subset W^-(p)$ be a cell neighbourhood of p and let U be an arbitrarily small neighbourhood of q such that the set $B_1 = U \cap W^-(p)$

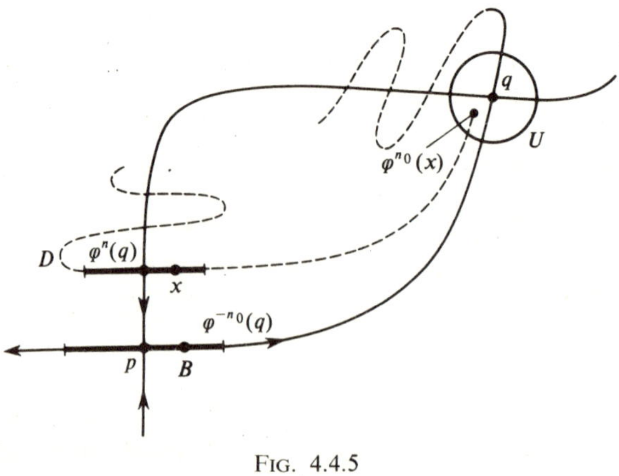

FIG. 4.4.5

is a cell. Since $\varphi^{-1}|_{W^-(p)}$ is a contraction (to p), there is an $n_0 \in Z^+$ such that $\varphi^{-n_0}(B_1) \subset B$. According to the λ-Lemma, for any $\varepsilon > 0$ there exists a cell $D \subset W^-(p)$ such that B is ε-C^1-close to D. We may additionally assume that $\varphi^n(q) \in D$ and $\varphi^n(B_1) \supset D$ (this follows from the proof of the λ-Lemma; see Fig. 4.4.5).

Let $D_1 \subset D$ be a cell such that D_1 is ε-C^1-close to $\varphi^{-n_0}(B_1)$. If ε is small enough, we have $\varphi^{n_0}(D_1) \cap U \neq \emptyset$ which in view of the inclusion $\varphi^{-n}(D_1) \subset U$ yields $\varphi^{n+n_0}(U) \cap U \neq \emptyset$. Since U can be arbitrarily small, it follows that q is non-wandering. Hence also each $\varphi^n(q)$, $n \in Z^+$, is a non-wandering point.

We have

$$\lim_{n\to\infty} \varphi^n(q) = p, \qquad \varphi^n(q) \neq p \ (n \in Z^+);$$

consequently

$$\operatorname{Card}\Omega(\varphi) \geqslant \operatorname{Card}\{\varphi^n(q)\colon\ n \in Z^+\},$$

which is infinite.

Remark 4.4.1 Corollaries 4.4.1 and 4.4.2 are true without changes for hyperbolic periodic points of a diffeomorphism $\varphi \in \operatorname{Diff}^r(M)$; they also retain validity in the case of flows generated by vector fields $\Phi \in \mathscr{F}^r(M)$, $r \geqslant 1$.

4.5. Properties of Morse–Smale Systems

Let M be a compact manifold and let $\varphi \in s^r(M)$. Let $\gamma_1, \ldots, \gamma_N$ denote all the periodic orbits of φ.

Definition 4.5.1 We say that an orbit γ_i is *preceded* by γ_k (in symbols, $\gamma_i \leqslant \gamma_k$) iff $W^+(\gamma_i) \cap W^-(\gamma_k) \neq \emptyset$.

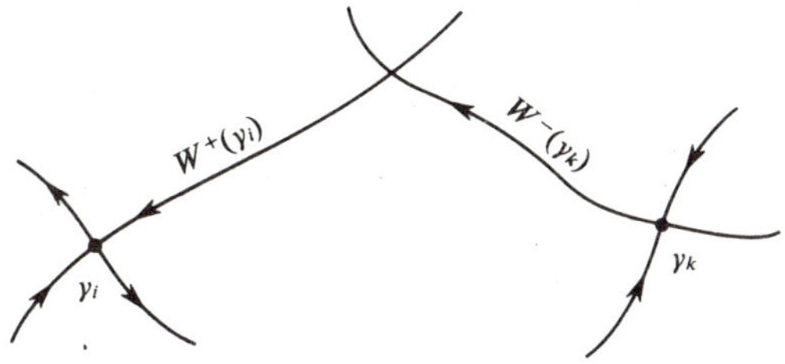

FIG. 4.5.1

Theorem 4.5.1 *The relation $\leqslant$ just introduced defines a partial order in the set* $\{\gamma_1, \ldots, \gamma_N\}$.

Proof Since $W^+(\gamma_i) \cap W^-(\gamma_i) \supset \hat{\gamma}_i \neq \emptyset$, we have $\gamma_i \leqslant \gamma_i$. Transitivity follows from Corollary 4.4.1. Assume that $\gamma_i \leqslant \gamma_k$ and $\gamma_k \leqslant \gamma_i$. If $\gamma_i \neq \gamma_k$, then (see Corollary 4.4.1) $W^+(\gamma_i) \cap W^-(\gamma_i) - \hat{\gamma}_i \neq \emptyset$, whence by Corollary 4.4.2 the set $\Omega(\varphi)$ is infinite, contrary to the definition of a Morse–Smale diffeomorphism.

Definition 4.5.2 The set $\{\gamma_1, \ldots, \gamma_N\}$ together with the order $\leqslant$ is called the *phase diagram* of φ.

Remark 4.5.1 An analogous relation is introduced also in the set of all critical elements of a vector field $\Phi \in S^r(M)$: by definition, $\gamma_i \leqslant \gamma_k$ iff $W^+(\gamma_i) \cap W^-(\gamma_k) \neq \emptyset$. Also in this case $\leqslant$ is a partial order (see Exercise 4.5.4) and the set of all critical elements equipped with this order relation is called the *phase diagram* of Φ.

Exercises

4.5.1.A. Let Φ be a vector field satisfying conditions (1) and (2) of Definition 4.4.1; show that for each trajectory γ the sets $\omega(\gamma)$ and $\alpha(\gamma)$ are critical elements.

Hint. Every point belonging either to $\omega(\gamma)$ or to $\alpha(\gamma)$ is non-wandering.

4.5.2.A. Let Φ be a Morse–Smale vector field. Let $\gamma_1 \leqslant \gamma_2 \leqslant \ldots \leqslant \gamma_k$ be any maximal chain of critical elements. Show that γ_1 is a sink (i.e., that $W^-(\gamma_1) = \emptyset$) and γ_k is a source ($W^+(\gamma_k) = \emptyset$).

Hint. If the conclusion were not true, the chain could not be maximal; see Palis (1969).

4.5.3.B. Suppose that a manifold M has the following property: there exists a vector field $\Phi \in S^r(M)$ which has exactly two critical elements, both of them being single points. Prove that M is homeomorphic to a sphere.

Hint. By the conclusion of Exercise 4.5.2, for one of the two critical points (denote it by x_1) the manifold $W^+(x_1)$ is homeomorphic to

$\boldsymbol{R}^m$, $m = \dim M$, and it is compactified by the other critical point x_2. Hence M is homeomorphic to S^m.

4.5.4.A. Let Φ be a Morse–Smale field on a manifold M. Show that the relation defined in Remark 4.5.1 is a partial order in the set of all critical elements of Φ.

Hint. See the proof of Theorem 4.5.1.

4.5.5.B. Let γ be a critical element of a Morse–Smale field. Prove that $W^+(\gamma)$ and $W^-(\gamma)$ are submanifolds of M (in the induced topology).

Hint. $W^+(\gamma)$ results by an injective embedding of $\boldsymbol{R}^k$: $W^+(\gamma) = i(\boldsymbol{R}^k)$, i: $\boldsymbol{R}^k \to M$, and $\psi_t = i^{-1} \circ \varphi_t \circ i$ is a contraction of $\boldsymbol{R}^k$ to 0. Hence $\boldsymbol{R}^k = \bigcup_{n=-\infty}^{+\infty} \psi_n(D)$, where D is some annulus, and $i(D)$ is a submanifold of M. Every point $x \in W^+(\gamma)$ has a neighbourhood U in M such that $U \cap W^+(\gamma)$ is the union of a finite number of $\psi_n(D)$.

A non-trivial example of a collection of trajectories of a certain Morse–Smale diffeomorphism is shown in Fig. 4.5.2.

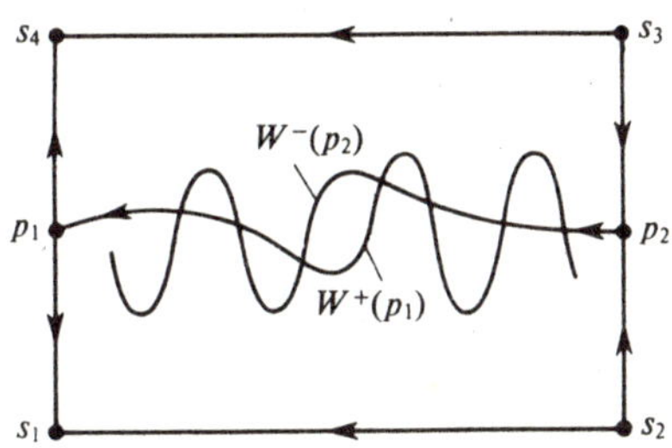

FIG. 4.5.2

We now present the theorem of Palis (1969) on the openness of $S^r(M)$ and $s^r(M)$.

Theorem 4.5.2 (1) *$S^r(M)$ is an open subset of $\mathscr{F}^r(M)$; similarly, $s^r(M)$ is an open subset of* $\mathrm{Diff}^r(M)$, $r \geqslant 1$.

(2) *If a vector field Ψ is sufficiently close to a vector field $\Phi \in S^r(M)$, then Ψ and Φ have isomorphic phase diagrams; a similar statement holds for diffeomorphisms.*

Proof We prove the theorem for $r = 1$ only; our proof is based on the Closing Lemma.

We consider only the vector field case.

Let $\Phi \in S^1(M)$ and let $\gamma_1, \ldots, \gamma_N$ be all the critical elements of Φ. Let U be a neighbourhood of a hyperbolic critical point x_0; U is called *hyperbolic* if there exists a larger neighbourhood $U_1 \supset U$ such that the system (U_1, φ_t) is topologically conjugate to the local system $(\tilde{U}_1, L_t)$, where $\tilde{U}_1 \subset T_{x_0}(M)$ and L_t is the flow generated by the vector field $\dot{\Phi}(x_0)x$ in the tangent space $T_{x_0}(M)$. Now let U be a tubular neighbourhood of a hyperbolic closed orbit γ; U is called *hyperbolic* iff, for every cross-section S of γ the set $S \cap U$ is a hyperbolic neighbourhood of the point $S \cap \gamma$ relative to the Poincaré map Θ_S.

Suppose that $U_1, \ldots, U_N$ are disjoint hyperbolic neighbourhoods of $\gamma_1, \ldots, \gamma_N$. Let V_1 be a neighbourhood of Φ in $\mathscr{F}^1(M)$ such that every vector field $\Psi \in V_1$ has exactly one critical element in each U_i, $i = 1, \ldots, N$ (see Theorem 3.4.2 on p. 157).

In virtue of Theorem 3.3.1 (p. 148) the mapping

$$G\colon \mathscr{F}^1(M) \times M \times \boldsymbol{R} \to M$$

given by

$$G(\Phi, x, t) = \varphi_t(x)$$

is of class C^1 at any point (Φ, x, t).

The proof of Theorem 4.5.2 is carried out in four steps.

Step I. We shall show that there exists a neighbourhood V_2 of Φ such that every $\Psi \in V_2$ has just as many critical elements as Φ.

Assume the contrary. Then there exists a sequence of vector fields Ψ_n, $n = 1, 2, \ldots$, C^1-convergent to Φ and such that each Ψ_n has N_n critical elements, where $N_n \neq N$. We may assume that $\Psi_n \in V_1$ for all n, which means that each Ψ_n has one critical element in every U_i, and thus $N_n > N$. Since Ψ_n has as many critical points as Φ, it follows that Ψ_n has more closed orbits.

For any n denote by β_n such an extra orbit, i.e., one not contained within any U_i. Pick a point x_n in each $\hat{\beta}_n$ so as to obtain a sequence (x_n) such that

$$x_n \to x_0 \notin \overline{\bigcup_{i=1}^{N} U_i}$$

(we may pass to a subsequence if necessary). Since $\Psi_n \to \Phi$ uniformly, we have

$$\Psi_n(x_n) \to \Phi(x_0).$$

Let l_1 denote the trajectory of Φ passing through x_0 (see Fig. 4.5.3).

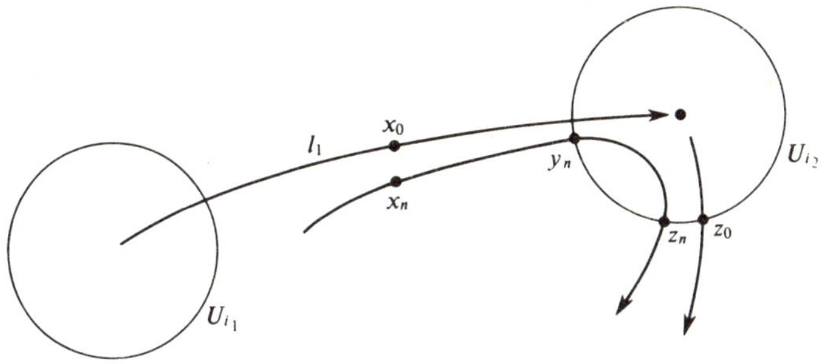

FIG. 4.5.3

By the conclusion of Exercise 4.5.1 on p. 212 the sets $\alpha(l_1)$ and $\omega(l_1)$ coincide with some critical elements of Φ. Let γ_{i_1} and γ_{i_2} be these elements (respectively). In view of the continuity of $\varphi_t(x)$, both in x and in the Φ-variable, each trajectory β_n (for large n) intersects U_{i_2} and U_{i_1}. Denote by y_n the earliest point on β_n, later than x_n, at which β_n enters U_{i_2}. We may assume that

$$y_n \to y_0 \in \hat{l}_1 \cap \mathrm{Fr}\, U_{i_2}.$$

Now, β_n is a closed orbit not entirely contained in U_{i_2}. Thus there is a point at which β_n leaves U_{i_2}; denote by z_n the earliest such point, later than y_n. Again we may assume that

$$z_n \to z_0 \in \mathrm{Fr}\, U_{i_2}.$$

Let l_2 denote the trajectory of Φ passing through z_0, $l_2(0) = z_0$. Then the "half-trajectory" l_2^- defined as l_1 restricted to $t < 0$ is wholly contained in U_{i_2}. Indeed: assume $l(t_1) \notin \overline{U_{i_2}}$ for some $t_1 < 0$; this, by the continuity of $G(\Phi, x, t)$, implies $\beta_n(t_n) \notin U_{i_2}$ for large n and for t_n near t_1, showing that the piece of β_n limited by y_n and z_n intersects the boundary of U_{i_2}, contrary to the construction of y_n and z_n.

The trajectory l_2 is approached by the trajectories β_n and its ω-set also coincides with some critical element γ_{i_3}. By repetition of the previous argument we obtain a sequence of critical elements

$$\gamma_{i_1} \geqslant \gamma_{i_2} \geqslant \gamma_{i_3} \geqslant \ldots$$

Since Φ has only finitely many critical elements, one of the elements γ_i must enter the sequence more than once, and this, in view of Theorem

4.5.1, implies

$$\gamma_{i_1} = \gamma_{i_2} = \gamma_{i_3} = \dots$$

Consequently l_1 is contained in $W^+(\gamma_{i_1}) \cap W^-(\gamma_{i_1})$. Hence by Corollary 4.4.2 (in the formulation involving flows) all points of l_1 are non-wandering and some of them do not belong to the union of $\hat{\gamma}_i$, $i = 1, \dots, N$. This contradicts the assumption that Φ is a Morse–Smale field.

Step II. Vector fields which are C^1-close to Φ have Property $\mathscr{G}_4$. Suppose that Φ can be C^1-approached by fields Ψ_n such that

$$\Gamma_{\Psi_n} := \overline{\operatorname{Per} \Psi_n} \neq \Omega(\Psi_n).$$

Thus let $x_n \in \Omega(\Psi_n) - \Gamma_{\Psi_n}$. Without loss of generality we may assume that $x_n \notin \overline{\bigcup_{i=1}^{N} U_i}$; otherwise the whole trajectory $\gamma^{\Psi_n}(x_n)$ passing through x_n would lie in some U_i, whereas the only trajectory of Ψ_n (for large n) contained in U_i is the closed orbit near to γ_i – a contradiction, because $\gamma^{\Psi_n}(x_n)$ cannot be periodic, by the choice of x_0.

By the Closing Lemma each Ψ_n can be C^1-approached by a field Ψ_n', whose trajectory through x_n is periodic. Choose Ψ_n' so as to have $\Psi_n' \to \Phi$ in C^1-topology. For n large enough each Ψ_n' has exactly one critical element in every U_i and one closed orbit passing through $x_n \in \overline{\bigcup U_i}$, i.e., has at least $N+1$ critical elements, contrary to the conclusion of Step I. Hence follows the claim.

Step III. Now we show that if Ψ is C^1-close to Φ, then Φ and Ψ have isomorphic phase diagrams. That means, we show that then

$$W^+(\gamma_j^\Phi) \cap W^-(\gamma_k^\Phi) \neq \emptyset \qquad \text{iff } W^+(\gamma_j^\Psi) \cap W^-(\gamma_k^\Psi) \neq \emptyset;$$

here γ_i^Φ and γ_i^Ψ $(i = j, k)$ denote the unique critical elements of Φ and Ψ, respectively, contained in the neighbourhood U_i.

Thus assume that $W^+(\gamma_j^\Phi) \cap W^-(\gamma_k^\Phi) \neq \emptyset$. Let $\Phi_n \to \Phi$ in C^1-topology. Let $i_n^{+,j}$: $P_j^+ \to M$ be an immersion that carries P_j^+, the underlying manifold, onto $W^+(\gamma_j^{\Phi_n})$, $n = 0, 1, 2, \dots$, where Φ_0 stands for Φ. Clearly, P_j^+ is the same for all Φ_n, provided that n is sufficiently large. In fact, P_j^+ is equal either to $\boldsymbol{R}^s$ or to $\boldsymbol{R}^s \times S^1$ or to $\boldsymbol{R}^s \dot{\times} S^1$. According to Theorem 1.6.2, we may assume that $i_n^{+,j} \to i_0^{+,j}$ in C^1-topology on any set of the form

$$K_j^+(a) := \{p \in P_j^+ : \|p\| \leqslant a\}.$$

Quite analogously, we have a sequence of immersions $i_n^{-,k}: P_k^- \to M$. $i_n^{-,k}(P_k^-) = W^-(\gamma_k^{\Phi_n})$, such that $i_n^{-,k} \to i_0^{-,k}$ in C^1-topology on any set of the form

$$K_k^-(a) := \{p \in P_k^- : \|p\| \leqslant a\},$$

where P_k^- denotes a manifold of one of the three types.

Let $q \in W^+(\gamma_j^{\Phi_0}) \cap W^-(\gamma_k^{\Phi_0})$. Then there exists a number $a > 0$ such that

$$q \in i_0^{+,j}(K_j^+(a)) \cap i_0^{-,k}(K_k^-(a)).$$

Since

$$W^+(\gamma_j^{\Phi_0}) \pitchfork_q W^-(\gamma_k^{\Phi_0})$$

and since $(i_n^{+,j})$, $(i_n^{-,k})$ converge C^1-uniformly on $K_j^+(a)$, $K_k^-(a)$, we infer that for every n, sufficiently large, there exists a point q_n such that

$$W^+(\gamma_j^{\Phi_n}) \pitchfork_{q_n} W^-(\gamma_k^{\Phi_n}).$$

To obtain the converse implication, assume that $W^+(\gamma_j^{\Phi}) \cap W^-(\gamma_k^{\Phi}) = \emptyset$. Repeating the argument of Step I, we show that there is no sequence (Φ_n) C^1-convergent to Φ and such that $W^+(\gamma_j^{\Phi_n}) \cap W^-(\gamma_k^{\Phi_n}) \neq \emptyset$ for all n.

Step IV. It remains to prove that if Ψ is sufficiently close to Φ, then Ψ satisfies the transversality condition (condition (3) of Definition 4.4.1 on p. 202).

Again the proof is indirect. Suppose that there exists a sequence (Φ_n) of vector fields C^1-convergent to Φ which do not satisfy the transversality condition. As shown in Step I, each Φ_n has just as many critical elements as Φ. By assumption, $W^+(\gamma_j^{\Phi_n})$ intersects $W^-(\gamma_k^{\Phi_n})$ non-transversally at some point q_n. We may assume that $q_n \to q_0$. Then $q_0 \in W^+(\gamma_j^{\Phi}) \cap W^-(\gamma_k^{\Phi})$. Put

$$E_n = T_{q_n}(W^+(\gamma_j^{\Phi_n})) + T_{q_n}(W^-(\gamma_k^{\Phi_n})).$$

Since $E_n \neq T_{q_n}(M)$, we have $\dim E_n < \dim M$. According to Theorem 1.6.2,

$$T_{q_n}(W^\pm(\gamma_i^{\Phi_n})) \to T_{q_0}(W^\pm(\gamma_i^{\Phi})) \qquad \text{for } i = j, k.$$

Hence

$$E_n \to T_{q_0}(W^+(\gamma_j^{\Phi})) + T_{q_0}(W^-(\gamma_k^{\Phi})) = T_{q_0}(M).$$

This is the required contradiction: $\dim T_{q_0}(M) = \dim M$ and $\dim E_n < \dim M$.

The theorem is thus proved.

A significant result obtained by Palis and Smale in 1968 establishes structural stability of Morse–Smale systems. We only formulate the result; the proof is long and difficult and therefore omitted here.

Theorem 4.5.3 Every diffeomorphism $\varphi \in s^r(M)$ is C^r-structurally stable. Every vector field $\Phi \in S^r(M)$ is C^r-structurally stable.

This result is not the strongest possible with regard to structural stability. In Section 4.12 we formulate a theorem (Theorem 4.12.2 on p. 264) of which Theorem 4.5.3 is a particular case.

4.6. Structural Stability of Flows on Two-Dimensional Manifolds

In the case of M being a two-dimensional manifold, the Morse–Smale conditions (1)–(3) (see Section 4.4, p. 202) admit an alternative restatement:

Theorem 4.6.1 For $\Phi \in \mathscr{F}^r(M)$, $\dim M = 2$, conditions (1)–(3) are equivalent to the following ones:

(1′) *for every trajectory γ the sets $\alpha(\gamma)$ and $\omega(\gamma)$ coincide with some critical elements,*

(2′) *all critical elements are hyperbolic,*

(3′) *there is no trajectory joining two saddles.*

Proof (1)–(3) imply (1′)–(3′): (1′) is the conclusion of Exercise 4.5.1; (2) and (2′) are identical; (3′) results from the following observation: if two saddles, call them x_1 and x_2, were connected by a trajectory γ, then $\hat{\gamma} \subset W^+(x_1) \cap W^-(x_2)$ and consequently this intersection would not be transversal.

(1′)–(3′) imply (1)–(3): (2) does not differ from (2′), (3) is an immediate consequence of (3′). It remains to prove (1).

To begin with, we show that Φ has only finitely many critical elements. Since all of them are hyperbolic, there cannot be infinitely many critical points. Thus assume that Φ has infinitely many closed orbits; denote them by $\beta_1, \beta_2, \ldots$ By compactness, the sets $\hat{\beta}_n$ cluster about some point x_0. According to condition (1′), the sets $\omega(x_0)$ and

$\alpha(x_0)$ are critical elements. They must be saddles; to see this, notice that otherwise the dimension of the stable (unstable) manifold of $\omega(x_0)$ or $\alpha(x_0)$ would be equal to 2, and so either the set $\omega(x_0)$ or the set $\alpha(x_0)$ would attract its vicinity (either at $+\infty$ or at $-\infty$), which is

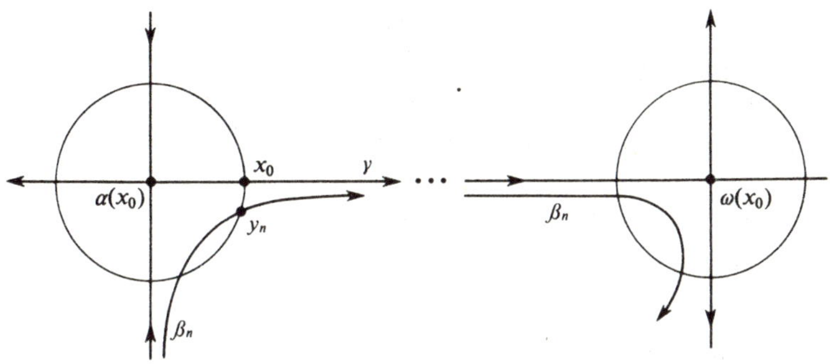

FIG. 4.6.1

impossible, because every neighbourhood of $\omega(x_0)$ or of $\alpha(x_0)$ is intersected by closed orbits β_n. Thus the trajectory passing through x_0 connects the two saddles: $\alpha(x_0)$ and $\omega(x_0)$, contrary to condition (3′).

We have yet to show that every non-wandering point belongs to some critical element. Assume that $x \in \Omega(\Phi)$ and that $\Phi(x) \neq 0$. The sets $\alpha(x)$ and $\gamma(x)$ are critical elements. If $\alpha(x) \neq \omega(x)$, both sets are saddles (otherwise x has a "wandering neighbourhood") and are connected by the trajectory passing through x; this contradicts (3′). Therefore $\omega(x)$ equals $\alpha(x)$ and it is not a saddle. Since all critical elements are hyperbolic, $\omega(x)$ is a closed orbit passing through x. This ends the proof.

Exercise

4.6.1.B. Show that if $M = S^2$, then condition (1′) is a consequence of (2′) and (3′).

Hint. On account of Schwartz' theorem (Theorem 2.2.1) the set $\omega(\gamma)$ contains a critical element. If the stable manifold of that element is of dimension 2, then $\omega(\gamma)$ is equal to that element. If $\omega(\gamma)$ contains saddle points (one or more) but is itself not a saddle, then (S^2 being simply connected) there must exist a trajectory joining two saddles or a trajectory $\tilde{\gamma}$ such that $\omega(\tilde{\gamma}) = \alpha(\tilde{\gamma}) =$ a saddle.

Example

4.6.1. Consider the vector field Φ_c defined in Exercise 1.4.4 on p. 37; we already came across that field in the discussion of Example 4.1.3 on p. 189. We recall that Φ_c results by projecting the gravitation field onto a tilted torus. In the conclusion of Example 4.1.3 we have stated that Φ_c is structurally stable. Now we show that it is a Morse Smale field.

Φ_c is a gradient field. Hence all of its trajectories begin and end at the critical points x_1, x_2, x_3, x_4. As we have already ascertained there is no trajectory joining the saddles x_2 and x_3. It remains to verify that the critical points are hyperbolic.

We will use the following parametrization of T^2:

$$x = (a+b\cos s)\cos t,$$
$$y = (a+b\cos s)\sin t,$$
$$z = b\sin s,$$

where a, b, t, s are as in Fig. 4.6.2; $a > b$.

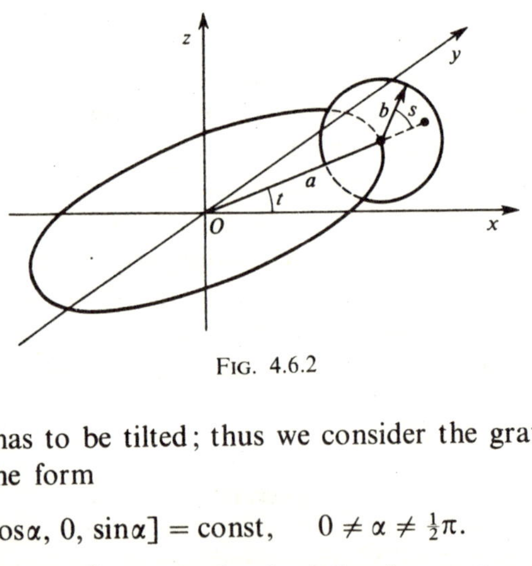

FIG. 4.6.2

The torus has to be tilted; thus we consider the gravitation field in $\boldsymbol{R}^3$ to be of the form

$$\Psi(x) = [\cos\alpha, 0, \sin\alpha] = \text{const}, \qquad 0 \neq \alpha \neq \tfrac{1}{2}\pi.$$

Projecting Ψ onto T^2 we obtain the following vector field (written in the coordinates t, s; we omit the calculations):

$$\Phi(t, s) = \left[-\frac{\sin t\cos\alpha}{a+b\cos s}, \frac{-\sin s\cos t\cos\alpha+\cos s\sin\alpha}{b}\right].$$

The critical points are:

$$x_4 = (0, \alpha), \quad x_3 = (0, \alpha+\pi), \quad x_2 = (\pi, -\alpha), \quad x_4 = (\pi, \pi-\alpha).$$

The Hessian is given by

$$\dot{\Phi}(t, s) = \begin{bmatrix} -\dfrac{\cos t \cos \alpha}{a + b\cos s} & -\dfrac{\sin s \sin t \cos \alpha}{b} \\ \dfrac{\sin s \sin t \cos \alpha}{b} & \dfrac{-\cos s \cos t \cos \alpha - \sin s \sin \alpha}{b} \end{bmatrix}.$$

At the critical points the Hessian takes on the form

$$\dot{\Phi}(x_i) = \begin{bmatrix} \dfrac{\pm\cos \alpha}{a \pm b\cos \alpha} & 0 \\ 0 & \pm\dfrac{1}{b} \end{bmatrix}, \qquad i = 1, 2, 3, 4.$$

It is readily seen that the eigenvalues of $\dot{\Phi}(x_i)$ are real and non-zero. Thus the four critical points are hyperbolic.

In virtue of the Palis–Smale theorem the field under consideration is structurally stable.

Observe that the above conclusion is a result of the general theory and an elementary calculus. A straightforward proof would by no means be easier than general considerations, carried out at least with regard to gradient fields.

The Palis–Smale theorem gives a sufficient condition for structural stability. As we shall see later on (see p. 264), this condition is not necessary in general. However, it is necessary in the case of two-dimensional orientable manifolds.

Theorem 4.6.2 (*Peixoto, 1962*) *Let M be a two-dimensional oriented manifold and let $\Phi_0 \in \mathscr{F}^r(M)$, $r \geqslant 1$. The field Φ_0 is C^r-structurally stable if and only if it is a Morse–Smale field.*

The "if" part is established by Theorem 4.5.3, which asserts the sufficiency of Morse–Smale conditions for stability in the general case.

As regards the necessity (the "only if" part), we shall present a proof only for the case of $r = 1$; our proof will resort to the Closing Lemma and to Smale's Theorem 4.3.4. A proof for the general case can be found in Palis and de Melo (1978).

We will need some auxiliary facts.

Let M be compact, connected, two-dimensional, and oriented.

Lemma 4.6.1 *Suppose that a vector field $\Phi \in \Sigma^r(M)$ has infinitely many* *closed orbits:* $\beta_1, \beta_2, \ldots$ *Write* $F = \overline{\bigcup \hat{\beta}_n} - \bigcup \hat{\beta}_n$ *and let* $x_0 \in F$, $\Phi(x_0) \neq 0$. *Then:*

1° *the sets* $\omega(x_0)$ *and* $\alpha(x_0)$ *are reduced to saddle points,*

2° *the trajectory* γ *passing through* x_0 *is a submanifold of* M *(in the topology of* M*).*

Proof Assume that there is a point $x_1 \in \omega(x_0)$ such that $\Phi(x_1) \neq 0$. Choose a rectifying chart (U, α), $U \ni x_1$. The trajectories of Φ in U are parallel straight line segments (see Fig. 4.6.3). Let I_0 denote the piece of

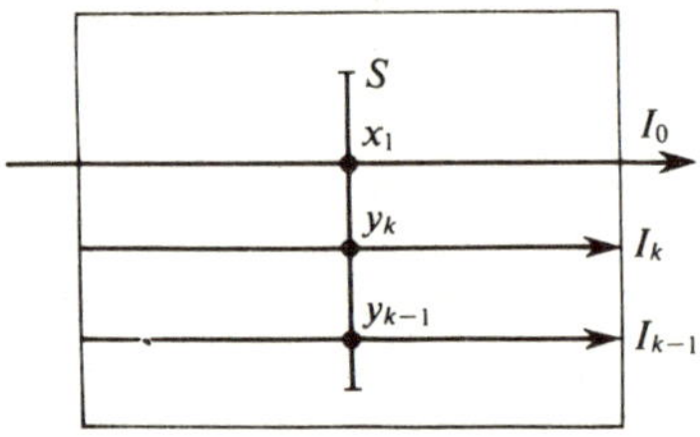

FIG. 4.6.3

the trajectory through x_1 contained within U, $x_1 \in I_0$. Let I_k, $k = 1, 2, \ldots$, be a sequence of segments of γ, converging to I_0. Draw a cross-section S through x_1, perpendicular in U to I_0 and denote by y_k the cross-point of S and I_k. It can be assumed that the sequence (y_k) converges to x_1 monotonically.

Each I_k is the limit of a sequence of pieces of β_n. Denote by z_n the point of the (finite) set $\hat{\beta}_n \cap S$ which lies nearest to x_1. Let (β_{n_i}) be a subsequence of (β_n) such that each z_{n_i} belongs to an interval (y_{k_i}, y_{k_i+1}), the sequence (y_{k_i}) being monotone.

It is well known (see e.g. Theorem 7.1 in Wallace, 1968) that M is homeomorphic to the sphere S^2 with a finite number of "handles"; denote this number (the genus of M) by p. There can exist at most p closed curves on M not disconnecting M (see Wallace, 1968).

Consider the curves $\beta_{n_1}, \ldots, \beta_{n_{p+1}}$. In view of what has just been said, one of these curves, say β_{n_s}, is a common boundary of two connected components of M. Since M is oriented, we may distinguish the two sides of β_{n_s}. The segments I_{k_s} and I_{k_s+1} lie on the opposite sides of β_{n_s} (see Fig. 4.6.4) and thus they belong to different components. But I_{k_s} and I_{k_s+1} are pieces of the same trajectory γ. Thus it is possible to pass from I_{k_s} to I_{k_s+1} along γ without crossing any of the curves β_n. We have arrived at a contradiction; this shows that $\omega(x_0)$ does not contain any non-critical point.

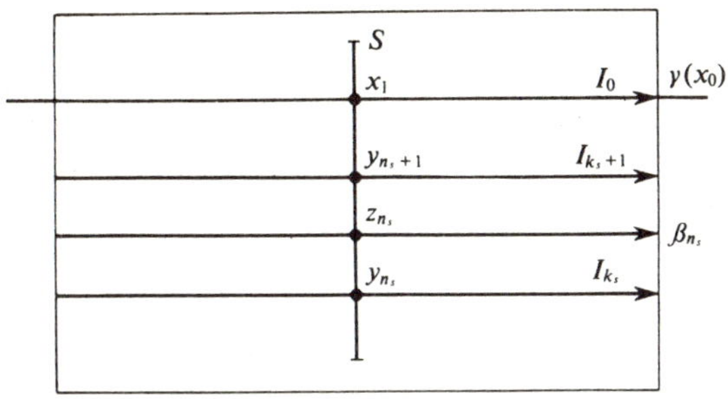

FIG. 4.6.4

The set $\omega(x_0)$ is connected (see Section 1.11). Thus it must be a singleton, because it consists entirely of critical points (as we have just shown) and all critical points are hyperbolic – here we use Theorem 4.3.1, namely the case considered in Step I of its proof (not involving condition (ε)).

The unique point in $\omega(x_0)$ is a saddle; otherwise it would attract some neighbourhood of x_0, and this is impossible, since the closed orbits β_n cluster about x_0.

This proves assertion 1° for $\omega(x_0)$; the reasoning for $\alpha(x_0)$ is the same.

Assertion 2° is a direct consequence of the proof of 1°: any segment of γ contained in U is isolated from any other such segment, and hence γ is a one-dimensional submanifold of M.

Lemma 4.6.1 implies

Corollary 4.6.1 *If $x_0 \in M$ is a recurrent point of a field $\Phi \in \mathscr{F}^r(M)$, M being compact, two-dimensional, orientable, then x_0 is periodic.*

From now on we restrict our considerations to $r = 1$.

Lemma 4.6.2 *If $\Phi \in \Sigma^1(M)$, then Φ has finitely many closed orbits.*

Proof Assume that Φ has infinitely many closed orbits $\beta_1, \beta_2, \ldots$. By Theorem 4.3.1 with Remark 4.3.1, all critical elements are hyperbolic. Let F be as in Lemma 4.6.1. Certainly there exists a point $x_0 \in F$ with $\Phi(x_0) \neq 0$; if there were no such points, F would consist only of critical points which are hyperbolic and thus cannot be cluster points of the union of closed orbits. In view of Lemma 4.6.1 the orbit $\gamma = \gamma(x_0)$ either is closed or connects two saddles. The first eventuality is not possible, since all closed orbits are hyperbolic (they are either attractive or repulsive, and so $\hat{\gamma}$ cannot contain cluster points of β_n). The second eventuality is contradicted by Theorem 4.3.4 (Smale's): if γ connects two saddle-points, then $\hat{\gamma}$ is a subset of the stable manifold of one of these points and of the unstable manifold of the other.

Proof of Theorem 4.6.2 (for $r = 1$) Assume that Φ_0 is C^1-structurally stable. In virtue of Theorem 4.3.3 Φ_0 has Property $\mathscr{G}_4$ (here is the Closing Lemma involved; the proof of Theorem 4.3.3 is based on it). Thus $\Omega(\Phi_0) = \overline{\operatorname{Per}\Phi_0}$. Since Φ_0 has only finitely many critical points and closed orbits (Lemma 4.6.2), we have $\overline{\operatorname{Per}\Phi_0} = \operatorname{Per}\Phi_0$. This proves the Morse–Smale condition (1) of Definition 4.4.1. Condition (2) is fulfilled in virtue of Theorem 4.3.1 with Remark 4.3.1, condition (3) follows from Theorem 4.3.4.

As we have seen, structurally stable vector fields on two-dimensional compact orientable manifolds have finitely many critical elements. This is not necessarily true in higher dimensions, e.g. the so-called Anosov flows (see Definition 4.8.2 on p. 242) on tori of dimension not less than 3 are structurally stable but have infinitely many closed orbits (see Anosov, 1967). The same remark applies to diffeomorphisms.

Another interesting property of structurally stable vector fields on two-dimensional orientable manifolds is expressed by the following theorem:

Theorem 4.6.3 (Peixoto, 1962) If M is a compact oriented two-dimensional manifold, then $\Sigma^r(M)$ is a dense open subset of $\mathscr{F}^r(M)$, $r \geqslant 1$.

Proof Again we restrict ourselves to the case of $r = 1$, referring to Palis and de Melo (1978) for the general case.

In virtue of Theorem 4.6.2 structurally stable fields are precisely the Morse–Smale fields, and the latter constitute an open subset of $\mathscr{F}^1(M)$, by Theorem 4.5.2.

Now we prove that $\mathscr{G}_3^1(M) \cap \mathscr{G}_4^1(M) \subset \Sigma^1(M)$. Let $\Phi \in \mathscr{G}_3^1(M) \cap \mathscr{G}_4^1(M)$. It follows from the definition of Property $\mathscr{G}_3$ and from Theorem 4.6.1 that then Φ satisfies the Morse–Smale conditions (2) and (3) (see p. 202). Hence by Corollary 4.6.2 Φ has only finitely many closed orbits, and thus finitely many critical elements in general. Consequently and by Property $\mathscr{G}_4$, we have $\Omega(\Phi) = \operatorname{Per}\Phi$. This shows that Φ fulfils also condition (1), and so $\Phi \in \Sigma^1(M)$.

Now we apply the Kupka–Smale theorem (Theorem 3.7.1) and Pugh's theorem (Corollary 3.8.1). According to those theorems, $\mathscr{G}_3^1(M)$ and $\mathscr{G}_4^1(M)$ are residual subsets of $\mathscr{F}^1(M)$. Since the set $\Sigma^1(M)$ contains the common part of $\mathscr{G}_3^1(M)$ and $\mathscr{G}_4^1(M)$, it is also residual, hence it is dense in $\mathscr{F}^1(M)$. This ends the proof.

Remark 4.6.1 It is not known whether the Closing Lemma is true for C^r-flows on compact oriented two-dimensional manifolds. If this is the case, the above proofs work without any change in the general situation (concerning C^r-flows and C^r-stability, $r \geqslant 2$).

4.7. Hyperbolic Sets

It has already been emphasized that Morse–Smale conditions are only sufficient for structural stability. In the next section we shall present a class of structurally stable diffeomorphisms which need not satisfy the Morse–Smale conditions – the so-called Anosov systems. To define and discuss those systems we shall need the concept of a hyperbolic set. Before introducing this concept, let us briefly recall some facts of differential geometry.

Let M be a Riemannian manifold with scalar product $(\cdot, \cdot)_x$, $x \in M$. Let f: $T(M) \to T(M)$. We call f a *bundle map* iff f preserves the fibres.

A bundle map f is called:

1° *contracting* iff there exist constants $c > 0$ and λ, $0 < \lambda < 1$, such that

$$\|f^n(v)\| \leqslant c\lambda^n \|v\| \qquad \text{for } v \in T(M),\ n \in Z^+;$$

2° *expanding* iff there exist constants $d > 0$ and $\mu > 1$ such that

$$\|f^n(v)\| \geqslant d\mu^n \|v\| \qquad \text{for } v \in T(M),\ n \in Z^+.$$

Theorem 4.7.1 If M is compact and if f is contracting then f is also contracting in any other Riemannian metric; a similar statement holds for expanding maps.

Proof On the fibre $T_x(M)$ any two norms $\|\cdot\|_x$ and $|\cdot|_x$ are equivalent, since M is finite-dimensional. Thus, for every $x \in M$ there exist numbers $a(x)$ and $b(x)$ such that

$$a(x)|\cdot|_x \leqslant \|\cdot\|_x \leqslant b(x)|\cdot|_x.$$

The norms $|\cdot|_x$ and $\|\cdot\|_x$ depend continuously on x and M is compact. Also, we have $\dim T_x(M) = \dim M$ for all x. Consequently there are constants $a > 0$ and $b < +\infty$ such that

$$a \leqslant a(x) \qquad \text{and} \qquad b(x) \leqslant b \qquad \text{for all } x \in M.$$

Suppose that f is $\|\cdot\|$-contracting. Then

$$|f^n(v)| \leqslant \frac{1}{a}\|f^n(v)\| \leqslant \frac{c}{a}\lambda^n\|v\| \leqslant \frac{bc}{a}\lambda^n|v|$$

and so f is $|\cdot|$-contracting. For expanding mappings the argument is exactly the same.

Exercise

4.7.1.A. Let f and M be as above, M being compact. Show that if f is contracting, then there exists a scalar product $(\cdot, \cdot)_x$ in each $T_x(M)$ such that the induced norm satisfies the inequality $\|f^n(v)\| \leqslant \lambda^n\|v\|$ for all $v \in T(M)$ and for some λ, $0 < \lambda < 1$. Show that an analogous statement is valid for expanding f.

Hint. See the proof of Lemma 1.5.1.

Suppose that φ: $M \to M$ is a diffeomorphism. Then $d\varphi$ is an automorphism of the tangent bundle $T(M)$.

Examples

4.7.1. Let φ: $S^1 \to S^1$ be given by $\varphi(z) = z^N$, $N > 1$. Then $d\varphi$ is an expanding mapping of $T(S^1)$: we have $d\varphi(z)v = Nv$ for $v \in T_z(S^1)$.

4.7.2. Let $M = \boldsymbol{R}^2$ and let φ be given by a matrix $\begin{bmatrix} \lambda & 0 \\ 0 & \lambda \end{bmatrix}$, where $|\lambda| < 1$. Then $d\varphi$ is contracting:

$$\|d\varphi(x)v\| = |\lambda|\,\|v\| \qquad \text{for } v \in T_x(\boldsymbol{R}^2).$$

Let φ: $M \to M$ be a diffeomorphism of a compact manifold M. Let $\Lambda \subset M$ be a closed invariant set: $\varphi(\Lambda) = \Lambda$.

Definition 4.7.1 We say that Λ has *hyperbolic structure* or that Λ is a *hyperbolic set* iff the tangent bundle $T(M)$ restricted to Λ (denote it by

$T_\Lambda(M)$) admits an invariant decomposition

$$T_\Lambda(M) = E^+ \oplus E^-,$$

where each of the spaces E^+, E^- is mapped by $d\varphi$ into itself, the mapping $d\varphi|_{E^+}$ is contracting and $d\varphi|_{E^-}$ is expanding.

Examples

4.7.3. Every hyperbolic periodic trajectory of a diffeomorphism is a hyperbolic set in the sense just defined.

4.7.4. Let $\varphi = \begin{bmatrix} m & n \\ p & q \end{bmatrix}$ be an algebraic automorphism of the torus and suppose that the spectrum of φ is disjoint with the unit circle. Then the entire torus T^2 has hyperbolic structure: indeed, let λ and μ be the eigenvalues of φ, $|\lambda| < 1$, $|\mu| > 1$, and let a and b be the corresponding eigenvectors: $\varphi(a) = \lambda a$, $\varphi(b) = \mu b$; the required decomposition of the tangent bundle $T_x(T^2) = \boldsymbol{R}^2$ is the decomposition into the direct sum of the eigenspaces: $\boldsymbol{R}^2 = [a] \oplus [b] = \operatorname{span}(a) \oplus \operatorname{span}(b)$.

Let us now pass to the flow case. Let f_t be a one-parameter group of bundle maps f_t: $T(M) \to T(M)$. The flow f_t is called:

1° *contracting* iff there exist constants $c > 0$ and $\lambda > 0$ such that

$$\|f_t(v)\| \leqslant ce^{-\lambda t}\|v\| \qquad \text{for } v \in T(M),\ t \in \boldsymbol{R};$$

2° *expanding* iff the flow f_{-t} is contracting.

Let M be a compact manifold and let φ_t be the flow generated by a vector field $\Phi \in \mathscr{F}^1(M)$. Then $d\varphi_t$ is a one-parameter group of mappings of $T(M)$. Let $\Lambda \subset M$ be a closed invariant set: $\varphi_t(\Lambda) = \Lambda$ for $t \in \boldsymbol{R}$.

Definition 4.7.2 We say that Λ has *hyperbolic structure* or that Λ is a *hyperbolic set* iff the tangent bundle $T(M)$ restricted to Λ admits a decomposition

$$T_\Lambda(M) = E^+ \oplus E^- \oplus E^0,$$

where E^+, E^-, E^0 are invariant under $d\varphi_t$, the flow $d\varphi_t|_{E^+}$ is contracting, $d\varphi_t|_{E^-}$ is expanding and E^0 is the one-dimensional bundle resulting by differentiation of φ_t with respect to t; i.e., $E^0_x = [\Phi(x)] = \operatorname{span}(\Phi(x))$.

E.g. every hyperbolic closed orbit γ has hyperbolic structure in the sense just defined:

$$T_\gamma(M) = T(W^+(\gamma)) \oplus T(W^-(\gamma)) \oplus [\Phi(\cdot)].$$

Exercise

4.7.2.A. Let $\Lambda \subset M$ be a hyperbolic set (relative to a diffeomorphism φ, $\varphi(\Lambda) = \Lambda$) and let $T_\Lambda(M) = E^+ \oplus E^-$ be a decomposition satisfying the conditions of Definition 4.7.1; it induces a decomposition of each fibre:

$$T_x(M) = E_x^+ \oplus E_x^-, \qquad x \in \Lambda.$$

Show that E_x^+, E_x^- depend continuously on x.

The space

$$T^m := S^1 \times \ldots \times S^1 \qquad (m \text{ factors})$$

is called the *m-torus*. It is easy to see that $T^m = \mathbf{R}^m/\Gamma$, where Γ denotes the subgroup of $\mathbf{R}^m$ consisting of all integer vectors. Thus T^m can be alternatively defined as follows:

$$T^m = \{(x^1, \ldots, x^m)(\text{mod } 1)\colon \ (x^1, \ldots, x^m) \in \mathbf{R}^m\}.$$

Let $A = (a_{ij})_{i,j=1,\ldots,m}$ be a matrix with integer entries and with determinant $\det A = \pm 1$. Since $A(\Gamma) \subset \Gamma$, A defines a mapping of T^m into itself. Since also A^{-1} has integer entries, we have $A^{-1}(\Gamma) \subset \Gamma$ and thus A defines a diffeomorphism of T^m. Mappings of that type for $m = 2$ have been introduced in Exercises 1.2.4 and 1.2.5 on p. 23 and have been repeatedly encountered throughout the text.

Definition 4.7.3 Any mapping of T^m of the type just described is called an *algebraic automorphism* of the *m*-torus.

Exercise

4.7.3.A. Show that the set of periodic points of an algebraic automorphism A of T^m is dense in T^m: $\overline{\text{Per}\, A} = T^m$.

Hint. Rational points are periodic.

4.8. Anosov Diffeomorphisms

Let $\varphi: M \to M$ be a diffeomorphism of a compact Riemannian manifold M.

Definition 4.8.1 We call φ an *Anosov diffeomorphism* iff the entire M has hyperbolic structure.

The system considered in Exercise 1.2.5 on p. 23 can serve as an example of an Anosov diffeomorphism.

Example

4.8.1. More generally, every algebraic automorphism A of the m-torus T^m such that the spectrum of A is disjoint with the unit circle is an Anosov diffeomorphism: the space $\boldsymbol{R}^m$ splits into the direct sum of two invariant subspaces E_0^+ and E_0^- corresponding to the eigenvalues lying inside and outside the unit circle; this yields the decomposition

$$T_x(T^m) = E_x^+ \oplus E_x^-, \qquad E_x^\pm = E_0^\pm \qquad \text{for all } x,$$

so that

$$\mathrm{d}A(x)(E_x^\pm) = E_{Ax}^\pm.$$

Evidently, $A|_{E^+}$ is contracting and $A|_{E^-}$ is expanding.

Anosov diffeomorphisms constitute a class with interesting properties. They need not be Morse–Smale diffeomorphisms, e.g. the algebraic automorphisms of tori have infinitely many periodic points (see Exercise 4.7.3); nevertheless, as shown by Anosov (1967), they are structurally stable.

Before presenting a proof of this fact, we show that they form an open subset of $\mathrm{Diff}^1(M)$.

Exercise

4.8.1*. A. Show that if $\varphi: M \to M$ is an Anosov diffeomorphism, then there is a constant $\vartheta > 0$ such that for any $x, y \in M$, $x \neq y$, we have

$$\varrho(\varphi^n(x), \varphi^n(y)) \geqslant \vartheta \qquad \text{for some } n \in Z.$$

Hint. Locally, the decomposition $T_x(M) = E_x^+ \oplus E_x^-$ implies the existence of subspaces W_x^+ and W_x^- such that $T_x(W_x^\pm) = E_x^\pm$, $x \in M$. The manifolds W_x^+ and W_x^- locally constitute a coordinate system;

consider the metric $\varrho_1^2 := \varrho_+^2 + \varrho_-^2$, where $\varrho_\pm$ denote the manifold metric restricted to $W_x^\pm$. Then ϱ_1 is equivalent to the original metric ϱ (see Bowen, 1975; cf. also Theorem 4.12.1).

Theorem 4.8.1 Let M be a compact Riemannian manifold. Then the set of all Anosov diffeomorphisms is open in $\mathrm{Diff}^1(M)$.

Proof Let $\varphi_0 \in \mathrm{Diff}^1(M)$ be an Anosov diffeomorphism and suppose that $\varphi \in \mathrm{Diff}^1(M)$ is C^1-close to φ_0. Define mappings F, F_0: $\mathscr{F}^1(M) \to \mathscr{F}^1(M)$ by:

$$F(\Phi)(x) = \mathrm{d}\varphi(\varphi^{-1}(x))(\Phi(\varphi^{-1}(x))),$$
$$F_0(\Phi)(x) = \mathrm{d}\varphi_0(\varphi_0^{-1}(x))(\Phi(\varphi_0^{-1}(x))).$$

The tangent bundle $T(M)$ admits a hyperbolic decomposition, invariant under $\mathrm{d}\varphi_0$:

$$T(M) = E^+ \oplus E^-, \qquad T_x(M) = E_x^+ \oplus E_x^- \qquad \text{for } x \in M. \tag{4.8.1}$$

The mappings F and F_0 are linear; let us write them in the matrix form corresponding to the decomposition (4.8.1):

$$F = \begin{bmatrix} F_{++} & F_{+-} \\ F_{-+} & F_{--} \end{bmatrix}, \qquad F_0 = \begin{bmatrix} F_{++}^0 & F_{+-}^0 \\ F_{-+}^0 & F_{--}^0 \end{bmatrix}.$$

This means that for

$$\Phi(x) = (\dot{x}^+(x), \dot{x}^-(x)) = (\dot{x}^+, \dot{x}^-)$$

we ought to have

$$F(\Phi)(x) = (F_{++}\dot{x}^+ + F_{+-}\dot{x}^-, F_{-+}\dot{x}^+ + F_{--}\dot{x}^-),$$

and similarly for F_0. Since the decomposition (4.8.1) is $\mathrm{d}\varphi_0$-invariant, and hence F_0-invariant, we have $F_{+-}^0 = F_{-+}^0 = 0$. Applying the result of Exercise 4.7.1 (slightly modified), we can introduce a metric on M such that

$$\|F_{++}^0 v^+\| \leqslant \lambda_1 \|v^+\| \qquad \text{for } v^+ \in E^+,$$
$$\|(F_{--}^0)^{-1} v^-\| \leqslant \lambda_1 \|v^-\| \qquad \text{for } v^- \in E^-,$$

where λ_1 is some number, $0 < \lambda_1 < 1$. Consequently, φ being C^1-close to φ_0, there exists a number λ, $0 < \lambda < 1$, such that

$$\|F_{++} v^+\| \leqslant \lambda \|v^+\| \qquad \text{for } v^+ \in E^+,$$
$$\|F_{--}^{-1} v^-\| \leqslant \lambda \|v^-\| \qquad \text{for } v^- \in E^-.$$

We will show that there exists a number $\eta > 0$ depending only on λ and F_0 (i.e., φ_0) and such that if $\|F_{-+}\| < \eta$, $\|F_{+-}\| < \eta$, then φ is Anosov; that is to say, we wish to have a $d\varphi$-invariant decomposition $T(M) = \tilde{E}^+ \oplus \tilde{E}^-$, such that $d\varphi$, restricted to $\tilde{E}^+$ and $\tilde{E}^-$, is respectively contracting and expanding.

We shall try to find $\tilde{E}^+$ among subspaces of the form

$$\{(v^+, Av^+):\ v^+ \in E^+\},$$

where A operates on fibres as follows:

$$A(x):\ E_x^+ \to E_x^-,\ x \in M,$$

each $A(x)$ being linear. I.e., $A = (A(x),\ x \in M)$ has to be a bundle map of E^+ into E^-; the space of all such maps will be denoted by $\mathscr{L}(E^+, E^-)$ (cf. Definition 3.1.8). Since the scalar product $(\cdot, \cdot)_x$ depends continuously on x, we have

$$\|A\| = \sup_{x \in M} \sup_{\|v^+\|_x = 1} \|Av^+\| < +\infty.$$

Since, further,

$$F\begin{bmatrix} v^+ \\ Av^+ \end{bmatrix} = \begin{bmatrix} F_{++} & F_{+-} \\ F_{-+} & F_{--} \end{bmatrix}\begin{bmatrix} v^+ \\ Av^+ \end{bmatrix}$$
$$= \big(F_{++}v^+ + F_{+-}(Av^+),\ F_{-+}v^+ + F_{--}(Av^+)\big), \tag{4.8.2}$$

the postulate that $\tilde{E}^+$ should be F-invariant takes on the form:

$$AF_{++} + AF_{+-}A = F_{-+} + F_{--}A, \tag{4.8.3}$$

which can be rewritten as

$$A = F_{--}^{-1}(-F_{-+} + AF_{++} + AF_{+-}A).$$

Thus define:

$$G(A) = F_{--}^{-1}(-F_{-+} + AF_{++} + AF_{+-}A);$$

what we need is to find an A which is a fixed point of the operation G.

If $\|F_{+-}\| < \eta$, $\|F_{-+}\| < \eta$, where η is small enough, then we have for $\|A\| \leqslant 1$

$$\|G(A)\| \leqslant \|F_{--}^{-1}\|\big(\eta + \|F_{++}\|\,\|A\| + \|A\|^2\eta\big) < 1.$$

Further, for $\|A\| \leqslant 1$, $\|B\| \leqslant 1$ we have

$$\begin{aligned}\|G(A)-G(B)\| &\leqslant \|F_{--}^{-1}\|\left(\|A-B\|\,\|F_{++}\|+\|AF_{+-}A-BF_{+-}B\|\right)\\ &\leqslant \lambda\left(\lambda\|A-B\|+\|AF_{+-}(A-B)\|+\|(A-B)F_{+-}B\|\right)\\ &\leqslant (\lambda^2+2\eta\lambda)\,\|A-B\|.\end{aligned}$$

Hence, if $\lambda^2+2\eta\lambda<1$, then G is a contraction on the unit ball in $\mathscr{L}(E^+, E^-)$ and thus there exists an A (exactly one) such that

$$G(A)=A.$$

This means that equation (4.8.6) has a unique solution. It is easy to verify that $d\varphi = F$ contracts $\tilde{E}^+$ (this is seen from (4.8.2)).

The space $\tilde{E}^-$ is constructed in exactly the same way: it is found among subspaces of the form

$$\{(Bv^-, v^-) \colon v^- \in E^-\},$$

where $B\colon E^- \to E^+$ is of the type analogous to A. As before, the method consists in finding a solution of a certain equation by applying the fixed point theorem.

The operators A and B found in this way have norms less than 1: $\|A\|<1$, $\|B\|<1$. Obviously, $\dim \tilde{E}_x^\pm = \dim E_x^\pm$. It remains to show that $\tilde{E}_x^+ \cap \tilde{E}_x^- = \{0\}$. But this is immediate: if $v \in \tilde{E}_x^+ \cap \tilde{E}_x^-$, $v^- \neq 0$, then

$$v^- = Av^+ = ABv^-$$

and so

$$\|v^-\| \leqslant \|A\|\,\|B\|\,\|v^-\| < \|v^-\|,$$

which is absurd. Thus $v^-=0$; hence also $v^+ = Bv^- = 0$, ending the proof.

We now proceed to the main goal of this section: we prove that every Anosov diffeomorphism is structurally stable. Anosov's original proof is very complicated; it has been essentially simplified by Moser (1969).

The proof which we present is reproduced from Robbin's work (1971); in fact, it is fundamentally Moser's proof with certain minor technical improvements.

Theorem 4.8.2 (*Anosov*) *Every Anosov diffeomorphism of a compact Riemannian manifold M is C^1-structurally stable. More precisely: if $\varphi \in \mathrm{Diff}^1(M)$ is an Anosov diffeomorphism (M compact), then for every* *neighbourhood* N *of the identity in* $C^0(M, M)$ *there exists a neighbourhood* V *of* φ *in* $\mathrm{Diff}^1(M)$ *such that any* $\psi \in V$ *is conjugated with* φ *by a homeomorphism* $h \in N$: $\psi \circ h = h \circ \varphi$.

The scheme of the proof is as follows: we try to find h among mappings of the form[17]

$$h(x) = \exp_x \Phi(x), \qquad \text{where } \Phi \in \mathscr{F}^0(M).$$

The postulate that h should conjugate ψ with φ is translated into certain conditions imposed on Φ. Then the vector field Φ is found as the fixed point of a contraction of a suitable closed subset of $\mathscr{F}^0(M)$.

In the proof we shall need several facts which we now formulate as lemmas.

Lemma 4.8.1 *If* $\varphi \in \mathrm{Diff}^r(M)$, $r \geqslant 0$ (*for* $r = 0$ *this just means that* φ *is a homeomorphism*), *and if* φ *is sufficiently* C^r*-close to identity, then* φ *is isotopic to identity, and moreover, there exists exactly one vector field* $\Phi \in \mathscr{F}^r(M)$ *such that* $\exp_x \Phi(x) = \varphi(x)$.

Proof If $\varphi(x)$ is close to x, then x and $\varphi(x)$ can be locally connected by a geodesic line $\gamma_x(t)$, $\gamma_x(0) = x$, $\gamma_x(1) = \varphi(x)$, in a unique way. Now we put $\Phi(x) = \dot{\gamma}_x(0)$. The curve $\gamma_x(t)$ is a solution of a certain second order differential equation with the initial condition $\gamma_x(0) = x$, $\dot{\gamma}_x(0) = \Phi(x)$.

Write $F(x, v) = \exp_x(v)$. The Riemannian metric on M can be chosen so as to be of arbitrarily high class of regularity C^{r_0}, $r_0 \geqslant \max(r, 1)$. Hence F is a C^{r_0}-map in the variables x and v. Since for x fixed the mapping $F(x, \cdot)$ is a diffeomorphism of some ball $B(0, a) \subset T_x(M)$ onto a neighbourhood of x, it follows that $\mathrm{d}_v F(x, v)$ is a surjection of $T_x(M)$ onto $T_{\gamma(1)}(M)$ and the equation $F(x, v) = \bar{x}$ can be solved with respect to v, provided that $\bar{x}$ is close enough to x. Denote this solution by $v = G(x, \bar{x})$. Thus we have $F(x, G(x, \bar{x})) = \bar{x}$ for $\bar{x}$ close to x, and G is a C^{r_0}-map. Setting $\bar{x} = \varphi(x)$, we obtain $\Phi(x) = G(x, \varphi(x))$; since φ is of class C^r, $r = 1, 2, \ldots$, Φ is of the same class.

The isotopy between φ and the identity is given by: $f(x, t) = \exp_x t\Phi(x)$. The uniqueness of Φ follows from the fact that, locally, two points can be joined by a geodesic line in a unique way. This proves the lemma.

According to Lemma 4.8.1, homeomorphisms h: $M \to M$ close to identity stand in a one-to-one correspondence with C^0-vector fields Ψ, via the formula $h = \exp \Psi$. Let $\varphi \in \mathrm{Diff}^1(M)$; we define a mapping $F\colon \mathscr{F}^0(M) \to \mathscr{F}^0(M)$ as follows:

$$\exp F(\Psi) = \varphi^{-1} \circ \exp \Psi \circ \varphi. \tag{4.8.5}$$

Consider also the map $\varphi^{\#}\colon \mathscr{F}^0(M) \to \mathscr{F}^0(M)$ given by

$$\varphi^{\#}(\Psi) = (\mathrm{d}\varphi^{-1} \circ \varphi)(\Psi \circ \varphi), \tag{4.8.6}$$

i.e.,

$$\varphi^{\#}(\Psi)(x) = \mathrm{d}\varphi^{-1}(\varphi(x))\Psi(\varphi(x)) \qquad \text{for } x \in M.$$

Finally, define $Q\colon \mathscr{F}^0(M) \to \mathscr{F}^0(M)$ by

$$F(\Psi) = \varphi^{\#}(\Psi) + Q(\Psi).$$

Lemma 4.8.2 For every $\varepsilon > 0$ *there exists* $\delta > 0$ *such that for any* Ψ, $\Psi' \in \mathscr{F}^0(M)$ *with* $\|\Psi\|_0 \leqslant \delta$, $\|\Psi'\|_0 \leqslant \delta$ *we have*

$$\|Q(\Psi)\|_0 \leqslant \varepsilon \|\Psi\|_0$$

and

$$\|Q(\Psi) - Q(\Psi')\|_0 \leqslant \varepsilon \|\Psi - \Psi'\|_0.$$

(Here and in what follows $\|\cdot\|_r$ denotes the norm in $\mathscr{F}^r(M)$; thus $\|\cdot\|_0$ is the C^0-norm.)

Proof Since $Q(0) = 0$, the first of the inequalities asserted results from the second one simply by putting $\Psi' = 0$.

F is a C^1-map (with respect to Ψ): indeed, exp is of class C^∞, φ is of class C^1, the composition is of class C^1. Obviously, $F(0) = 0$. Calculate $\mathrm{d}F(0)$ by differentiating (4.8.5) with respect to Ψ; we get (omitting Ψ in notation):

$$\mathrm{d}\exp_x F(0) \circ \mathrm{d}F(0) = \mathrm{d}\varphi^{-1}(\varphi(x)) \circ \mathrm{d}\exp_{\varphi(x)}(0).$$

Since $\mathrm{d}\exp_x(0) =$ identity, this yields

$$\mathrm{d}F(0) = \mathrm{d}\varphi^{-1}(\varphi(x)) \circ \mathrm{id}_{\varphi(x)},$$

i.e.,

$$\mathrm{d}F(0)(\Psi) = \mathrm{d}\varphi^{-1}(\varphi(x))\Psi(\varphi(x)).$$

Thus

$$\mathrm{d}F(0) = \varphi^{\#}.$$

Consequently $Q(0) = 0$ and $dQ(0) = 0$, which, by continuity, leads to the estimate

$$\|dQ(\Psi)\|_0 \leqslant \varepsilon \qquad \text{for } \|\Psi\|_0 \leqslant \delta.$$

The assertion of the Lemma now follows immediately from the Mean Value Theorem:

$$\|Q(\Psi) - Q(\Psi')\|_0 \leqslant \|dQ(\bar{\Psi})\|_0 \|\Psi - \Psi'\|_0 \leqslant \varepsilon \|\Psi - \Psi'\|_0,$$

where $\bar{\Psi}$ belongs to the straight line segment (in the space $\mathscr{F}^0(M)$) joining Ψ with Ψ'. The lemma is proved.

Now let Φ, $\Psi \in \mathscr{F}^0(M)$ be fields with small norms. Then the mappings $\exp\Phi$, $\exp\Psi$ are close to the identity in $C^0(M, M)$. Hence there exists a unique field in $\mathscr{F}^0(M)$, denote it by $C_\Phi(\Psi)$, such that

$$\exp C_\Phi(\Psi) = \exp\Phi \circ \exp\Psi. \tag{4.8.7}$$

We define one more mapping P_Φ: $\mathscr{F}^0(M) \to \mathscr{F}^0(M)$ (given Φ) as follows:

$$C_\Phi(\Psi) = \Psi + P_\Phi(\Psi). \tag{4.8.8}$$

Lemma 4.8.3 For every $\varepsilon > 0$ there exists $\delta > 0$ such that for any $\Phi \in \mathscr{F}^1(M)$, Ψ, $\Psi' \in \mathscr{F}^0(M)$ with $\|\Phi\|_1 \leqslant \delta$, $\|\Psi\|_0 \leqslant \delta$, $\|\Psi'\|_0 \leqslant \delta$ we have

$$\|P_\Phi(\Psi)\|_0 \leqslant \|\Phi\|_0 + \varepsilon\|\Psi\|_0$$

and

$$\|P_\Phi(\Psi) - P_\Phi(\Psi')\|_0 \leqslant \varepsilon\|\Psi - \Psi'\|_0.$$

Proof The map P_Φ is of class C^1 in Ψ in a neighbourhood of 0 in $\mathscr{F}^0(M)$. Equalities (4.8.7), (4.8.8) and the equalities resulting from them by differentiation with respect to Ψ show that the expressions $P_\Phi(\Psi)$ and $\partial_\Psi P_\Phi(\Psi)$ are jointly continuous in $(\Phi, \Psi) \in \mathscr{F}^1(M) \times \mathscr{F}^0(M)$. Easy calculations show that the latter expression is zero if $\Phi = 0$ and $\Psi = 0$. Hence

$$\|\partial_\Psi P_\Phi(\Psi)\|_0 \leqslant \varepsilon \qquad \text{for } \|\Phi\|_1 \leqslant \delta, \ \|\Psi\|_0 \leqslant \delta. \tag{4.8.9}$$

Applying the Mean Value Theorem, we obtain, as before,

$$\begin{aligned}\|P_\Phi(\Psi) - P_\Phi(\Psi')\|_0 &\leqslant \|\partial_\Psi P_\Phi(\bar{\Psi})\|_0 \|\Psi - \Psi'\|_0 \\ &\leqslant \varepsilon\|\Psi - \Psi'\|_0,\end{aligned}$$

i.e., the second inequality of the assertion of the lemma (here again $\bar{\Psi}$ denotes some element of the segment joining Ψ with Ψ' in $\mathscr{F}^0(M)$). The first inequality results from the second by putting $\Psi' = 0$, in view of the fact that $P_\Phi(0) = \Phi$, which follows immediately from (4.8.7).

Proof of Theorem 4.8.2 Let $\psi \in \mathrm{Diff}^1(M)$ be C^1-close to φ. Then $\varphi^{-1} \circ \psi$ is C^1-close to the identity. By Lemma 4.8.1 there exists a unique vector field $\Phi \in \mathscr{F}^1(M)$ such that

$$\varphi^{-1} \circ \psi = \exp \Phi.$$

The equation of conjugacy

$$\psi \circ h = h \circ \varphi$$

can be rewritten in the following equivalent form:

$$\varphi^{-1} \circ \psi \circ h = \varphi^{-1} \circ h \circ \varphi.$$

We wish to find a solution h near the identity; in view of Lemma 4.8.1, we may restrict our attention to mappings of type

$$h = \exp \Psi, \qquad \Psi \in \mathscr{F}^0(M).$$

Our equation then takes on the form

$$\exp \Phi \circ \exp \Psi = \varphi^{-1} \circ \exp \Psi \circ \varphi.$$

Using the notation introduced in the lemmas above, we rewrite this as follows:

$$\Psi + P_\Phi(\Psi) = \varphi^{\#}(\Psi) + Q(\Psi),$$

i.e.,

$$(I - \varphi^{\#})\Psi = R_\Phi(\Psi), \tag{4.8.10}$$

where

$$R_\Phi(\Psi) = Q(\Psi) - P_\Phi(\Psi).$$

We shall show that the operator $I - \varphi^{\#}$ has an inverse on $\mathscr{F}^0(M)$. By assumption φ is an Anosov diffeomorphism and thus it induces an invariant decomposition $T(M) = E^+ \oplus E^-$, where E^+ is contracted, E^- expanded by $\mathrm{d}\varphi$. This, in turn, induces the decomposition of $\mathscr{F}^0(M)$ into an analogous direct sum; for simplicity, we use the same symbols E^+ and E^-:

$$\mathscr{F}^0(M) = E^+ \oplus E^-.$$

E^+ and E^- are invariant under $d\varphi$, hence under φ^*. Furthermore, we may assume that φ^* and $(\varphi^*)^{-1}$ are contractions on E^+ and E^- respectively:

$$\|\varphi^*(\Psi)\|_0 \leqslant \lambda\|\Psi\|_0 \qquad \text{for } \Psi \in E^+,$$

$$\|(\varphi^*)^{-1}(\Psi)\|_0 \leqslant \mu\|\Psi\|_0 \qquad \text{for } \Psi \in E^-,$$

where $\lambda, \mu \in (0, 1)$. This follows from the fact that we can choose the Riemannian metric so as to have

$$\|d\varphi(x)v\| \leqslant \lambda\|v\| \qquad \text{for } v \in E^+,$$

$$\|d\varphi^{-1}(x)v\| \leqslant \mu\|v\| \qquad \text{for } v \in E^-.$$

The desired inverse operator $(I-\varphi^*)^{-1}$ can be written explicitly on the subspaces E^+, E^-:

$$(I-\varphi^*)^{-1} = \sum_{n=0}^{\infty} (\varphi^*)^n \qquad \text{on } E^+,$$

$$(I-\varphi^*)^{-1} = -(\varphi^*)^{-1}\sum_{n=0}^{\infty} (-1)^n(\varphi^*)^{-n} \qquad \text{on } E^-.$$

These formulas define $(I-\varphi^*)^{-1}$ as a continuous linear operator on the whole of $\mathscr{F}^0(M)$.

We use the opportunity to inform the reader that also conversely, if the operator φ^* defined by (4.8.6) is such that $(I-\varphi^*)^{-1}$ exists on the entire $\mathscr{F}^0(M)$, then φ is an Anosov diffeomorphism; this result is due to Mather (1968).

Now equation (4.8.10) obtains the form

$$\Psi = (I-\varphi^*)^{-1} \circ R_\Phi(\Psi).$$

Fix $\varepsilon > 0$ and choose $\delta > 0$ in accordance with Lemmas 4.8.3 and 4.8.4. We shall show that the operator $(I-\varphi^*)^{-1} \circ R_\Phi$ takes the ball $\|\Psi\|_0 \leqslant \delta$ into itself, whenever $\|\Phi\|_0 \leqslant \varepsilon\delta$ and ε is small enough:

$$\begin{aligned}
&\|(I-\varphi^*)^{-1} \circ R_\Phi(\Psi)\|_0 \\
&\quad\leqslant \|(I-\varphi^*)^{-1}\|(\|Q(\Psi)\|_0 + \|P_\Phi(\Psi)\|_0) \\
&\quad\leqslant \|(I-\varphi^*)^{-1}\|(\varepsilon\|\Psi\|_0 + \|\Phi\|_0 + \varepsilon\|\Psi\|_0) \\
&\quad\leqslant \|(I-\varphi^*)^{-1}\| \cdot 3\varepsilon\delta,
\end{aligned}$$

and this number is not greater than δ if ε is chosen so small that $3\varepsilon\|(I-\varphi^*)^{-1}\| \leqslant 1$.

Further, we show that if ε, δ and Φ are as above, then the operator under consideration is a contraction in the ball $\|\Psi\|_0 \leqslant \delta$:

$$\begin{aligned}
&\|(I-\varphi^{\#})^{-1} \circ R_{\Phi}(\Psi)-(I-\varphi^{\#})^{-1} \circ R_{\Phi}(\Psi')\|_0 \\
&\quad \leqslant \|(I-\varphi^{\#})^{-1}\| \left(\|Q(\Psi)-Q(\Psi')\|_0+\|P_{\Phi}(\Psi)-P_{\Phi}(\Psi')\|_0\right) \\
&\quad \leqslant \|(I-\varphi^{\#})^{-1}\| \cdot 2\varepsilon \|\Psi-\Psi'\|_0 \\
&\quad \leqslant \tfrac{2}{3}\|\Psi-\Psi'\|_0;
\end{aligned}$$

the estimates follow from Lemmas 4.8.2 and 4.8.3 and from the choice of ε.

Thus $(I-\varphi^{\#})^{-1} \circ R_{\Phi}$ is indeed a contraction on $\{\Psi\colon \|\Psi\|_0 \leqslant \delta\}$. Consequently there exists a Ψ belonging to that ball and satisfying equation (4.8.10), which means that the mapping $h = \exp\Psi$ fulfils the equation

$$\psi \circ h = h \circ \varphi; \tag{4.8.11}$$

h can be arbitrarily close to identity because δ can be arbitrarily small. It remains to prove that h is a homeomorphism.

We first show that h is one-to-one. Assume that $h(\bar{x}) = h(\bar{y})$, $\bar{x}, \bar{y} \in M$. From (4.8.11) we get by induction

$$\psi^n \circ h = h \circ \varphi^n \qquad \text{for all } n \in Z.$$

Hence

$$h \circ \varphi^n(\bar{x}) = \psi^n \circ h(\bar{x}) = \psi^n \circ h(\bar{y}) = h \circ \varphi^n(\bar{y}).$$

If $\bar{x} \neq \bar{y}$, then by Exercise 4.8.1 we can find $n \in Z$ such that

$$\varrho\left(\varphi^n(\bar{x}), \varphi^n(\bar{y})\right) \geqslant \vartheta > 0,$$

ϑ denoting a constant depending on φ only.

Since h may be arbitrarily close to identity, we may assume that

$$\varrho\left(h(x), x\right) \leqslant \tfrac{1}{3}\vartheta \qquad \text{for all } x \in M.$$

Then

$$\begin{aligned}
\vartheta &\leqslant \varrho\left(\varphi^n(\bar{x}), \varphi^n(\bar{y})\right) \\
&\leqslant \varrho\left(\varphi^n(\bar{x}), h \circ \varphi^n(\bar{y})\right)+\varrho\left(h \circ \varphi^n(\bar{y}), \varphi^n(\bar{y})\right) \\
&= \varrho\left(\varphi^n(\bar{x}), h \circ \varphi^n(\bar{x})\right)+\varrho\left(h \circ \varphi^n(\bar{y}), \varphi^n(\bar{y})\right) \\
&\leqslant \tfrac{2}{3}\vartheta,
\end{aligned}$$

which is nonsense. Thus $\bar{x} = \bar{y}$ and h is one-to-one.

By a familiar theorem on open mappings [5] we hence conclude that $h(M)$ is an open set and h maps M homeomorphically onto $h(M)$. On the other hand, $h(M)$ is compact since M is so. Consequently $h(M)$ is a non-empty closed-open subset of M. Thus $h(M) = M$, provided that M is connected. If M is not, the situation is none the worse: h is close to identity and therefore it carries components onto themselves. Thus also in this case $h(M) = M$.

This completes the proof.

We now state one more interesting and important theorem concerning Anosov diffeomorphisms.

Theorem 4.8.3 (*Manning, 1974*) *Every Anosov diffeomorphism on the m-torus T^m is topologically conjugate to an algebraic automorphism of T^m.*

This result was first obtained by Franks (1970) for the case of T^2; in 1974 Manning extended it to a fairly broad class of manifolds, the so called *infranilmanifolds*; this class comprises, in particular, the tori of all dimensions. We are not going to deal with such manifolds in this book, referring the readers interested in the topic to Nitecki (1971).

Manning's result was presented at the International Congress of Mathematics 1974 in Vancouver. The proof is difficult; it can be found in Manning (1974).

The question arises whether there exist any Anosov diffeomorphisms other than torus automorphisms. The answer is affirmative (Shub, 1969). Here we show an example of an Anosov diffeomorphism on a differentiable manifold not homeomorphic to a torus.

Consider the following set in $\boldsymbol{R}^4$:

$$\Gamma := \{v = (\alpha, \beta, \bar{\alpha}, \bar{\beta});\ \alpha, \beta \in Z[\sqrt{3}]\},$$

where $Z[\sqrt{3}]$ denotes the ring $\{a + b\sqrt{3}:\ a, b \in Z\}$ and $\bar{\alpha}$ stands for $a - b\sqrt{3}$ if $\alpha = a + b\sqrt{3}$.

Put $\omega = \frac{1}{2} + \frac{1}{2}\sqrt{3}$. Then $\omega \notin Z[\sqrt{3}]$, but $2\omega \in Z[\sqrt{3}]$. We define a mapping B: $\boldsymbol{R}^4 \to \boldsymbol{R}^4$ as follows:

$$Bx = (-x^1, x^2, -x^3, x^4) \qquad \text{for } x = (x^1, x^2, x^3, x^4) \in \boldsymbol{R}^4.$$

Further, writing $g_0 = (0, \omega, 0, \bar{\omega})$ we define a new mapping B_1:

$$B_1 x = g_0 + Bx \qquad \text{for } x \in \boldsymbol{R}^4.$$

Let G be the least group of isometries of $\boldsymbol{R}^4$ containing all translations by vectors $v \in \Gamma$ and the mapping B_1. It is easy to verify that

$$G = \Gamma \cup \{B_1^n:\ n \in Z\}$$

(we identify a vector $v \in \Gamma$ with the translation by v). G is non-commutative: in general, we have

$$v + B_1 x \neq B_1(v+x) \qquad \text{for } v \in \Gamma,\ x \in \boldsymbol{R}^4.$$

It is not difficult to see that every $g \in G$ is of the form

$$g(x) = v + ng_0 + B^\sigma x \qquad \text{for } x \in \boldsymbol{R}^4,$$

where $v \in \Gamma$, $\sigma = 0, 1$ and n and σ are simultaneously even or odd. Hence g can be written as

$$g(x) = v + \sigma g_0 + B^\sigma x \qquad \text{for } x \in \boldsymbol{R}^4, \tag{4.8.12}$$

where $v \in \Gamma$ and $\sigma = 0, 1$ (since $2g_0 \in \Gamma$).

We will show that the following inequality holds for all $x \in \boldsymbol{R}^4$ and for any $g \in G$ different from the identity:

$$\|x - g(x)\| \geqslant \sqrt{2}. \tag{4.8.13}$$

Insert $v = (\alpha, \beta, \bar{\alpha}, \bar{\beta})$, $\alpha = a + b\sqrt{3}$, $\beta = c + d\sqrt{3}$, $a, b, c, d \in Z$, into (4.8.12) and calculate the square of the norm of $x - g(x)$: for $\sigma = 0$ we have

$$\begin{aligned}\|x - g(x)\|^2 &= \|v\|^2 \\ &= (a+b\sqrt{3})^2 + (c+d\sqrt{3})^2 + (a-b\sqrt{3})^2 + (c-d\sqrt{3})^2 \\ &= 2(a^2+c^2) + 6(b^2+d^2) \geqslant 2\end{aligned}$$

(a, b, c, d cannot all be zero, because g is not the identity); for $\sigma = 1$ and $x = (x^1, x^2, x^3, x^4) \in \boldsymbol{R}^4$ we have

$$\begin{aligned}\|x - g(x)\|^2 &= \|x - v - g_0 - Bx\|^2 \\ &= \|(2x^1 - \alpha,\ -\beta - \omega,\ 2x^3 - \bar{\alpha},\ -\bar{\beta} - \bar{\omega})\|^2 \\ &\geqslant (\beta+\omega)^2 + (\bar{\beta}+\bar{\omega})^2 = 2(c+\tfrac{1}{2})^2 + 6(d+\tfrac{1}{2})^2 \geqslant 2.\end{aligned}$$

This proves (4.8.13).

Now we establish another property of the group G. Write

$$G(0) = \{x \in \boldsymbol{R}^4:\ x = g(0),\ g \in G\}.$$

We claim that

$$\operatorname{dist}(x, G(0)) < 2+\sqrt{12} \qquad \text{for } x\in \boldsymbol{R}^4. \tag{4.8.14}$$

To show this take any point $x\in \boldsymbol{R}^4$ and denote by $\bar{x} = (n, m, k, l)$ the point with even integer coordinates which is nearest to x; clearly, $\|x-\bar{x}\| < 2$. Consider the point $x_0 = (\alpha, \beta, \bar{\alpha}, \bar{\beta})$, where

$$\alpha = a+b\sqrt{3}, \qquad \beta = c+d\sqrt{3},$$

$$a = \frac{n+k}{2}, \qquad b = E\left(\frac{n-k}{2\sqrt{3}}\right), \qquad c = \frac{m+1}{2}, \qquad d = E\left(\frac{m-1}{2\sqrt{3}}\right).$$

We have

$$0 < n-\alpha = \frac{n-k}{2} - E\left(\frac{n-k}{2\sqrt{3}}\right)\sqrt{3} < \sqrt{3},$$

$$0 < k-\bar{\alpha} = \frac{k-n}{2} - E\left(\frac{n-k}{2\sqrt{3}}\right)\sqrt{3} < \sqrt{3},$$

and similarly,

$$0 < m-\beta < \sqrt{3}, \qquad 0 < l-\bar{\beta} < \sqrt{3}.$$

Thus $\|\bar{x}-x_0\| < \sqrt{12}$ and (4.8.14) follows, because $x_0\in G(0)$.

We shall need the following lemma:

Lemma 4.8.4 *Suppose that G is a discrete group of isometries of $\boldsymbol{R}^m$ with the following property: there exist constants $\vartheta > 0$ and $r < +\infty$ such that*

$$\|x-g(x)\| \geqslant \vartheta \qquad \textit{for } x\in \boldsymbol{R}^m \textit{ and } g\in G,\ g \neq \mathrm{id}, \tag{4.8.15}$$

$$\operatorname{dist}(x, G(0)) < r \qquad \textit{for } x\in \boldsymbol{R}^m, \tag{4.8.16}$$

where $G(0) = \{x\in \boldsymbol{R}^m\colon x = g(0), g\in G\}$. *Then* $\boldsymbol{R}^m/G$ *is a compact manifold.*

Proof The points of the quotient space $\boldsymbol{R}^m/G$ are cosets $G(x) = \{g(x)\colon g\in G\}$. From (4.8.15) we see that if U is a sufficiently small neighbourhood of x in $\boldsymbol{R}^m$, then the sets $g(U)$, $g\in G$, are pairwise disjoint in $\boldsymbol{R}^m$. The set $V = \{G(x)\colon x\in U\}$, is, by definition, a neighbourhood of

$G(x)$ in $\boldsymbol{R}^m/G$ and the map $x \mapsto G(x)$ is a diffeomorphism and a local isometry between U and V.

Let (z_n) be any sequence in $\boldsymbol{R}^m/G$. Each z_n corresponds to a coset $G(x_n)$, where $x_n \in \boldsymbol{R}^m$. In view of (4.8.16) the points x_n can be chosen in the ball $B(0, r) \subset \boldsymbol{R}^m$. Thus the sequence (x_n) contains a convergent subsequence (x_{n_i}), $x_{n_i} \to x_0$, and so the sequence (z_{n_i}) converges to some $z_0 \in \boldsymbol{R}^m/G$. The lemma is proved.

Now let G be as before. The estimates (4.8.13) and (4.8.14) show that G satisfies the conditions of Lemma 4.8.4. Thus $\boldsymbol{R}^4/G$ is a compact manifold. Since G is non-commutative, the fundamental group $\pi_1(\boldsymbol{R}^4/G)$ is non-commutative. On the other hand, $\pi_1(T^4)$ is a commutative group. This shows that $\boldsymbol{R}^4/G$ and T^4 are not homeomorphic.

We define a certain Anosov diffeomorphism on $\boldsymbol{R}^4/G$. Write $\lambda = 2+\sqrt{3}$. Then $\bar{\lambda} = 2-\sqrt{3}$, $\lambda\bar{\lambda} = 1$ and $\lambda\omega = (2+\sqrt{3})(\frac{1}{2}+\frac{1}{2}\sqrt{3}) = (2+\sqrt{3})+(\frac{1}{2}+\frac{1}{2}\sqrt{3}) = \lambda+\omega$. Denote by A the following matrix:

$$A = \begin{bmatrix} \lambda & & & 0 \\ & \lambda & & \\ & & \bar{\lambda} & \\ 0 & & & \bar{\lambda} \end{bmatrix}.$$

The mapping defined by A is an automorphism of G: we have $A^{-1}GA = G$. Indeed, let g be given by (4.8.12); then

$$\begin{aligned} A^{-1}gA &= A^{-1}v+\sigma A^{-1}g_0+A^{-1}B^\sigma A \\ &= A^{-1}v+\sigma(0, \lambda\omega, 0, \bar{\lambda}\bar{\omega})+B^\sigma \\ &= A^{-1}v+\sigma(0, \lambda, 0, \bar{\lambda})+\sigma g_0+B^\sigma, \end{aligned}$$

which is an element of G, since $A^{-1}v$ and $\sigma(0, \lambda, 0, \bar{\lambda})$ are in Γ. Consequently, A induces a mapping $\hat{A}$: $\boldsymbol{R}^4/G \to \boldsymbol{R}^4/G$. Now, A is a hyperbolic linear map in $\boldsymbol{R}^4$: $\operatorname{sp}A = \{\lambda, \bar{\lambda}\}$, $\lambda > 1$, $0 < \bar{\lambda} < 1$, and thus A is an Anosov diffeomorphism on $\boldsymbol{R}^4$. Since $\boldsymbol{R}^4/G$ and $\boldsymbol{R}^4$ are locally isometric, $\hat{A}$ is an Anosov diffeomorphism on the manifold $\boldsymbol{R}^4/G$. As we have seen, $\boldsymbol{R}^4/G$ and T^4 are not homeomorphic. Therefore the system $(\boldsymbol{R}^4/G, \hat{A})$ is not topologically conjugate to any torus diffeomorphism.

The concept of an Anosov system is extended to the case of a flow as follows.

Let M be a compact manifold.

Definition 4.8.2 The flow φ_t generated by a vector field $\Phi \in \mathscr{F}^1(M)$ is called an *Anosov flow* if the entire M has hyperbolic structure (in the sense of Definition 4.7.2).

The Anosov theorem for diffeomorphisms has its counterpart for flows:

Theorem 4.8.4 (Anosov) Every Anosov flow generated by a vector field $\Phi \in \mathscr{F}^1(M)$, where M is compact, is C^1-structurally stable.

We conclude our considerations concerning Anosov systems by mentioning two open problems.

Problem 1 Let M be a Riemannian manifold and suppose that there exists an Anosov diffeomorphism $\varphi: M \to M$. Is $\boldsymbol{R}^m$ necessarily a covering space for M ($m = \dim M$)?

Problem 2 If $\varphi: M \to M$ is an Anosov diffeomorphism, need the set of periodic points of φ be dense in M?

4.9. Smale's Counter-Example. The Existence of Structurally Stable Dynamical Systems

As we have seen in the previous section, C^r-structurally stable vector fields are dense in the space $\mathscr{F}^r(M)$ if M is a two-dimensional oriented manifold. In Section 4.2 we proved that on the circle, C^r-structurally stable diffeomorphisms are dense in the space of all C^r-diffeomorphisms. It is natural to expect that these are particular cases of a general principle and that, for any compact manifold M, $\dim M < +\infty$, structurally stable dynamical systems are dense in the corresponding space of all dynamical systems considered: $\sigma^r(M)$ is a dense subset of $\text{Diff}^r(M)$ and $\Sigma^r(M)$ is a dense subset of $\mathscr{F}^r(M)$. Such a conjecture was posed and examined in the early sixties. Yet it turned out to be false. S. Smale (1966) showed that the set of C^1-structurally stable diffeomorphisms of the 3-torus T^3 is not dense in $\text{Diff}^1(T^3)$. Smale found an example of a diffeomorphism φ of T^3 with the following property: there is a neighbourhood V_0 of φ in $\text{Diff}^3(T^3)$ such that, for any $\psi \in V_0$, every neighbourhood V of ψ contains a diffeomorphism which is not structurally stable. Consequently, no diffeomorphism $\psi \in V_0$ is structurally stable, and hence $\Sigma^1(T^3)$ is not dense in $\text{Diff}^1(T^3)$.

We now discuss that example. Consider the algebraic automorphism of T^2 given by $A_0 = \begin{bmatrix} 2 & 1 \\ 1 & 1 \end{bmatrix}$. Since $\operatorname{sp} A = \{\frac{1}{2}(3-\sqrt{5}), \frac{1}{2}(3+\sqrt{5})\}$, A_0 is an Anosov diffeomorphism. Denote by A the matrix

$$A = \begin{bmatrix} 2 & 1 & 0 \\ 1 & 1 & 0 \\ 0 & 0 & a_0 \end{bmatrix},$$

where a_0 is a small positive number. Regard the torus T^3 as the rectangular parallelepiped $\langle -1, 1\rangle \times \langle -1, 1\rangle \times \langle -3, 3\rangle$ with the usual identifications. The first step in the construction is to introduce a mapping $\varphi_0\colon T^3 \to T^3$ which is subject to the following conditions (see Fig. 4.9.1):

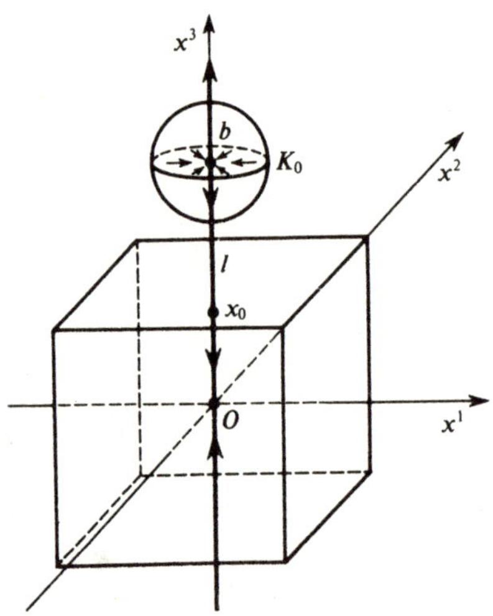

FIG. 4.9.1

1° φ_0 is linear in the ball K_0 with centre b and radius $\frac{1}{2}$, where $b = (0, 0, 2)$ is a hyperbolic fixed point of φ_0 such that

$$W^+(b) \supset \{(x^1, x^2, 0)\colon x^1, x^2 \in \langle -1, 1\rangle\} \cap K_0,$$

$$W^-(b) \supset l := \{(0, 0, t)\colon 0 \leqslant t \leqslant 2\}.$$

2° $\varphi_0(x) = Ax$ in the set
$\{x = (x^1, x^2, x^3)\colon |x^1| \leqslant 1, |x^2| \leqslant 1, |x^3| \leqslant 1\}$.

3° In the remaining portion of T^3, φ_0 is defined in any manner making it a C^1 diffeomorphism of T^3.

Let $x_0 = (0, 0, \frac{3}{4})$ and let $G_0(x_0, \alpha)$ be the ball with centre at x_0 and radius α, where α is some small fixed positive number. Write $G = \varphi_0^{-1}(G_0)$, $x_1 = \varphi_0^{-1}(x_0)$.

We define the desired diffeomorphism φ as follows:

$$\varphi(x) = \varphi_0(x) + \eta f(x) e_1,$$

where $\eta > 0$ is a small number, f is a C^∞-function with support contained in G, having a peak (a non-degenerate maximum) at x_1, $f(x_1) = 1$, and $e_1 = (1, 0, 0)$. Obviously, for small η the diffeomorphism φ is C^1-close to φ_0.

Now, the set $T^2 \times \{0\}$ is φ-invariant and we have $\varphi|_{T^2 \times \{0\}} = A_0$. The set $\{q_i\}$ of all periodic points of A_0 is dense in T^2 (see Exercise 4.7.3) and each of these points is hyperbolic. In the region

$$\{x = (x^1, x^2, x^3): |x^1| \leqslant 1, |x^2| \leqslant 1, |x^3| \leqslant 1\}$$

the stable manifolds $W^+(q_i)$ are planes perpendicular to $T^2 \times \{0\}$, and the sets $W^+(q_i) \cap (T^2 \times \{0\})$ are the stable manifolds of q_i in T^2.

Examine the unstable manifold $W_\varphi^-(b)$ and denote by $\bar{l}$ the piece of it which goes "downwards" from b; it is marked by a bold line in Fig. 4.9.2. Till the moment of penetrating into G_0 the curve $\bar{l}$ coincides with

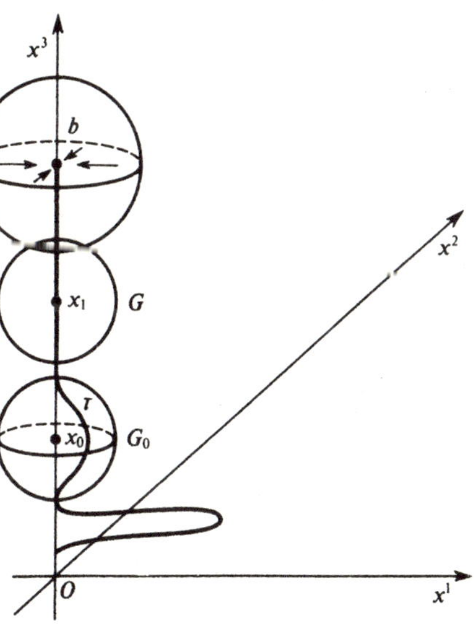

FIG. 4.9.2

l. In G_0, $\bar{l}$ has a "deviation" in the direction of e_1, then it returns to l (consequently it then has another "deviation" from l, and so on).

We will show that there is a neighbourhood V_0 of φ in $\mathrm{Diff}^1(T^3)$ such that, for any $\psi \in V_0$, there exists a diffeomorphism ψ_1 arbitrarily close to ψ, without Property $\mathscr{G}_3$ (transversality of stable and unstable manifolds). Such a ψ_1 cannot be structurally stable.

In our further considerations we shall need the following fact, which we leave to the reader as an exercise:

Exercise

4.9.1*.A. Let ψ be a diffeomorphism sufficiently C^1-close to φ. Prove that there exists a torus T_1^2 which is ψ-invariant and is C^1-close to $T^2 \times \{0\}$.

Hint. Apply the technique used in the proof of the Hadamard–Perron theorem: define T_1^2 to be the graph of a suitably chosen C^1-function $x^3 = g(x^1, x^2)$; a function g with all needed properties is found with the use of the Fixed Point Theorem.

Let ψ, T_1^2 and g be as in this exercise. Projecting ψ from T_1^2 onto T^2 we get a mapping ψ_0: $T^2 \to T^2$; ψ_0 acts as follows:

$$(x^1, x^2) \mapsto (x^1, x^2, g(x^1, x^2)) \mapsto \psi(x^1, x^2, g(x^1, x^2))$$
$$= (\bar{x}^1, \bar{x}^2, \bar{x}^3) \mapsto (\bar{x}^1, \bar{x}^2).$$

If ψ is close enough to φ, then g is close to 0 in C^1-topology, and so ψ_0 is C^1-close to A_0. Since A_0 is Anosov, ψ_0 and A_0 are topologically conjugate by Theorem 4.8.2. The set of all periodic points of A_0 is dense in T^2; hence the same can be said about ψ_0. The systems (T^2, ψ_0) and (T_1^2, ψ) are topologically conjugate by construction (the conjugacy being established by orthogonal projection of T_1^2 onto T^2). It follows that the set of periodic points of $\psi|_{T_1^2}$ is dense in T_1^2 and that T_1^2 is a hyperbolic set. Then, as it has been shown* by M. Hirsch and C. Pugh (1970), the stable and unstable manifolds of nearby periodic points are "C^1-close on their compact subsets" (i.e., on subsets resulting by immersion from compact subsets of a suitable Euclidean space).

---

* The theorem which we quote is formulated in much greater generality, both in the original paper and in Nitecki (1971); the version given here is adapted to what we actually need.

Thus if $q_1, q_2 \in T_1^2$ are periodic points of ψ sufficiently close to each other, then $W^+(q_1)$ and $W^-(q_2)$ are C^1-close on their compact subsets. Also, if ψ is C^1-close to φ, $q \in T_1^2$ and $p \in T^2$ are periodic points of ψ and φ, lying close enough, then $W_\psi^+(q)$ is C^1-close to $W_\varphi^-(p)$, again by the Hirsch–Pugh theorem.

Denote by $\bar{b}$ the fixed point of ψ which corresponds to the fixed point b of φ ($\bar{b}$ close to b) and consider the unstable manifold $W_\psi^-(\bar{b})$. It is a curve which is C^1-close to $\bar{l}$ within the piece from b to the point of escape from G_0. That piece is a compact subset of $W_\varphi^-(b)$; denote by $\bar{l}$ the corresponding piece of $W_\psi^-(b)$.

Since $\bar{l}$ intersects some of the stable manifolds $W_\varphi^+(p)$, where $p \in T^2$ are periodic points of φ close to 0, then $\bar{l}$ intersects certain manifolds $W_\psi^+(q_i)$, where $q_i \in T_1^2$ are periodic points of ψ, also close to 0. The manifolds $W_\psi^+(q_i)$ intersect the x^1-axis $0x^1$; let us consider those which intersect $\bar{l}$. Consider the set

$$A = \{\alpha \in 0x^1 : \alpha \in W_\psi^+(q_i) \text{ where } W_\psi^+(q_i) \cap \bar{l} \neq \emptyset\}.$$

This set is bounded; write $\alpha_0 = \sup A$.

If $\alpha_0 \in A$, then there is a manifold $W_\psi^+(q_{i_0})$ passing though α_0, such that $W_\psi^+(q_{i_0}) \cap \bar{l} \neq \emptyset$, and tangent to $\bar{l}$ at the meeting point (which is the point of maximal deviation of $\bar{l}$ in the direction of e_1). This means that $W_\psi^+(q_{i_0})$ intersects $\bar{l}$ non-transversally, and so ψ does not have Property $\mathscr{G}_3$.

Now suppose that $\alpha_0 \notin A$. The periodic points q_i of ψ lie densely on T_1^2. Perturbing ψ a little (in C^1-topology) we make it into a diffeomorphism ψ_1 for which $\alpha_0 \in A$; as has been shown, such a ψ_1 does not have Property $\mathscr{G}_3$.

We readily admit that the above reasoning is not rigorous enough to constitute a formal proof. But any further formalization would completely obscure the geometric idea contained in it. Therefore we prefer to appeal to the reader's geometrical imagination.

Remark 4.9.1 Using this example, we can easily construct a flow on T^4 with a neighbourhood containing no structurally stable flow. It suffices to pass to suspension in T^4.

The following question arises: does every compact manifold admit a structurally stable flow? The theorem which follows, together with Theorems 4.5.3 and 3.7.2, gives a positive answer to this question (see Smale, 1961; Palis and Smale, 1970).

Theorem 4.9.1 Every gradient vector field with Property $\mathscr{G}_3$ is a Morse–Smale field.

Proof Let $\Phi \in \mathrm{grad}^r(M) \cap \mathscr{G}_3^r(M)$. The field Φ has no closed orbits (because it is a gradient field) and has only finitely many critical points (Property $\mathscr{G}_1$). All that remains is to check that the set of non-wandering points coincides with the set of critical points (see Definition 4.4.1).

Let $\Phi = \mathrm{grad} f$. Take an $x_0 \in \Omega(\Phi)$ and suppose, contrary to the assertion, that $\Phi(x_0) \neq 0$. Fix a time-instant $s > 0$ and write $y = \varphi_s(x_0)$. Since f is strictly increasing along trajectories not reducing to fixed points, we have

$$f(y) > f(x_0). \tag{4.9.1}$$

Let ε be a number satisfying $0 < \varepsilon < \frac{1}{2}(f(y) - f(x_0))$ and let U_1 be a neighbourhood of y such that

$$|f(x) - f(y)| < \varepsilon \qquad \text{for } x \in U_1. \tag{4.9.2}$$

Further, let U be a neighbourhood of x_0 such that $\varphi_s(U) \subset U_1$ and

$$|f(x) - f(x_0)| < \varepsilon \qquad \text{for } x \in U. \tag{4.9.3}$$

Since x_0 is a non-wandering point, there are an instant $t > s$ and a point $x \in U$ such that $\varphi_t(x) \in U$. From (4.9.2) and (4.9.3) we get

$$f(y) - \varepsilon < f(\varphi_s(x)) < f(\varphi_t(x)) < f(x_0) + \varepsilon,$$

whence

$$f(y) < f(x_0) + 2\varepsilon < f(y),$$

which is absurd.

Corollary 4.9.1 Structurally stable gradient vector fields constitute a residual subset of $\mathrm{grad}^r(M)$.

This is a straightforward consequence of Theorems 4.9.1 and 3.7.2.

4.10. Geodesic Flows

Let M be a compact Riemannian manifold. Consider the set

$$W = \{(q, v) \colon q \in M,\ v \in T_q(M),\ \|v\| = 1\}.$$

W is called the *sphere bundle* over M. If $\dim M = m$, W is a manifold of dimension $2m-1$. Any flow φ_t: $W \to W$, $t \in \boldsymbol{R}$, is called a *tangent flow*; the underlying manifold M is termed the *configuration space* of the flow φ_t. Denote by π the natural projection of W onto M: $\pi(q, v) = q$.

Certain flows induced by second order differential equations on M can be reduced to tangent flows. Let

$$\ddot{q} = f(q, \dot{q})$$

be a second order equation on M. It induces the flow on $T(M)$:

$$\dot{q} = v, \qquad \dot{v} = f(q, v).$$

If this flow has the property that $\|v(t)\| = 1$ whenever $\|v(0)\| = 1$, it is a tangent flow.

The most important type of tangent flows are the so-called *geodesic flows*:

Definition 4.10.1 Let W be the sphere bundle over M and let φ_t: $W \to W$, $t \in \boldsymbol{R}$, be the tangent flow acting as follows: given a pair $w = (q, v) \in W$, shift the point q by the distance t along the geodesic line[16] sensed according to v; denote the resulting point by q_t and the unit vector tangent to the geodesic line at q_t by v_t, and define φ_t by $\varphi_t(w) := (q_t, v_t)$. Then φ_t is called the *geodesic flow* on M.

It follows from this definition that the set $\{\pi(\varphi_t(w)) \colon t \in \boldsymbol{R}\}$ is a geodesic line on M.

Geodesic flows constitute an important class of dynamical systems on account of their application in mechanics. In fact, we have the following theorem.

Theorem 4.10.1 *By a suitable change of variable on the time axis, every holonomic dynamical system reduces to a geodesic flow.*

Proof In mechanics, a dynamical system is called *holonomic* iff the configuration space is a Riemannian manifold M, $\dim M = m$, with kinetic energy ($= (1/2) \times$ Riemannian metric)

$$T(q, \dot{q}) = \tfrac{1}{2} \sum_{i,j=1}^{m} a_{ij}(q)\dot{q}^i \dot{q}^j, \qquad q \in M, \ \dot{q} \in T_q(M),$$

the motion being described by the Lagrange equations (see Arnold, 1979):

$$\frac{\mathrm{d}}{\mathrm{d}t}\left(\frac{\partial L}{\partial \dot{q}}\right) = \frac{\partial L}{\partial q}, \qquad L = T - V, \tag{4.10.1}$$

where $V(q)$ is a certain scalar-valued function, called the *potential*. Physically, the function $T(\cdot, \cdot)$ denotes the kinetic energy of the system, and $V(\cdot)$ the potential energy. The phase space is the tangent bundle $T(M)$.

Introduce the vector

$$p = \frac{\partial L}{\partial \dot{q}} = \frac{\partial T}{\partial \dot{q}}$$

(differentiation in charts) and consider the system in the coordinates q, p. The components p_i of p, $i = 1, \ldots, m$, are called the *generalized momenta* and are expressed by:

$$p_i = \sum_{j=1}^{m} a_{ij}(q)\dot{q}^j.$$

Then p acts as a functional on the space $T_q(M)$:

$$(\dot{q}, p)_q = 2T(q, \dot{q}),$$

hence p is an element of the co-tangent bundle $T^*(M)$. The Lagrange equations (4.10.1) then transform into the system of Hamilton canonical equations:

$$\frac{\mathrm{d}q}{\mathrm{d}t} = \frac{\partial H}{\partial p}, \qquad \frac{\mathrm{d}p}{\mathrm{d}t} = -\frac{\partial H}{\partial q}, \tag{4.10.2}$$

where

$$H(p, q) = T + V = \tfrac{1}{2} \sum_{i,j=1}^{m} a^{ij}(q)p_i p_j + V,$$

$(a^{ij}(q))$ denoting the inverse matrix to $(a_{ij}(q))$. The function H is called the *Hamiltonian function* of the system (4.10.2); it expresses the total energy of the system. It is a first integral of the system under consideration, that is to say, the motion occurs on the submanifolds $M_h = \{(p, q)\colon H(p, q) = h\}$.

According to the Maupertuis variational principle (*principle of least action*), the motion trajectories are the curves minimalizing the functional

$$\int_a^b p\mathrm{d}q, \tag{4.10.3}$$

where a, b are given points and

$$p\mathrm{d}q = \sum_{i=1}^{m} p_i \mathrm{d}q^i.$$

More precisely, consider all smooth curves connecting a with b on M_h. The space of those curves can be identified with a certain Banach space of vector fields on M_h: namely, the space of those fields which possess an integral curve joining a with b. For any such curve γ we have the quantity

$$F(\gamma) = \int_\gamma p\,\mathrm{d}q.$$

Then the motion is actually accomplished along those curves γ which are critical points of the functional F.

For any curve γ in M_h with parametrization $q(t)$, $t \in \langle \alpha, \beta \rangle$, $q(\alpha) = a$, $q(\beta) = b$, we have

$$\begin{aligned} F(\gamma) &= \int_\alpha^\beta p(t)\dot{q}(t)\,\mathrm{d}t \\ &= \int_\alpha^\beta \frac{\partial T}{\partial \dot{q}}(t)\dot{q}(t)\,\mathrm{d}t \\ &= \int_\alpha^\beta 2T\big(q(t), \dot{q}(t)\big)\,\mathrm{d}t \\ &= \sqrt{2}\int_\alpha^\beta \sqrt{h - V\big(q(t)\big)}\sqrt{\sum_{i,j=1}^{m} a_{ij}\big(q(t)\big)\dot{q}^i(t)\dot{q}^j(t)}\,\mathrm{d}t, \end{aligned}$$

which yields

$$\int_a^b p\,\mathrm{d}q = \sqrt{2}\int_a^b \sqrt{h - V}\sqrt{\sum_{i,j=1}^{m} a_{ij}(q)\,\mathrm{d}q^i\mathrm{d}q^j}. \tag{4.10.4}$$

We now define a new Riemannian metric on M by putting

$$b_{ij}(q) = (h - V(q))a_{ij}(q),$$

i.e.,

$$\sum_{i,j=1}^{m} b_{ij}(q)\mathrm{d}q^i\mathrm{d}q^j = (h - V(q)) \sum_{i,j=1}^{m} a_{ij}(q)\,\mathrm{d}q^i\mathrm{d}q^j. \tag{4.10.5}$$

Consider only those points $q \in M_h$ for which $h - V(q) \neq 0$. The set $\{q \in M_h\colon h = V(q)\}$ is closed, and every trajectory of (4.10.2) meets it in a nowhere dense set; for otherwise in some time-interval we would have $dq/dt = \partial H/\partial p = 0$, i.e., $q(t) = \text{const}$.

The length element ds assumes in the new metric the form

$$\mathrm{d}s^2 = (h - V(q)) \sum_{i,j=1}^{m} a_{ij}(q)\,\mathrm{d}q^i\mathrm{d}q^j. \tag{4.10.6}$$

It is readily seen from (4.10.4) and (4.10.6) that the curves of motion are the critical points of the functional

$$F(\gamma) = \int_a^b p\,dq$$

$$= \sqrt{2} \int_a^b \sqrt{h - V(q)} \sqrt{\sum_{i,j=1}^{m} a_{ij}(q)\,\mathrm{d}q^i\mathrm{d}q^j} = \sqrt{2} \int_a^b \mathrm{d}s.$$

Clearly enough, this occurs if and only if the length of the curve is minimal (provided that a and b lie close enough to each other). Thus the curves of motion are just the geodesic lines in the metric (4.10.6), i.e., the motion in the set $\{q \in M_h\colon V(q) < h\}$ is accomplished along the geodesics.

Since

$$h - V(q) = T(q, \dot{q}) = \frac{1}{2} \sum_{i,j=1}^{m} a_{ij}(q)\dot{q}^i\dot{q}^j,$$

we have

$$\mathrm{d}t^2 = \frac{1}{2}\frac{1}{h - V(q)} \sum_{i,j=1}^{m} a_{ij}(q)\,\mathrm{d}q^i\mathrm{d}q^j.$$

Dividing the left and the right hand side of (4.10.6) by the corresponding sides of the last equality, we get

$$\frac{ds}{dt} = \sqrt{2(h - V(q))},$$

which means that the speed of motion along the geodesic (in the metric (4.10.6)) equals $\sqrt{2(h-V(q))}$.

Introduce the new time variable τ according to the formula

$$d\tau = \sqrt{2(h-V(q))}\,dt.$$

Then

$$\frac{ds}{d\tau} = \frac{ds}{dt}\frac{dt}{d\tau} = \sqrt{2(h-V(q))}\frac{1}{\sqrt{2(h-V(q))}} = 1$$

and the system turns out to be a geodesic flow. The proof is finished.

Obviously, the converse theorem is also true: every geodesic flow is induced by some holonomic system; it suffices simply to take $V=0$, i.e., to consider the motion in the field of forces equal to zero (uniform motion—constant scalar velocity along the geodesic).

Now we present a theorem which is known under several names, depending on how it is formulated and interpreted. It might rightly be called the Anosov–Cartan–Hadamard–Lobachevsky theorem.

Theorem 4.10.2 If M is a compact Riemannian manifold of negative curvature [19], *then the geodesic flow is Anosov.*

We will prove this only in the case of $\dim M = 2$, under the additional assumption that M has constant curvature equal to -1.

An instance of such a manifold is the Lobachevsky plane $\mathscr{L}_2$. In Poincaré's model $\mathscr{L}_2$ is described as follows:

$$\mathscr{L}_2 = \{z \in \mathbf{C}\colon \operatorname{Im} z > 0\},$$

$$ds^2 = \frac{1}{y^2}(dx^2 + dy^2).$$

We shall need certain properties of $\mathscr{L}_2$ and we invite the reader to supply the missing proofs:

Exercises

4.10.1.A. Let $z_0 \in \mathscr{L}_2$, $v \in T_{z_0}(\mathscr{L}_2)$. If v is not parallel to the axis $\{z\colon \operatorname{Re} z = 0\}$, then the geodesic line passing through z_0 in the direction of v is the semicircle with centre on the axis $\{z\colon \operatorname{Im} z = 0\}$ passing through z_0 and tangent at z_0 to v. If v is parallel to $\{z\colon \operatorname{Re} z = 0\}$, then the geodesic is the half-line $\{z_0 + tv\colon t \in \boldsymbol{R}\} \cap \mathscr{L}_2$.

Hint. First show that any transformation $F(z) = (az+b)/(cz+d)$ preserves the Riemannian metric ds^2; then show that vertical lines are geodesics and use the fact that F is a conformal mapping taking straight lines and circles into straight lines and circles.

4.10.2.A. Every orientation preserving isometry F of $\mathscr{L}_2$ is of the form: $F(z) = (az+b)/(cz+d)$, where

$$\begin{bmatrix} a & b \\ c & d \end{bmatrix} \in \mathrm{SL}(2, \boldsymbol{R}) = \{A \in \mathscr{L}(\boldsymbol{R}^2, \boldsymbol{R}^2)\colon \det A = 1\}.$$

The group of orientation preserving isometries of $\mathscr{L}_2$ can be identified with $\mathrm{SL}(2, \boldsymbol{R})/Z_2$, where Z_2 is the cyclic group generated by $\begin{bmatrix} -1 & 0 \\ 0 & -1 \end{bmatrix}$.

Hint. Use the same argument as in the preceding exercise (see the hint): transform the geodesics involved into vertical lines.

Now, let $w_0 = (q_0, v_0) \in T(\mathscr{L}_2)$, v_0 being non-vertical, and let $\|v_0\| = 1$. Denote by γ^+ the geodesic line through q_0 in the direction of v_0 and by γ^- the geodesic through q_0 in the direction of $-v_0$. Then $\hat{\gamma}^+ \cup \hat{\gamma}^-$ is a semi-circle passing through q_0 and orthogonal to the real axis. Let q_+ and q_- denote the points at which γ^+ and γ^- meet the line $\{z\colon \operatorname{Im} z = 0\}$. Consider the family of all semicircles passing through q_+, orthogonal to the real line, and the family of all semicircles passing through q_-, orthogonal to the real line. Denote these families by S^+ and S^-, respectively. S^+ and S^- are thus certain families of geodesics in $\mathscr{L}_2$.

Definition 4.10.2 Let $H^+ \subset T(\mathscr{L}_2)$ be any set such that $\pi(H^+) = \{q\colon (q, v) \in H^+\}$ is a circle tangent to $\{z\colon \operatorname{Im} z = 0\}$ at q_+ and, for every $(q, v) \in H^+$, v is the unit normal vector to $\pi(H^+)$, sensed to the interior of $\pi(H^+)$. Any such set H^+ is called a *positive* (or *contracting*) *horocycle* of the family S^+.

Putting everywhere the sign $-$ in place of $+$ and demanding that v be outward normal to $\pi(H^-)$ we define a *negative* (*expanding*) *horocycle* of the family S^-.

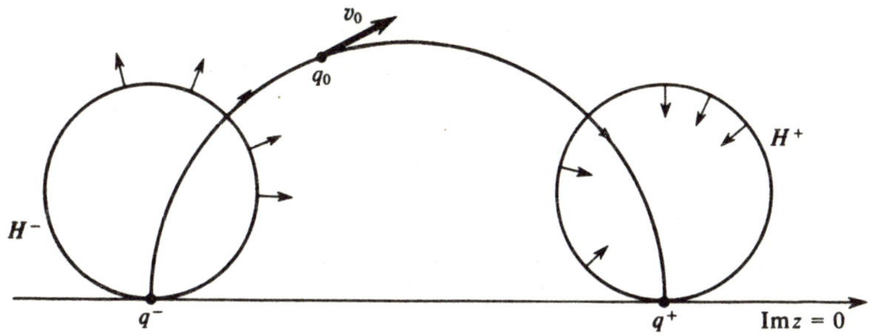

FIG. 4.10.1

Fig. 4.10.1 shows a positive and a negative horocycle of S^+ and S^-.

Exercises

4.10.3.A. Let S^+ and H^+ be as above. Pushing each element $w \in H^+$ by a fixed distance t along the geodesic (determined by w), we obtain a set $H_t^+ \subset T(\mathscr{L}_2)$, which is also a horocycle of S^+.

Hint. Apply an inversion with centre at q_+.

4.10.4.A. Let $w', w'' \in H^+$ and denote by w_t', w_t'' the elements of H_t^+ (see the preceding exercise) resulting from w', w'' by translation along their geodesics. Let l_t denote the length of the arc of $\pi(H_t^+)$ limited by $\pi(w_t')$ and $\pi(w_t'')$. Show that there exist constants $\alpha, \beta > 0$ such that

$$\alpha e^t \leqslant \frac{l_0}{l_t} \leqslant \beta e^t.$$

Hint. Calculation becomes simple if we apply a suitable inversion, as in the preceding exercise.

Proof of Theorem 4.10.2 We first prove the theorem in the case of $M = \mathscr{L}_2$, the Lobachevsky plane.

Let W be the sphere bundle over $\mathscr{L}_2$. For each $w = (q, v) \in W$ the tangent space $T_w(W)$ admits the splitting

$$T_w(W) = [a] \oplus [b] \oplus [c];$$

in this formula, a is a vector (in $T_w(W)$, i.e. 3-dimensional) tangent at w to H^+ (the positive horocycle passing through w viewed as a line in W), b is a vector tangent at w to H^- (the negative horocycle), and c is a vector tangent to the trajectory of the geodesic flow passing through

w. The above decomposition is invariant under the action of the geodesic flow (Exercise 4.10.3). As follows from the result of Exercise 4.10.4 the norm $\|d\varphi_t(w)a\|$ is exponentially decreasing and the norm $\|d\varphi_t(w)b\|$ is exponentially increasing as $t \to +\infty$. Thus the theorem is true for $M = \mathscr{L}_2$.

Now let M be a compact two-dimensional manifold of constant curvature equal to -1. Then $\mathscr{L}_2$ is a covering space for M and the covering projection is locally isometric (see Springer, 1957). Let W and W_2 denote the sphere bundles over M and $\mathscr{L}_2$, respectively. It follows that the decomposition of $T(W_2)$ into the contracting part (E^+), the expanding part (E^-) and the neutral part (E^0) can be projected onto $T(W)$, yielding the corresponding decomposition for the geodesic flow over M.

The proof in the general case (that of a compact Riemannian manifold with negative curvature at each point) is much more difficult; it can be found in Anosov (1967).

4.11. Expanding Mappings

An interesting class of mappings of smooth manifolds is furnished by the so called *expanding mappings*.

Let M be a Riemannian manifold; M need not be compact.

Definition 4.11.1 A differentiable mapping $\varphi: M \to M$ is called *expanding* iff there exist constants $\alpha > 0$, $\lambda > 1$ such that

$$\|d\varphi^n(x)v\| \geqslant \alpha\lambda^n\|v\| \qquad \text{for all } x \in M,\ v \in T_x(M),\ n = 1, 2, \ldots$$

The class of all expanding C^r-mappings will be denoted by the symbol $\mathscr{E}^r(M)$.

Examples

4.11.1. Let $M = S^1$, $\varphi(z) = z^N$, $N \in Z^+$, $N > 1$; φ is expanding: $\|d\varphi^n(z)v\| = N^n\|v\|$.

4.11.2. Let $M = S^1$, $\varphi(e^{2\pi i x}) = e^{2\pi i f(x)}$, f: $\boldsymbol{R} \to \boldsymbol{R}$, $f \in C^1(\boldsymbol{R})$, $f(x+1) = N + f(x)$, $N \in Z^+$, $N > 1$ and let $f'(x) \geqslant \lambda > 1$ for all x; φ is expanding: for $z = e^{2\pi i x}$, $v \in T_z(S^1)$ we have $\|d\varphi(z)v\| \geqslant f'(x)\|v\| \geqslant \lambda\|v\|$, whence $\|d\varphi^n(z)v\| \geqslant \lambda^n\|v\|$.

4.11.3. Let $M = \boldsymbol{R}^m$, $\varphi(x) = Ax$, $A \in \mathscr{L}(\boldsymbol{R}^m, \boldsymbol{R}^m)$, $\operatorname{sp} A \subset \{|\lambda|: |\lambda| > 1\}$. By Lemma 1.5.1 there is a norm $|\cdot|$ in $\boldsymbol{R}^m$ such that $\vartheta := |A^{-1}| < 1$. Then we have for each $x \in \boldsymbol{R}^m$

$$|x| \leqslant |A^{-n}| \cdot |A^n x| \leqslant \vartheta^n |A^n x|,$$

and so

$$|A^n x| \geqslant \lambda^n |x|,$$

where $\lambda = 1/\vartheta > 1$. Since in $\boldsymbol{R}^m$ all norms are equivalent, there are constants a, b such that $a\|\cdot\| \leqslant |\cdot| \leqslant b\|\cdot\|$, $\|\cdot\|$ denoting the standard norm. Hence

$$\|A^n x\| \geqslant \frac{1}{b}|A^n x| \geqslant \frac{1}{b}\lambda^n |x| \geqslant \frac{a}{b}\lambda^n \|x\|.$$

Exercises

4.11.1.A. Let M be compact. Show that if $\varphi \in \mathscr{E}^r(M)$ in some Riemannian metric then $\varphi \in \mathscr{E}^r(M)$ in every Riemannian metric.

Hint. Cp. Example 4.11.3.

4.11.2.A. Let M be compact. Show that if φ is expanding, then there exists a C^r-smooth Riemannian metric $\|\cdot\|_1$ and a constant $\lambda_1 > 1$ such that

$$\|\mathrm{d}\varphi(x)v\|_1 \geqslant \lambda_1 \|v\|_1 \qquad \text{for all } x \in M,\ v \in T_x(M).$$

Hint. See the proof of Lemma 1.5.1; cf. also Exercise 4.7.1.

4.11.3.A. Let M be a Riemannian manifold, compact or not. Show that if φ is expanding, then φ is locally a diffeomorphism[2]. If M is connected and compact, then the sets $\varphi^{-1}(x)$, $x \in M$, are of constant finite cardinality, equal to $\deg \varphi$.

Hint. That φ is a local diffeomorphism follows from the Inverse Mapping Theorem. The function $x \mapsto \operatorname{card} \varphi^{-1}(x)$ assumes finite values (by the compactness of M) and is continuous.

The conclusion of Exercise 4.11.2 yields:

Corollary 4.11.1 *For M compact, $\mathscr{E}^r(M)$ is an open subset of $C^r(M, M)$.*

Proof Let $\|\cdot\|_1$ and λ_1 be as in Exercise 4.11.2. If ψ is sufficiently close to φ, then we have for some λ_2 with $1 < \lambda_2 < \lambda_1$: $\|d\psi(x)v\|_1 \geqslant \lambda_2 \|v\|_1$ for $x \in M$, $v \in T_x(M)$; hence $\|d\psi^n(x)v\|_1 \geqslant \lambda_2^n \|v\|_1$.

According to the statement in Exercise 4.11.3, every expanding map of a compact manifold is a covering.

Before proceeding to further considerations let us survey certain useful facts concerning coverings. Their proofs (or, at least, suitable bibliographical indications) can be found in Spanier (1966), Chapter II.

Let M be a manifold. The symbol $\bar{M}$ will denote the universal covering space[3] of M. Let π: $\bar{M} \to M$ be the covering map and let D_M denote the group of all homeomorphisms j: $\bar{M} \to \bar{M}$ such that $\pi \circ j = \pi$; we call D_M the *group of covering transformations.*

Let M and N be two manifolds. The following statements are true:

1° Every continuous map f: $M \to M$ has a continuous lift $\bar{f}$: $\bar{M} \to \bar{M}$: $\bar{f}$ is unique in the sense that if $\bar{f}_1$ and $\bar{f}_2$ are two such lifts (for the same f), then $\bar{f}_1 = j \circ \bar{f}_2$ for some $j \in D_M$.

2° If a mapping f: $M \to M$ is covering, then its lift $\bar{f}$: $\bar{M} \to \bar{M}$ is a homeomorphism of $\bar{M}$.

3° A mapping $\bar{f}$: $\bar{N} \to \bar{M}$ is a lift of some mapping f: $N \to M$ iff there exists a mapping h: $D_N \to D_M$ such that $\bar{f} \circ j = h(j) \circ \bar{f}$ for every $j \in D_N$. Such an h will be denoted by the symbol $\bar{f}^{\#}$; it is a group homomorphism $D_N \to D_M$.

4° A set $C \subset \bar{M}$ is called a *covering domain* iff $C = \overline{\operatorname{int} C}$ and $\pi(C) = M$. For every compact M there exist compact covering domains.

Theorem 4.11.1 (Shub, 1969) Let M be a compact connected Riemannian manifold and let φ: $M \to M$ be an expanding map. Then

(1) *φ has a fixed point,*

(2) *the universal covering space of M is diffeomorphic to $\boldsymbol{R}^m$, $m = \dim M$,*

(3) *for every open set $U \subset M$ there is an n such that $\varphi^n(U) \supset M$,*

(4) *the periodic points of φ are dense in M.*

Proof (1) Since π is locally a diffeomorphism, we can endow $\bar{M}$ (the universal covering space of M) with the Riemannian structure by simply importing the metric from M to $\bar{M}$: $\bar{M}$ is complete in the resulting metric.

According to 1° and 2°, the lift $\bar{\varphi}$: $\bar{M} \to \bar{M}$ is a diffeomorphism of $\bar{M}$; since φ is expanding, so is $\bar{\varphi}$. We may assume (re-defining the metric if necessary – see Exercise 4.11.2) that

$$\|d\bar{\varphi}(x)v\| \geqslant \lambda \|v\| \qquad \text{for } v \in T_x(\bar{M})$$

holds with some $\lambda > 1$; hence

$$\|d\bar{\varphi}^{-1}(x)v\| \leqslant \frac{1}{\lambda}\|v\|.$$

If γ is a smooth path connecting two given points a and b in $\bar{M}$, $\gamma(0) = a$, $\gamma(1) = b$, then

$$\text{the length of } \bar{\varphi}^{-1}(\hat{\gamma}) = \int_0^1 \|d\bar{\varphi}^{-1}(\gamma(t))\dot{\gamma}(t)\|dt \leqslant \frac{1}{\lambda}\int_0^1 \|\dot{\gamma}(t)\|dt.$$

The distance between two points in $\bar{M}$ is equal to the infimum of the lengths of all connecting paths; hence

$$\varrho(\bar{\varphi}^{-1}(a), \bar{\varphi}^{-1}(b)) \leqslant \frac{1}{\lambda}\varrho(a, b),$$

which means that $\bar{\varphi}^{-1}$ is a contraction of $\bar{M}$. Thus $\bar{\varphi}$ has exactly one fixed point in $\bar{M}$.

(2) As shown above, $\bar{\varphi}^{-1}$ is a contraction of $\bar{M}$. It hence follows without much difficulty that $\bar{M}$ is diffeomorphic to $\boldsymbol{R}^m (m = \dim M)$; to see this, apply an argument as in the proof of Theorem 1.6.1.

(3) Since M is compact, $\operatorname{diam} M < +\infty$ and any two points in M can be connected by a smooth curve of length not exceeding d, where d is some number, $d > \operatorname{diam} M$. Fix a point $x_0 \in M$; then take another point $x \in M$ and a curve γ of length $\leqslant d$ connecting x_0 with x (there are many such curves). Let $y_0 \in \pi^{-1}(x_0)$ and denote by $\bar{\gamma}$ the component of $\pi^{-1}(\hat{\gamma})$ which contains y_0. Define C to be the point union of all $\bar{\gamma}$ obtained in this fashion (for a fixed x_0). Then C is a covering domain; we leave the details to the reader. Obviously, $\operatorname{diam} C \leqslant 2d$.

Now let $U \subset M$ be an open set, let $U_1 \subset \pi^{-1}(U)$ and let $B(\bar{y}, r)$ be a ball contained in U_1. The diffeomorphism $\bar{\varphi}^{-1}$ is a contraction of $\bar{M}$, the Lipschitz constant being $1/\lambda$. Therefore

$$\varrho(\bar{\varphi}(x), \bar{\varphi}(y)) \geqslant \lambda\varrho(x, y) \qquad \text{for any } x, y \in \bar{M}. \tag{4.11.1}$$

Hence

$$\bar{\varphi}^n(B(\bar{y}, r)) \supset B(\bar{\varphi}^n(\bar{y}), \lambda^n r). \tag{4.11.2}$$

Suppose that C is a covering domain constructed as described above, the point y_0 coinciding with $\bar{\varphi}^n(\bar{y})$, where n is such that $\lambda^n r > 2d$. Since $\operatorname{diam} C \leqslant 2d$ and $y_0 = \bar{\varphi}^n(\bar{y}) \in C$, we see that

$$B(\bar{\varphi}^n(\bar{y}), \lambda^n r) \supset C,$$

whence by (4.11.2) we get

$$\varphi^n(U) \supset \pi(\bar{\varphi}^n(U_1)) \supset \pi(\bar{\varphi}^n(B(\bar{y}, r))) \supset \pi(C) = M.$$

(4) Let $B = B(x, r)$ be an arbitrary ball in M, $x \in M$, $r > 0$. Then there is an $n \in Z^+$ such that $\varphi^n(B) \supset M$, as we have just shown. Thus $\varphi^{-n}(B) \cap B \neq 0$. Choose a point $x_0 \in \varphi^{-n}(B) \cap B$. Then there exists a point $x_1 \in B$ such that $\varphi^n(x_1) = x_0$; similarly, there exists an $x_2 \in B$ such that $\varphi^n(x_2) = x_1$, and so on. Proceeding inductively, we find a sequence of points $x_k \in B$ such that

$$\varphi^n(x_{k+1}) = x_k, \qquad k = 0, 1, 2, \ldots$$

Using condition (4.11.1) we obtain

$$\varrho(x_{k+1}, x_k) \leqslant \frac{1}{\lambda}\varrho(x_k, x_{k-1}) \leqslant \ldots \leqslant \frac{1}{\lambda^k}\varrho(x_1, x_0),$$

whence it easily follows that the sequence x_k converges; clearly, the limit point is a fixed point for φ^n and lies in the closure of B.

Thus every ball in M contains a periodic point in its closure; hence the assertion.

Suppose that α: $N \to M$ is a continuous map of compact manifolds; as before, $\bar{\alpha}$: $\bar{N} \to \bar{M}$ stands for the lift of α to a mapping between the universal covering spaces of N and M. Consider the following family of mappings:

$$V_{\bar{\alpha}} = \{B\colon \bar{N} \to \bar{M},\ B \text{ is a lift of a continuous map } \beta\colon N \to M, \text{ and } B^{\#} = \bar{\alpha}^{\#}\}.$$

It is readily seen that $V_{\bar{\alpha}}$ becomes a complete metric space if the distance between $A, B \in V_{\bar{\alpha}}$ is defined by

$$d(A, B) = \sup_{x \in \bar{N}} \varrho(A(x), B(x))$$

(see Exercise 4.11.4 below).

Lemma 4.11.4 Let f: $M \to M$, g: $N \to N$ be expanding mappings and let α: $N \to M$ be a continuous map. If there exist lifts $\bar{f}$: $\bar{M} \to \bar{M}$, $\bar{g}$: $\bar{N}$

$\to \bar{N}$, $\bar{\alpha}$: $\bar{N} \to \bar{M}$ such that $\bar{f}^{\#} \circ \bar{\alpha}^{\#} = \bar{\alpha}^{\#} \circ \bar{g}^{\#}$, then $V_{\bar{\alpha}}$ contains a unique B satisfying the condition $\bar{f} \circ B = B \circ \bar{g}$.

Proof Write $T(A) = \bar{f}^{-1} \circ A \circ \bar{g}$ for $A \in V_{\bar{\alpha}}$. To begin with, we prove that $T(A)$ is a lift of some map $N \to M$. By the definition of $\#$, we have $\bar{f} \circ j = \bar{f}^{\#}(j) \circ \bar{f}$ for any $j \in D_M$ (cf. statement 3° on p. 258). Hence $j = \bar{f}^{-1} \circ \bar{f}^{\#}(j) \circ \bar{f}$, because $\bar{f}$ is a homeomorphism (statement 2°) and so $\bar{f}^{-1}$ does exist. Thus, if $\bar{f}^{\#}(j_1) = \bar{f}^{\#}(j_2)$, then $j_1 = j_2$, showing that $\bar{f}^{\#}$ is an invertible transformation of D_M. Therefore the following the definition is meaningful:

$$h(j) := (\bar{f}^{\#})^{-1} \circ A^{\#} \circ \bar{g}^{\#}(j) \qquad \text{for } j \in D_N.$$

In order to prove that $T(A)$ is a lift of some map $N \to M$, it is enough to show that

$$T(A) \circ j = h(j) \circ T(A) \qquad \text{for all } j \in D_N, \tag{4.11.3}$$

(again we use statement 3° on p. 258). Now, we have

$$A \circ \bar{g} \circ j = \left(A^{\#} \circ \bar{g}^{\#}(j)\right) \circ A \circ \bar{g} \qquad \text{for } j \in D_N$$

and

$$\bar{f}^{-1} \circ (\bar{f}^{\#}(j)) \circ \bar{f} = j \qquad \text{for } j \in D_M;$$

consequently, for any $j \in D_N$ we get

$$\begin{aligned}
T(A) \circ j &= \bar{f}^{-1} \circ A \circ \bar{g} \circ j \\
&= \bar{f}^{-1} \circ \left(A^{\#} \circ \bar{g}^{\#}(j)\right) \circ A \circ \bar{g} \\
&= \bar{f}^{-1} \circ \left(\bar{f}^{\#} \circ (\bar{f}^{\#})^{-1} \circ A^{\#} \circ \bar{g}^{\#}(j)\right) \circ \bar{f} \circ \bar{f}^{-1} \circ A \circ \bar{g} \\
&= \left((\bar{f}^{\#})^{-1} \circ A^{\#} \circ \bar{g}^{\#}(j)\right) \circ \bar{f}^{-1} \circ A \circ \bar{g} \\
&= h(j) \circ T(A),
\end{aligned}$$

i.e., (4.11.3).

We claim that $T(A) \in V_{\bar{\alpha}}$. It is already settled that $T(A)$ is a lift. It remains to show that $T(A)^{\#} = \bar{\alpha}^{\#}$, or, equivalently,

$$\left(T(A)^{\#}(j)\right) \circ \bar{\alpha} = \bar{\alpha} \circ j \qquad \text{for } j \in D_N.$$

The following identity is an easy consequence of the definition of the operation $\#$:

$$(\bar{f}^{-1} \circ A \circ \bar{g})^{\#} = (\bar{f}^{-1})^{\#} \circ A^{\#} \circ \bar{g}^{\#}.$$

The expression on the left is nothing else than $T(A)^*$. Thus, for $j \in D_N$,

$$(T(A)^*(j)) \circ \bar{\alpha} = ((\bar{f}^{-1})^* \circ A^* \circ \bar{g}^*(j)) \circ \bar{\alpha} = ((\bar{f}^{-1})^* \circ \bar{\alpha}^* \circ \bar{g}^*(j)) \circ \bar{\alpha}.$$

By hypothesis, $\bar{\alpha}^* \circ \bar{g}^* = \bar{f}^* \circ \bar{\alpha}^*$. Hence

$$(T(A)^*(j)) \circ \bar{\alpha} = ((\bar{f}^{-1})^* \circ \bar{f}^* \circ \bar{\alpha}^*(j)) \circ \bar{\alpha} = \bar{\alpha}^*(j) \circ \bar{\alpha} = \bar{\alpha} \circ j,$$

proving the claim.

The assignment $A \mapsto T(A)$ thus defines a mapping $T\colon V_{\bar{\alpha}} \to V_{\bar{\alpha}}$. As it is easy to guess, T is a contraction. Indeed, let $A, B \in V_{\bar{\alpha}}$. Since $\bar{f}^{-1}$ is contractive (with the contraction coefficient $1/\lambda$), we have

$$\begin{aligned} d(T(A), T(B)) &= \sup_{x \in \bar{N}} \varrho(\bar{f}^{-1} \circ A \circ \bar{g}(x), \bar{f}^{-1} \circ B \circ \bar{g}(x)) \\ &\leqslant \frac{1}{\lambda} \sup_{y \in \bar{N}} \varrho(A(y), B(y)) = \frac{1}{\lambda} d(A, B). \end{aligned}$$

By the Banach Fixed Point Theorem there exists a unique $B \in V_{\bar{\alpha}}$ satisfying the condition $T(B) = B$, i.e., $\bar{f} \circ B = B \circ \bar{g}$.

Theorem 4.11.2 (Shub, 1969) Let M be compact and connected. If $f, g\colon M \to M$ are expanding mappings and f is homotopic to g, then the systems (M, f) and (M, g) are topologically conjugate.

Proof Select a point $a \in M$. Let f_t, $0 \leqslant t \leqslant 1$, be a homotopy joining f with $g (f_0 = f, g_0 = g)$. The curve $\gamma(t) = f_t(a)$, $0 \leqslant t \leqslant 1$, connects the points $f(a)$ and $g(a)$.

Lift f to a mapping $\bar{f}\colon \bar{M} \to \bar{M}$. Choose $y_0 \in \pi^{-1}(a) \subset \bar{M}$ and write $y_1 = \bar{f}(y_0)$. Denote by $\bar{\gamma}$ the lift of γ whose one endpoint is y_1 and by $\bar{g}$ the lift of g such that $y_2 = \bar{g}(y_0)$ is the other endpoint of $\bar{\gamma}$. For any $j \in D_M$ we have $\bar{f}_t \circ j = \bar{f}_t^*(j) \circ \bar{f}_t$; the lifts $\bar{f}_t$ have to be chosen continuously with respect to t. It is evident that also $\bar{f}_t^*(j)$ depends continuously on t. On the other hand, the set D_M is discrete; hence

$$\bar{f}_t^*(j) = \text{const} = \bar{f}_0^*(j) = \bar{f}_1^*(j) = \bar{f}^*(j) = \bar{g}^*(j).$$

Now apply Lemma 4.11.1 to the mappings f, g and $\alpha = $ identity: there is a unique $B \in V_{\text{id}}$ such that $\bar{f} \circ B = B \circ \bar{g}$. Interchanging the roles of f and g and repeating the argument, we find an $A \in V_{\text{id}}$ satisfying $\bar{g} \circ A = A \circ \bar{f}$. Multiplying (in the sense of composition) the former equality by A and applying the latter, we get

$$\bar{f} \circ B \circ A = B \circ \bar{g} \circ A = B \circ A \circ \bar{f}.$$

Since $A^{\#} = \mathrm{id}^{\#} = \mathrm{id}$, $B^{\#} = \mathrm{id}^{\#} = \mathrm{id}$, we have $(B \circ A)^{\#} = \mathrm{id}$ and so $B \circ A \in V_{\mathrm{id}}$. By uniqueness (again in virtue of Lemma 4.11.1, now applied to the mappings f, f and $\alpha = \mathrm{id}$), $B \circ A = \mathrm{id}$. Resorting once more to symmetry, we see that also $A \circ B = \mathrm{id}$, and thus A and B are homeomorphisms.

Corollary 4.11.2 Every expanding mapping of a connected compact Riemannian manifold is structurally stable.

Proof Let f: $M \to M$ be an expanding map. If g: $M \to M$ is close to f, then g is also an expanding map (Corollary 4.11.1), homotopic to f (see e.g. the proof of Lemma 4.8.1). Hence by Theorem 4.11.2 the systems (M, f) and (M, g) are topologically conjugate.

Anosov diffeomorphisms and expanding mappings show certain similarities. Also, the two classes are subjected to a common extension to the class of so-called *Anosov endomorphisms*. We shall not formulate their precise definition; let us only mention that e.g. a mapping f: $T^m \to T^m$ is an Anosov endomorphism whenever f is given by a matrix $f = (a_{ij})$, $i, j = 1, \ldots, m$, $a_{ij} \in Z$, $|\det f| \geqslant 1$, $\mathrm{sp} f \cap \{\lambda\colon |\lambda| = 1\} = \emptyset$. In particular, if $|\det f| = 1$ then f is an Anosov diffeomorphism; if $\mathrm{sp} f \subset \{\lambda\colon |\lambda| > 1\}$, then f is an expanding map. As it has been shown by F. Przytycki (1976), an Anosov endomorphism is structurally stable iff it is either a diffeomorphism or an expanding map.

Exercises

4.11.4.A. Show that $d(A, B)$ is finite for all $A, B \in V_{\bar{\alpha}}$ (see the proof of Theorem 4.11.1) and defines a complete metric in $V_{\bar{\alpha}}$.

4.11.5.B. Let $f \in \mathscr{E}^1(S^1)$, $\deg f = N$. Show that there exist compact intervals $A_1, \ldots, A_N$ with the following properties:

(a) $\mathrm{int} A_i \cap \mathrm{int} A_j = \emptyset$ for $i \neq j$,

(b) $\bigcup_{i=1}^{N} A_i = S^1$,

(c) $f(A_i) = S^1$ for each i,

(d) f is one-to-one on each $\mathrm{int} A_i$,

(e) $\bigcup_{i=1}^{N} f^{-1}(\mathrm{Fr} A_i) \supset \mathrm{Fr} A_j$ for each j.

Hint. Lift f to $\boldsymbol{R}$; since f has a fixed point, we may assume that $\bar{f}(0) = 0$. Examine the inverse images $\bar{f}^{-1}(I_i)$, where $I_i = \langle i-1, i\rangle$.

4.12. Axiom A Diffeomorphisms. Remarks on Stability

We shall now discuss a certain specific condition, the so-called *Axiom A*, which is of vital importance in the theory of smooth dynamical systems.

Let M be a given manifold and let $\varphi \in \mathrm{Diff}^r(M)$.

Definition 4.12.1 We say that φ *satisfies Axiom A*, or that φ is an *Axiom A diffeomorphism*, iff:

(1) $\Omega(\varphi)$ is a hyperbolic set,

(2) periodic points are dense in $\Omega(\varphi)$.

For flows, Axiom A is formulated in a slightly different fashion; let $\Phi \in \mathscr{F}^r(M)$ and let φ_t be the flow generated by Φ.

Definition 4.12.2 We say that the flow φ_t *satisfies Axiom A*, iff:

(1) the fixed points of φ_t are hyperbolic,

(2) $\Omega(\Phi)$ is the union of the set of all critical points (denote it by F) and the closure of the set of all periodic points (denote this set by Λ): $\Omega(\Phi) = F \cup \bar{\Lambda}$,

(3) $\Lambda \cap F = \emptyset$,

(4) Λ is a hyperbolic set.

The property which follows is proved along the same lines as the Hadamard–Perron theorem.

Theorem 4.12.1 Let M be a compact Riemannian manifold. Suppose that $\varphi \in \mathrm{Diff}^r(M)$ is an Axiom A diffeomorphism. Then for every $x \in \Omega(\varphi)$ the sets

$$W^+(x) = \{y \in M\colon \lim_{n \to +\infty} \varrho(\varphi^n(x), \varphi^n(y)) = 0\},$$

$$W^-(x) = \{y \in M\colon \lim_{n \to -\infty} \varrho(\varphi^n(x), \varphi^n(y)) = 0\}$$

are immersed C^r-submanifolds.

The proof can be found e.g. in Hirsch and Pugh (1970) and Nitecki (1971).

In 1971 J. Robbin proved the following important theorem.

Theorem 4.12.2 Let M be compact and suppose that a diffeomorphism $\varphi \in \mathrm{Diff}^2(M)$ satisfies the conditions:

(1) Axiom A,

(2) for any $x, y \in \Omega(\varphi)$ the manifolds $W^+(x)$ and $W^+(y)$ intersect transversally.

Then φ is structurally stable.

It seems likely that the conditions (1) and (2) are also necessary for structural stability, but so far nobody has proved this conjecture. Thus we have:

Problem 4.12.1 Does every structurally stable diffeomorphism φ necessarily satisfy conditions (1) and (2)?

Remark 4.12.1 As has been shown by S. E. Newhouse (1970), the set of all Axiom A diffeomorphisms $\varphi \in \mathrm{Diff}^1(M)$ need not be dense in $\mathrm{Diff}^1(M)$. For instance, it is not dense even for the sphere $M = S^2$.

Let us point out another property of Axiom A diffeomorphisms which deserves attention.

Suppose that $\varphi\colon M \to M$ satisfies Axiom A.

Definition 4.12.3 By a *Markov partition* of $\Omega(\varphi)$ we mean any finite collection of sets $\xi = \{A_1, \dots, A_k\}$ such that:

(1) A_i are closed, $i = 1, \dots, k$,

(2) $A_i = \overline{\mathrm{int}\, A_i}$, $i = 1, \dots, k$,

(3) $\bigcup_{i=1}^{k} A_i = \Omega(\varphi)$,

(4) $A_i \cap A_j \subset \mathrm{Fr}\, A_i \cap \mathrm{Fr}\, A_j$ for $i \neq j$,

(5) if $x \in \mathrm{int}\, A_i \cap \varphi^{-1}(\mathrm{int}\, A_j)$, then

$$\varphi(A_i \cap W^-_{\mathrm{loc}}(x)) \supset A_j \cap W^-_{\mathrm{loc}}(\varphi(x)),$$

$$\varphi(A_i \cap W^+_{\mathrm{loc}}(x)) \subset A_j \cap W^+_{\mathrm{loc}}(\varphi(x));$$

here $W^{\pm}_{\mathrm{loc}}(x)$ denote the local submanifolds of the manifolds $W^{\pm}(x)$ defined in Theorem 4.12.1.

The concept of a Markov partition has been introduced by Sinai for the case of Anosov systems. We have already come across an example of such a partition in our course: recall the family of parallelograms on the torus described in Section 1.6.

Theorem 4.12.3 (Bowen, 1970a) If a system (M, φ) satisfies Axiom A, then it admits a Markov partition.

Markov partitions exist also in the case of expanding mappings; but their definition differs slightly from the one given above (see Krzyżewski, 1971b); the family of intervals on the circle considered in Exercise 4.11.5 can serve as an example.

By a certain modification of the definition of structural stability we arrive at the notion of so-called Ω-stability.

Definition 4.12.4 Given two dynamical systems (X, φ) and (Y, ψ), we say that they are *Ω-conjugate* iff the systems $(\Omega_X(\varphi), \varphi)$ and $(\Omega_Y(\psi), \psi)$ are topologically conjugate.

The symbols $\Omega_X(\varphi)$ and $\Omega_Y(\psi)$ denote the sets of non-wandering points of φ and ψ in X and Y, respectively.

Similarly, we say that two flows are Ω-conjugate iff they have equivalent phase portraits when restricted to the sets of non-wandering points.

There exist pairs of systems which are Ω-conjugate without being conjugate: e.g. the flows (T^2, φ_t^a) and (T^2, φ_t^c) generated by the vector fields Φ_a and Φ_c on the torus and described in Exercise 1.4.4 on p. 36 are Ω-conjugate (obvious) though they do not have equivalent phase portraits.

Definition 4.12.5 *Let* $\varphi_0 \in \mathrm{Diff}^r(M)$. The system (M, φ_0) is said to be *C^r-Ω-stable* iff there is a neighbourhood V of φ_0 in $\mathrm{Diff}^r(M)$ such that for any $\varphi \in V$ the systems (M, φ) and (M, φ_0) are Ω-conjugate.

Analougously we define C^r-Ω-stability for flows (i.e., for vector fields).

Certain necessary and sufficient conditions for Ω-stability can be found in Nitecki (1971).

4.13. Homoclinic Points

The starting point is an example known as *Smale's horseshoe*.

Let $Q = ABCD$ be a square in the plane $\boldsymbol{R}^2$:

$$Q = \{x\colon |x^i| \leqslant 1, i = 1, 2\}.$$

Let φ be a mapping of Q into $\boldsymbol{R}^2$ carrying Q into a horseshoe-like set as shown in Fig. 4.13.1.

Denote by Q_0 and Q_1 the components of $Q \cap \varphi(Q)$; write $\varphi(A) = A'$, $\varphi(B) = B'$, $\varphi(C) = C'$, $\varphi(D) = D'$. Denote by P_0 and P_1 the inverse images $\varphi^{-1}(Q_0)$ and $\varphi^{-1}(Q_1)$ (see Fig. 4.13.2).

Assume that:

1° φ is linear within P_0 and P_1,

2° φ is expanding on the sides AD and BC and contracting on the sides AB and CD (with coefficients greater than 1 and smaller than 1, respectively).

The image $\varphi(Q_0)$ is represented by the shaded area in Fig. 4.13.3. The image $\varphi(Q_1)$ is similar (a curvilinear strip "parallel" to $\varphi(Q_0)$). The set $Q \cap \varphi^2(Q)$ consists of 4 components, each of them being a rectangle (the shaded area in Fig. 4.13.4).

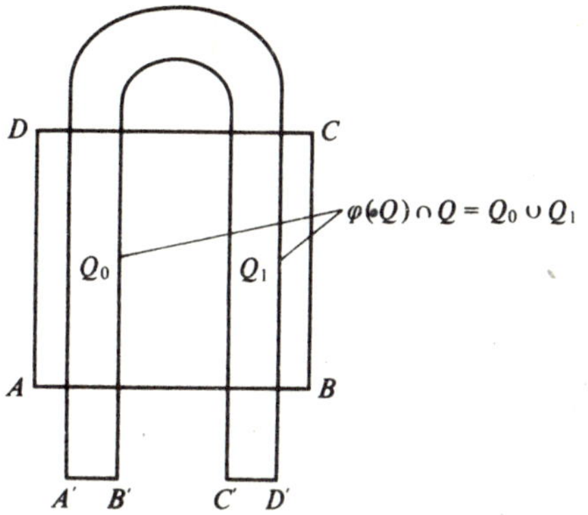

FIG. 4.13.1

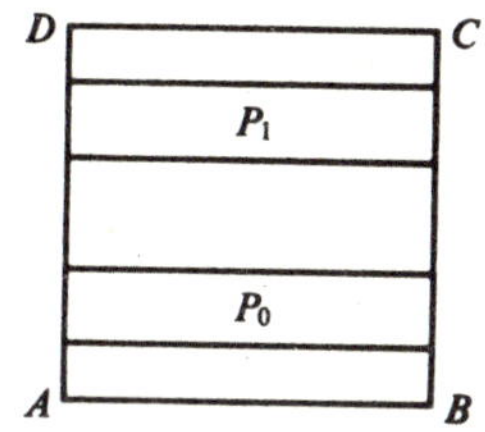

FIG. 4.13.2

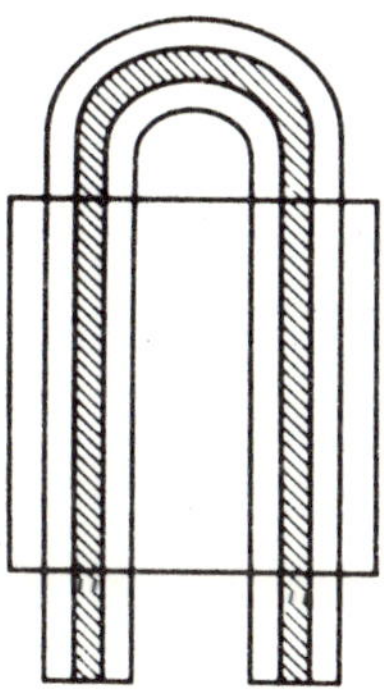

FIG. 4.13.3

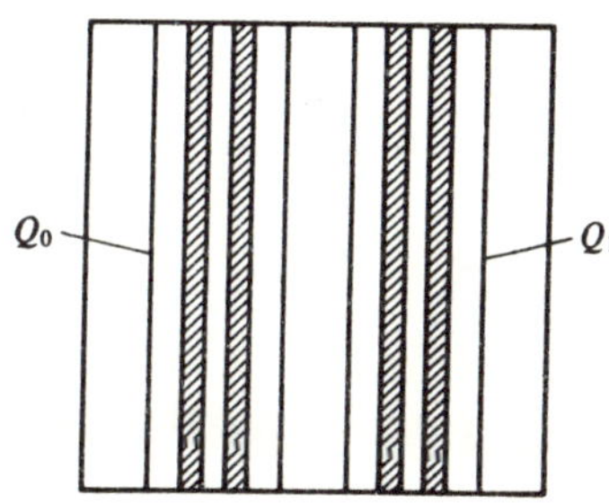

FIG. 4.13.4

Iterating, we obtain a sequence of sets; the n-th set $Q \cap \varphi^n(Q)$ splits into 2^n rectangular components, whose vertical sides are of constant length (equal to the length of AD), whereas the horizontal sides shorten to zero as $n \to +\infty$. Hence the set

$$\Lambda^+ := \bigcap_{n \geq 0} \varphi^n(Q)$$

is the product of a Cantor set $\Delta_1 \subset AB$ by the side AD.

Iterating backwards, we get the sets $Q \cap \varphi^{-n}(Q)$. Since $Q \cap \varphi^{-1}(Q) = P_0 \cup P_1$, the subsequent sets are just as those obtained for $n > 0$, rotated by 90°. Hence, as in the former case, the set

$$\Lambda^- := \bigcap_{n \leqslant 0} \varphi^n(Q)$$

is the product of the side AB by a Cantor set $\Delta_2 \subset AD$.

Now consider the set

$$\Lambda := \bigcap_{n \in Z} \varphi^n(Q) = \Lambda^+ \cap \Lambda^- = \Delta_1 \times \Delta_2.$$

This is the product of two Cantor sets; thus it is a closed nowhere dense 0-dimensional subset of Q.

With every point $x \in \Lambda$ we associate a two-tailed sequence $(\alpha_n(x))_{n=-\infty}^{+\infty}$ whose entries are 0 or 1, according to the rule:

$$\alpha_n(x) = \begin{cases} 0 & \text{if } \varphi^n(x) \in Q_0, \\ 1 & \text{if } \varphi^n(x) \in Q_1. \end{cases}$$

The resulting correspondence is one-to-one.

In the infinite product

$$X = \prod_{n=-\infty}^{+\infty} \{0, 1\}_n = \{x = (\alpha_n) \colon \alpha_n = 0 \text{ or } 1, n \in Z\} \tag{4.13.1}$$

we introduce the product topology (of coordinatewise convergence). This is a compact metric space with metric

$$\varrho(x, y) = \sum_{n=-\infty}^{+\infty} 2^{-|n|} |\alpha_n - \beta_n| \qquad \text{for } x = (\alpha_n),\ y = (\beta_n).$$

Denote by $s \colon X \to X$ the shift to the left:

$$\text{if } x = (\alpha_n) \text{ then } s(x) = (\beta_n) \text{ where } \beta_n = \alpha_{n+1}. \tag{4.13.2}$$

It is a homeomorphism of the space X; s^{-1} is the shift to the right.

Theorem 4.13.1 (1) *The set Λ is φ-invariant.*

(2) *The dynamical system (Λ, φ) is topologically conjugate to the system (X, s) given by* (4.13.1) *and* (4.13.2).

Proof (1) $\varphi(\Lambda) = \varphi(\bigcap_{n \in Z} \varphi^n(Q)) = \bigcap_{n \in Z} \varphi^{n+1}(Q) = \Lambda.$

(2) Every point $x = (x^1, x^2) \in \Lambda$ is fully determined by the numbers $\alpha_n(x)$, $n \in Z$; those with $n \geqslant 0$ define the coordinate x^1, and those with $n < 0$ define x^2. It is easy to check that the map $x \mapsto (\alpha_n(x))$ is a homeomorphism of Λ onto X. By the definition of $\alpha_n(x)$ we have $\alpha_n(\varphi(x)) = \alpha_{n+1}(x)$, and so the systems (Λ, φ) and (X, s) are conjugate, as asserted.

Remark 4.13.1 Let us modify the conditions imposed on φ in the horseshose example as follows:

(1) Replace the requirement that φ should be linear on $Q_0 \cup Q_1$ by the assumptions 1°–3° below:

1° the sets $Q \cap \varphi^n(AD)$ and $Q \cap \varphi^n(BC)$ are uniformly C^1-bounded with respect to $n \in Z^+$ (which means that those sets, regarded as graphs of functions of the variable x^2, have derivatives uniformly bounded with respect to n);

2° $Q \cap \varphi^n(AD)$ does not intersect $Q \cap \varphi^n(BC)$; the same applies to AB and CD;

3° φ is contracting on segments parallel to AB and expanding on segments parallel to AD.

Then the entire reasoning retains validity; the set $\Lambda := \bigcap_{n \in Z} \varphi^n(Q)$ is homeomorphic to X and Theorem 4.13.1 holds true for the system (Λ, φ).

(2) If the set $Q \cap \varphi(Q)$ consists of more than 2 components (say, of m components), then repeating the procedure with obvious modifications we find a φ-invariant set $\Lambda \subset Q$ defined as the product of two Cantor-like sets and such that the system (Λ, φ) is topologically conjugate to the system (X, s), where $X = \prod_{n=-\infty}^{+\infty} \{1, \ldots, m\}_n$ and s is the shift to the left.

(3) In the above, the set $\varphi(Q)$ may be variously shaped and variously situated relative to Q; see Fig. 4.13.5.

The observations contained in Remark 4.13.1 (1) are given a precise formulation in the following statement (see Nitecki, 1971, p. 149):

Theorem 4.13.2 The horseshoe system is C^1-Ω-stable; here $\Omega = \Omega(\varphi) \cap Q$. In other words, if ψ is a diffeomorphism C^1-close to φ within Q, then ψ restricted to the set of its non-wandering points contained in Q is topologically conjugate to the shift in the space of $\{0, 1\}$-sequences.

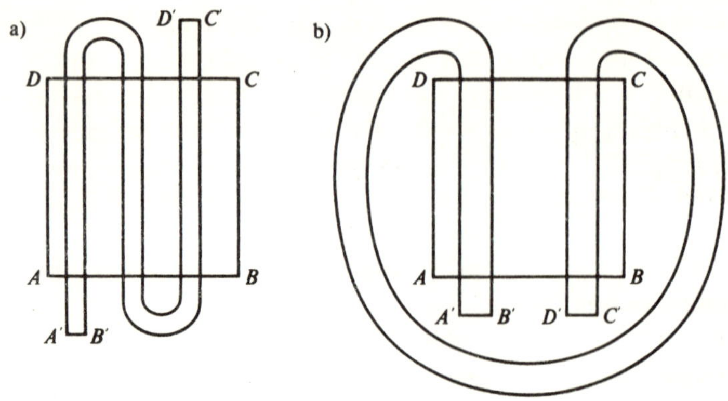

FIG. 4.13.5

The horseshoe phenomenon occurs in a natural way in many dynamical systems. Let us examine the problem more carefully.

Let M be a compact manifold and let $\varphi \in \mathrm{Diff}^r(M)$.

Definition 4.13.1 A point $x \in M$ is called a *homoclinic point* of φ iff $x \in W^+(p) \cap W^-(p)$ for some hyperbolic periodic point p different from x. If $W^+(p) \pitchfork_x W^-(p)$, then x is called a *transversal homoclinic point*.

We have already met such points when dealing with the Morse–Smale diffeomorphisms. The first to observe those points was Poincaré (in his works on celestial mechanics, 1892-99). He also noticed that the existence of points of that type greatly complicates the behaviour of orbits. Poincaré did not examine the phenomenon. It was studied later on (in the sixties) by Smale and others. Look at Fig. 4.13.6 to get an idea of the type of difficulties that arise as a consequence of the existence of a homoclinic point.

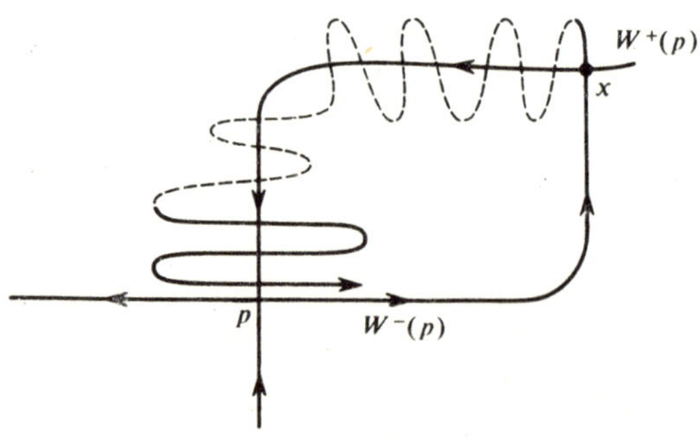

FIG. 4.13.6

If a system has a transversal homoclinic point, the horseshoe phenomenon necessarily occurs. Let us try to find its cause.

We confine our attention to the case of $\dim M = 2$. If $\dim M > 2$, the principal idea remains the same, but the technical details become very cumbersome.

Assume that a system (M, φ), $\varphi \in \mathrm{Diff}^1(M)$, $\dim M = 2$, has a transversal homoclinic point, i.e., there is a periodic point $p \in M$ such that $W^+(p)$ intersects $W^-(p)$ transversally at some point $x \neq p$. Without loss of generality we may assume that $\varphi(p) = p$ and (passing to a chart) that $p = 0$, the manifolds $W^-(p)$ and $W^+(p)$ coinciding locally with the coordinate axes $0x^1$ and $0x^2$, respectively. In virtue of the λ-Lemma, for any $\varepsilon > 0$ there is a set Q of the form

$$Q = \{(x^1, x^2)\colon x^1 \in \bar{B},\ x^1 \geqslant 0,\ 0 \leqslant x^2 \leqslant f(x^1)\},$$

where B denotes some cell neighbourhood of 0 in $W^-(0)$ (an interval in the present case) and f is a C^1-function, ε-C^1-close to 0, such that the set $D = \{(x^1, f(x^1))\colon x^1 \in B\}$ is contained in $W^-(0)$ and is a cell neighbourhood of the point $x_0 = (0, f(0))$ in $W^-(0)$; the cells B and D are ε-C^1-close to each other and x_0 is a homoclinic point (Fig. 4.13.7).

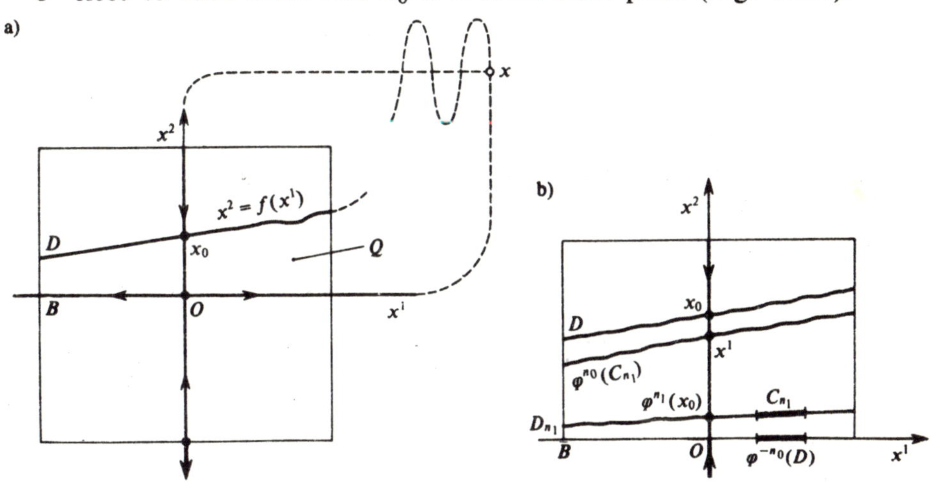

FIG. 4.13.7

Now, let n_0 be an integer such that $\varphi^{-n_0}(D) \subset B$. Take a rectangular neighbourhood U of 0, $U = B \times \langle -\eta, \eta \rangle$, where η is some fixed number. Define D_n to be the component of $\varphi^n(D) \cap U$ which contains $\varphi^n(x_0)$. On account of the λ-Lemma, the sequence of manifolds D_n approaches B in C^1-topology and each D_n contains a submanifold C_n which is C^1-close to $\varphi^{-n_0}(D)$. Therefore we can find an n_1 such that $\varphi^{n_0}(C_{n_1})$ is ε-C^1-close to D. Denote by x_1 the point of intersection of $\varphi^{n_0}(C_{n_1})$ with the $0x^2$-axis (Fig. 4.13.7).

Take an integer n_2 large enough for $\varphi^{n_2}(x_1)$ to fall into Q. Then $\varphi^{n_2}(\varphi^{n_0}(C_{n_1}))$ has points in common with Q. Write $k = n_1 + n_0 + n_2$ and consider the set Q_1 limited by the curves: the component of $\varphi^{n_2}(D) \cap Q$ containing $\varphi^{n_2}(x_0)$, the component of $\varphi^{n_2}(\varphi^{n_0}(C_{n_1})) \cap Q$ containing $\varphi^{n_2}(x_1)$, the segment of the $0x^2$-axis and the segment of the vertical line passing through the endpoint of B. Along with Q_1, consider the set Q_0 limited by: the interval B, the component of $\varphi^k(D) \cap Q$ containing $\varphi^k(x_0)$, the segment of the $0x^2$-axis and the segment of the vertical line passing through the endpoint of B (Fig. 4.13.8).

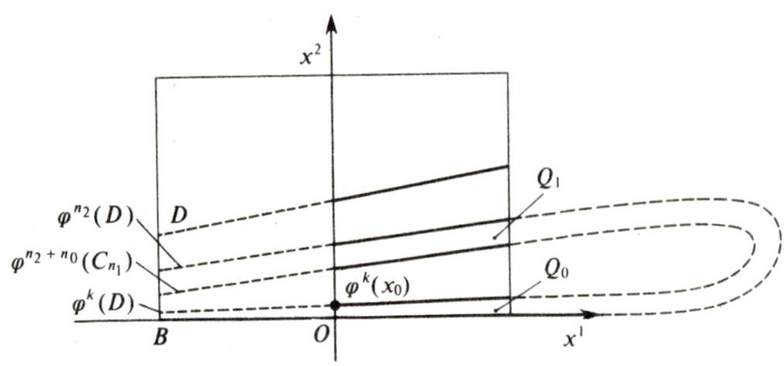

FIG. 4.13.8

It follows from the construction that for $\psi = \varphi^k$ the set $\psi(Q) \cap Q$ has at least two components (Q_0 and Q_1) and that the conditions 1°–3° of Remark 4.13.1 (1) are satisfied. Therefore the horseshoe phenomenon does occur. Theorem 4.13.1 implies

Corollary 4.13.1 *In the vicinity of p one can find a nowhere dense set Λ which is φ^k-invariant and such that the system (Λ, φ^k) is topologically conjugate to the shift on the space of $\{0, 1\}$-sequences.*

Corollary 4.13.2 *Every neighbourhood of p contains infinitely many periodic points of φ.*

Proof By Corollary 4.13.1 the systems (Λ, φ^k) and (X, s) are topologically conjugate, X and s being defined by (4.13.1) and (4.13.2). Evidently, the system (X, s) has infinitely many periodic points. Hence the assertion.

As we have mentioned, these facts are also true for manifolds of arbitrary (finite) dimension.

5

Ergodic Theory of Topological Dynamical Systems. Entropy

5.1. Measure Preserving Transformations

The object of study in ergodic theory are transformations which preserve measure. Before we pass to the investigation of invariant measures in the case of topological dynamical systems (and, in particular, smooth systems), we introduce and briefly discuss certain notions and facts of general ergodic theory. A detailed development of those topics can be found e.g. in Sinai (1970), Sinai (1976), Totoki (1970), and Halmos (1956).

Let $(X, \mathfrak{M}, \mu)$ be a normalized measure space; X is the space, $\mathfrak{M}$ is a σ-field of subsets of X and μ is a measure on $\mathfrak{M}$, $\mu(X) = 1$. All members of $\mathfrak{M}$ are considered modulo nullsets, that is to say, we regard two sets $A, B \in \mathfrak{M}$ as equal iff $\mu(A \dot{-} B) = 0$; the symbol $\dot{-}$ denotes symmetric difference: $A \dot{-} B = (A-B) \cup (B-A)$.

Let us consider a continuous semigroup of mappings

$$\varphi^\alpha: X \to X, \qquad \alpha \in \mathfrak{A},$$

and suppose that each φ^α preserves measure (is a measure preserving transformation, abbreviated as m.p.t.) i.e., we assume that $\mathfrak{A}$ is a metric Abelian semigroup with zero,

$$\varphi^\alpha \cdot \varphi^\beta = \varphi^{\alpha+\beta} \qquad \text{for } \alpha, \beta \in \mathfrak{A},$$

the φ^α satisfy the condition

$$\mu(\varphi^{-\alpha}(A)) = \mu(A) \qquad \text{for } \alpha \in \mathfrak{A},\ A \in \mathfrak{M},$$

where $\varphi^{-\alpha}$ stands for $(\varphi^{\alpha})^{-1}$, and we have

$$\lim_{n\to\infty} \mu(\varphi^{-\alpha_n}(A) \dot{-} A) = 0$$

for any sequence $\alpha_n \to 0$.

We will confine our attention to two types of semigroups: $\mathfrak{A} = Z^+$ or Z and $\mathfrak{A} = \boldsymbol{R}^+$ or $\boldsymbol{R}$. More general situations are also studied in the literature, e.g. $\mathfrak{A} = \boldsymbol{R}^m$.

Definition 5.1.1 Any system $(X, \mathfrak{M}, \mu, \varphi^{\alpha})$ satisfying the above conditions is called a *metric dynamical system.*

In the case of $\mathfrak{A} = Z$ or Z^+ we write simply: $(X, \mathfrak{M}, \mu, \varphi)$.

Examples

5.1.1. $X = S^1$, $\mathfrak{M} = \mathfrak{L}(S^1)$, the family of Lebesgue measurable sets (note that $\mathfrak{L}(S^1)$, considered modulo nullsets, can be identified with the σ-field of Borel sets), $\mu(\cdot) = \frac{1}{2\pi}|\cdot|$, $\varphi(z) = z_0 z$, $z_0 = e^{2\pi i\alpha}$; φ is the rotation by the angle $2\pi\alpha$. Obviously, φ preserves the measure μ.

5.1.2. Let $\mathscr{A} = \{a_1, \dots, a_m\}$ be some fixed set, which in the sequel will be called an *alphabet*. Let

$$X = \prod_{n=-\infty}^{+\infty} \mathscr{A}_n,$$

where $\mathscr{A}_n = \mathscr{A}$ for every $n \in Z$. Define $\mathfrak{M}$ as the least σ-field of subsets of X containing all cylindric sets, i.e., sets of the form

$$C = \{x = (x_n) \colon x_{i_s} = a_{i_s}, s = 1, \dots, N\},$$

where (i_s) is any increasing finite sequence of integers and (a_{i_s}) is any sequence (of the same length) of elements of $\mathscr{A}$. Let μ_0 be a normalized measure on $\mathscr{A}$:

$$\mu_0\{a_i\} = p_i, \qquad i = 1, \dots, m,$$
$$p_i > 0, \qquad \sum p_i = 1.$$

Define μ to be the product measure[38]

$$\mu = \prod_{n=-\infty}^{+\infty} \mu_n,$$

where $\mu_n = \mu_0$ for all n. The measure of a cylindric set C is given by

$$\mu(C) = \prod_{s=1}^{N} p_{i_s}.$$

Clearly, $\mu(X) = 1$. Consider the transformation $\varphi\colon X \to X$ defined as the shift by one notch to the left:

$$\varphi((x_n)) = (y_n), \qquad \text{where } y_n = x_{n+1},\ n \in Z.$$

Then φ preserves the measure of all cylindric sets and hence preserves μ on the whole of $\mathfrak{M}$. The dynamical system just defined is known as the *Bernoulli shift*.

5.1.3. Let $X = T^m$, $\mathfrak{M} = \mathfrak{L}(T^m)$, the family of Lebesgue measurable sets, $\mu(\cdot) = |\cdot|$, the Lebesgue measure on T^m, and let φ be an algebraic automorphism of T^m. Then $\det\varphi = \pm 1$, and so φ preserves the Lebesgue measure.

Exercises

5.1.1.B. Let Q be the square with vertices (0, 0), (1, 0), (0, 1), (1, 1) and define $\varphi\colon Q \to Q$ by:

$$\varphi(x, y) = \begin{cases} (2x, \tfrac{1}{2}y) & \text{if } 0 \leqslant x < \tfrac{1}{2}, \\ (2x-1, \tfrac{1}{2}y+\tfrac{1}{2}) & \text{if } \tfrac{1}{2} \leqslant x \leqslant 1. \end{cases}$$

Show that φ preserves the Lebesgue measure. (This is the so-called *baker's transformation*; it resembles dough rolling).

5.1.2.A. Let X, $\mathfrak{M}$ and φ be as in Example 5.1.2; let $P = (p_{ij})$ be a stochastic matrix (i.e. $p_{ij} \geqslant 0$, $\sum_{j=1}^{m} p_{ij} = 1$, $i = 1, \ldots, m$) and let $p_1, \ldots, p_m$ be a stationary Markov chain distribution with m states, P being the transition matrix. Define μ by the following condition:

$$\mu(C) = p_{a_{i_1}} p^{(i_2-i_1)}_{a_{i_1}, a_{i_2}} p^{(i_3-i_2)}_{a_{i_2}, a_{i_3}} \cdots p^{(i_N-i_{N-1})}_{a_{i_{N-1}}, a_{i_N}}$$

if C denotes the cylindric set determined by the N-tuples $(i_1, \ldots, i_N)$ and $(a_{i_1}, \ldots, a_{i_N})$; in the above formula the symbol $p^{(k)}_{a_i, a_j}$ denotes the probability of passing in k steps from state a_i to state a_j, and $p_{a_{i_1}}$ is the probability of being in state a_{i_1} at the moment 0. Show that φ preserves the measure μ. (This is the so-called *Markov shift*.)

Hint. μ is invariant on cylindric sets.

5.1.3.B. Imagine a billiard table B limited by a piecewise smooth curve K. A billiard ball moves on B without stops rebounding from the boundary in accordance with the usual rule: the angle of incidence equals the angle of reflexion; these angles are to be understood as the corresponding angles between the trajectory of the ball and the straight line normal to K at the reflection point; we assume that the ball never arrives at a point of non-differentiability of K (there are only finitely many such points). Consider the phase space $B \times S^1$ whose elements are pairs (x, ϑ), x denoting the position of the ball and ϑ the direction of the speed vector. Prove that the flow on $B \times S^1$ describing the motion of the ball preserves the Lebesgue measure.

Hint. Examine sets of the form $A = B_1 \times V$ where $B_1 \subset B$ and V is an open arc on S^1. Sets of this type generate the σ-field of measurable sets. Fix t and assume that for every $v \in V$ the set B_1 is reflected from just one "curvilinear side" of B (a smooth piece of K). Then apply the Fubini theorem.

Remark 5.1.1 The dynamical system described in Exercise 5.1.3 naturally extends to the one in which more than one ball run on the table. Also in such an extended system the flow defining motion preserves the Lebesgue measure in the phase space.

A physical model for such a billiard with several balls is supplied by the behaviour of an ideal gas in a closed vessel.

Metric dynamical systems have a property called the *recurrence property*; the fact which we now prove is known as the *recurrence theorem*.

Theorem 5.1.1 (Poincaré) Let $(X, \mathfrak{M}, \mu, \varphi)$ be a metric dynamical system and let $A \in \mathfrak{M}$, $\mu(A) > 0$. Then the set

$$B = \{x \in A \colon \varphi^n(x) \notin A,\ n = 1, 2, \dots\}$$

is of measure zero.

Proof We first show that $\varphi^{-n}(B) \cap B = \emptyset$ for $n = 1, 2, \dots$ Indeed, if we had an x in $\varphi^{-n}(B) \cap B$, then $x \in B$ and $\varphi^n(x) \in B \subset A$, contrary to the definition of B. Thus for any m, n with $m < n$ we have

$$\varphi^{-n}(B) \cap \varphi^{-m}(B) = \varphi^{-m}(\varphi^{-(n-m)}(B) \cap B) = \emptyset.$$

Since φ is measure preserving, $\mu(\varphi^{-n}(B)) = \mu(B)$. Hence

$$1 = \mu(X) \geqslant \mu\Big(\bigcup_{n=1}^{\infty} \varphi^{-n}(B)\Big) = \sum_{n=1}^{\infty} \mu(\varphi^{-n}(B)) = \sum_{n=1}^{\infty} \mu(B)$$

and consequently $\mu(B) = 0$, as asserted.

The recurrence theorem has given rise to a certain paradox in the kinetic theory of gas. The motion of gas in a closed vessel preserves the Lebesgue measure (cf. Remark 5.1.1). By the recurrence theorem, the system will, after some time, return to the initial state as close as we please. Suppose that the vessel is divided by a partition into two equal parts, one of them filled with chlorine and the other with helium; then the partition is removed. It must follow that, after some time, nearly all the chlorine will gather where it was at the outset and nearly all the helium will find itself in the remaining part. Physically, this is absurd. The paradox is explained as follows: the time it would take that astonishing phenomenon to occur is incredibly long, it exceeds thousands of times the life of the solar system. Our theory fails to work when applied to such enormous time periods. (Similarly, we cannot apply the laws of classical mechanics to the solar system for very long time periods and very high velocities if we want to arrive at conclusions consistent with experience.)

Exercise

5.1.4.B. Let $(X, \mathfrak{M}, \mu, \varphi)$ be a metric dynamical system, let $A \in \mathfrak{M}$, $\mu(A) > 0$, and let B be the set defined in Theorem 5.1.1. For every $x \in A - B$ define $n(x)$ to be the earliest moment of recurrence to A, i.e. the least integer $k > 0$ such that $\varphi^k(x) \in A$. Suppose that the system is ergodic (see Definition 5.2.1). Prove that then $\int_A n(x)\,\mathrm{d}\mu = 1$ (Kac' theorem).

Hint. Write $A_k = \{x \in A \colon n(x) = k\}$, $k = 1, 2, \ldots$ Observe that $\varphi^i(A_k) \cap \varphi^j(A_k) = \emptyset$ for $i < j < k$ and show that

$$\bigcup_{k=1}^{\infty} \bigcup_{i=0}^{k-1} \varphi^i(A_k) = X \text{ (modulo a nullset).}$$

5.2. Birkhoff's Theorem. Ergodicity

We devote this section to a theorem known as Birkhoff's Individual Ergodic Theorem. This is one of the most vital facts of mathematical analysis. The original proof is long and involved (see Halmos, 1956). The proof which we present below is a simplified version of Birkhoff's proof; we owe it to A. M. Garsia (1965); it can also be found in Totoki (1970). Though the theorem is not so very closely connected with the theory of smooth dynamical systems, we present it in full detail because of its great significance for the general theory.

Theorem 5.2.1 (Birkhoff's Individual Ergodic Theorem) Let $(X, \mathfrak{M}, \mu, \varphi)$ *be a metric dynamical system. Then for every function* $f \in \mathscr{L}^1(X, \mathfrak{M}, \mu)$ *the following limit exists almost everywhere:*

$$\lim_{n\to\infty} \frac{1}{n} \sum_{k=0}^{n-1} f\circ\varphi^k(x) = f^*(x); \tag{5.2.1}$$

moreover,

$$f^* \in \mathscr{L}^1(X, \mathfrak{M}, \mu), \qquad \int f^* \mathrm{d}\mu = \int f \mathrm{d}\mu, \qquad f^*\circ\varphi = f^*. \tag{5.2.2}$$

We first prepare a lemma.

Lemma 5.2.1 Set

$$S_n(x) = \sum_{k=0}^{n-1} f\circ\varphi^k(x), \qquad n \geqslant 1, \qquad S_0(x) = 0,$$

$$B = \{x\colon \sup_{n\geqslant 1} S_n(x) > 0\}.$$

Then

$$\int_{A\cap B} f \mathrm{d}\mu \geqslant 0$$

for every set $A \in \mathfrak{M}$ *which is* φ*-invariant.*

Proof Write

$$S_n^+(x) = \max_{0\leqslant k\leqslant n} S_k(x).$$

Then $S_n^+(x) \geqslant 0$ (since $S_0(x) = 0$) and

$$f(x) + S_n^+(\varphi(x)) \geqslant f(x) + S_k(\varphi(x)) = S_{k+1}(x) \qquad \text{for } k = 0, \ldots, n.$$

Thus for any $x \in B_n := \{x\colon S_n^+(x) > 0\}$ we have

$$f(x) + S_n^+(\varphi(x)) \geqslant \max_{1\leqslant k\leqslant n} S_k(x) = S_n^+(x).$$

Hence

$$\int_{A\cap B_n} f\,\mathrm{d}\mu \geqslant \int_{A\cap B_n} \left(S_n^+(x)-S_n^+(\varphi(x)\right)\mathrm{d}\mu(x)$$

$$= \int_{A\cap B_n} S_n^+(x)\,\mathrm{d}\mu(x) - \int_{A\cap B_n} S_n^+(\varphi(x))\,\mathrm{d}\mu(x)$$

$$\geqslant \int_A S_n^+(x)\,\mathrm{d}\mu(x) - \int_A S_n^+(\varphi(x))\,\mathrm{d}\mu(x) = 0,$$

since φ is measure-preserving and A is φ-invariant. Clearly, $B_n \subset B_{n+1}$ and $\bigcup B_n = B$, and so the last inequality immediately yields the assertion of the lemma.

Proof of Theorem 5.2.1 Write

$$B_\alpha = \left\{x\colon \sup_{n\geqslant 1} \frac{1}{n}\sum_{k=0}^{n-1} f\circ\varphi^k(x) > \alpha\right\}.$$

Applying Lemma 5.2.1 to the function $f(x)-\alpha$ we obtain

$$\int_{A\cap B_\alpha} f\,\mathrm{d}\mu \geqslant \alpha\mu(A\cap B_\alpha)$$

for every set $A\in\mathfrak{M}$ which is φ-invariant.

Further, set

$$\bar{f}(x) = \limsup_{n\to\infty} \frac{1}{n}\sum_{k=0}^{n-1} f\circ\varphi^k(x), \quad \underline{f}(x) = \liminf_{n\to\infty} \frac{1}{n}\sum_{k=0}^{n-1} f\circ\varphi^k(x),$$

$$E_{\alpha,\beta} = \left\{x\colon \underline{f}(x) < \beta, \bar{f}(x) > \alpha\right\}.$$

The set $E_{\alpha,\beta}$ is measurable, φ-invariant and contained in B_α. Hence, by Lemma 5.2.1,

$$\int_{E_{\alpha,\beta}} f\,\mathrm{d}\mu \geqslant \alpha\mu(E_{\alpha,\beta}).$$

Repeating the argument with f, α, β replaced by $-f, -\alpha, -\beta$, we obtain

$$\int_{E_{\alpha,\beta}} f\,\mathrm{d}\mu \leqslant \beta\mu(E_{\alpha,\beta}).$$

Hence $\mu(E_{\alpha,\beta}) = 0$ for $\beta < \alpha$. Consequently

$$\mu\{x: \underline{f}(x) < \bar{f}(x)\} = \mu\Big(\bigcup_{\substack{\beta<\alpha \\ \alpha,\beta \text{ rational}}} E_{\alpha,\beta}\Big) \leqslant \sum_{\substack{\beta<\alpha \\ \alpha,\beta \text{ rational}}} \mu(E_{\alpha,\beta}) = 0.$$

This proves the a.e. existence of the limit in (5.2.1).

We now pass to the proof of (5.2.2). Let $f = f^+ - f^-$ be the standard decomposition of f into the difference of non-negative functions. Limits analogous to (5.2.1) exist likewise for f^+ and f^-:

$$\frac{1}{n}\sum_{k=0}^{n-1} f^{\pm} \circ \varphi^k(x) \underset{\text{a.e.}}{\rightarrow} f^{*\pm}(x).$$

Introducing the truncated functions

$$f_m^{\pm}(x) = \begin{cases} f^{\pm}(x) & \text{if } f^{\pm}(x) \leqslant m, \\ 0 & \text{if } f^{\pm}(x) > m \end{cases}$$

(with m natural), we have also

$$\frac{1}{n}\sum_{k=0}^{n-1} f_m^{\pm} \circ \varphi^k(x) \underset{\text{a.e.}}{\rightarrow} f_m^{*\pm}(x)$$

and

$$\left|\frac{1}{n}\sum_{k=0}^{n-1} f_m^{\pm} \circ \varphi^k(x)\right| \leqslant m \qquad \text{for all } x.$$

By the Lebesgue Dominated Convergence Theorem ($\mu(X) = 1$) we obtain hence

$$\int f_m^{*\pm}\,\mathrm{d}\mu = \lim_{n\to\infty} \int \frac{1}{n}\sum_{k=0}^{n-1} f_m^{\pm} \circ \varphi^k\,\mathrm{d}\mu = \int f_m^{\pm}\mathrm{d}\mu,$$

because φ is measure preserving. The sequences $(f_m^{*\pm})$, $(f_m^{\pm})$ are monotonically convergent to $f^{*\pm}, f^{\pm}$, respectively. Thus

$$\int f^{*\pm}\mathrm{d}\mu = \int f^{\pm}\mathrm{d}\mu,$$

showing that

$$f^* \in \mathscr{L}^1(X, \mathfrak{M}, \mu) \qquad \text{and} \qquad \int f^*\mathrm{d}\mu = \int f\mathrm{d}\mu.$$

The equality

$$f^* \circ \varphi(x) = f^*(x)$$

is obvious. The proof is complete.

The Ergodic Theorem admits also the following version:

Theorem 5.2.2 (Statistical Ergodic Theorem) *Let $(X, \mathfrak{M}, \mu, \varphi)$ be a metric dynamical system. Then for every function $f \in \mathscr{L}^p(X, \mathfrak{M}, \mu)$, $p \geq 1$, the sequence $n^{-1} \sum_{k=0}^{n-1} f \circ \varphi^k$ converges to f^* in the $\mathscr{L}^p$-norm.*

A proof can be found in Dunford and Schwartz (1958), Ch. VIII, Section 5, Th. 9.

Analogous theorems hold also for flows:

Theorem 5.2.3 *Let $(X, \mathfrak{M}, \mu, \varphi_t)$ be a metric flow. Then for every function $f \in \mathscr{L}^1(X, \mathfrak{M}, \mu)$ the following limit exists almost everywhere:*

$$\lim_{t\to\infty} \frac{1}{t} \int_0^t f \circ \varphi_s(x)\, \mathrm{d}s = f^*(x);$$

moreover,

$$f^* \in \mathscr{L}^1(X, \mathfrak{M}, \mu), \qquad \int f^* \mathrm{d}\mu = \int f \mathrm{d}\mu, \qquad f^* \circ \varphi_t = f^*.$$

If $f \in \mathscr{L}^p(X, \mathfrak{M}, \mu)$, the convergence holds in $\mathscr{L}^p$-norm.

In the class of metric dynamical systems we distinguish the so-called ergodic systems.

Definition 5.2.1 A metric dynamical system $(X, \mathfrak{M}, \mu, \varphi^\alpha)$ is called *ergodic* iff every φ^α-invariant set is of measure either 0 or 1, i.e., iff the condition

$$\varphi^{-\alpha}(A) \subset A \qquad \text{for all } \alpha \in A$$

implies

$$\mu(A) = 0 \qquad \text{or} \qquad \mu(A) = 1.$$

Thus ergodicity means just as much as saying that the system in question does not contain properly any non-trivial subsystem.

Let $(X, \mathfrak{M}, \mu, \varphi)$ be a metric dynamical system and let $U\colon \mathscr{L}^p(X, \mathfrak{M}, \mu) \to \mathscr{L}^p(X, \mathfrak{M}, \mu)$, $p \geq 1$, be the operator defined by

$$Uf = f \circ \varphi, \qquad f \in \mathscr{L}^p(X, \mathfrak{M}, \mu).$$

We leave the proof of the following statement to the reader as an easy exercise:

Exercise

5.2.1.A. A system $(X, \mathfrak{M}, \mu, \varphi)$ is ergodic iff there are no measurable functions f satisfying the equation $Uf = f$ except $f = \text{const}$ a.e.

Hint. If $Uf = f$ and f is not equal a.e. to a constant, then there exists an α such that the set $\{x: f(x) < \alpha\}$ is invariant and is neither of measure 0 nor of measure 1.

Examples

5.2.1. Let $X = S^1$, $\mathfrak{M} = \mathfrak{L} = \mathfrak{L}(S^1)$, $\mu(\cdot) = \frac{1}{2\pi}|\cdot|$ (the normalized Lebesgue measure), $\varphi(z) = z_0 z$, where $z_0 = e^{2\pi i \alpha}$, α being irrational. The system $(S^1, \mathfrak{L}, \mu, \varphi)$ is ergodic. To show this, we apply Exercise 5.2.1. Suppose that f fulfils the equation $Uf = f$, i.e. $f(z_0 z) = f(z)$. Then also

$$f_m(z) = \begin{cases} f(z) & \text{if } |f(z)| \leqslant m, \\ 0 & \text{if } |f(z)| > m \end{cases}$$

fulfils the equation $Uf_m = f_m$. Therefore we may freely assume that f is, say, square-summable: $f \in \mathscr{L}^2(S^1, \mathfrak{L}, \mu)$. Then $f(z) = \sum_{n=-\infty}^{+\infty} a_n z^n$. Inserting this into the equation $Uf = f$ we get

$$\sum_{n=-\infty}^{+\infty} a_n z_0^n z^n = \sum_{n=-\infty}^{+\infty} a_n z^n$$

and, by uniqueness,

$$a_n z_0^n = a_n \qquad \text{for all } n.$$

Since α is irrational, $z_0^n \neq 1$ for each $n \neq 0$. Hence $a_n = 0$ for $n \neq 0$, and so $f(z) = a_0 = \text{const}$.

5.2.2. Let $X = T^m$, $\mathfrak{M} = \mathfrak{L} = \mathfrak{L}(T^m)$, $\mu(\cdot) = |\cdot|$, $A = (a_{ij})$ an algebraic isomorphism of T^m. We shall show that the system $(T^m, \mathfrak{L}, |\cdot|, A)$ is ergodic if an only if the spectrum of A contains no roots of unity.

We again use Exercise 5.2.1 and the theory of Fourier series. Suppose that f satisfies the equation

$$Uf(x) = f(Ax) = f(x). \tag{5.2.3}$$

As before, we may confine our attention to $f \in \mathscr{L}^2(T^m, \mathfrak{L}, |\cdot|)$ and write f as the sum of the Fourier series

$$f(x) = \sum_{n \in Z^m} a_n e^{2\pi i (n,x)}, \tag{5.2.4}$$

where $n = (n_1, \ldots, n_m)$, $n_j \in Z$, $j = 1, \ldots, m$, $x = (x^1, \ldots, x^m) \in T^m$. Inserting (5.2.4) into (5.2.3), we get

$$\sum_{n \in Z^m} a_n e^{2\pi i (n,x)} = \sum_{n \in Z^m} a_n e^{2\pi i (n,Ax)} = \sum_{n \in Z^m} a_n e^{2\pi i (A^* n,x)}.$$

The Fourier expansion is unique, and thus

$$a_n = a_{A^* n} = \ldots = a_{A^{*k} n} = \ldots$$

Fix an $n \neq 0$ and consider the sequence of indices $(A^{*k} n)$. Two situations can occur:

1° The terms of this sequence take on finitely many values. Then $A^{*k} n = n$ for some k, which means that $\operatorname{sp} A$ contains a value λ such that $\lambda^k = 1$.

2° The terms of this sequence are different from each other. Then, in view of the equality

$$\|f\|_{\mathscr{L}^2} = \sum_{n \in Z^m} a_n^2,$$

we have $a_n = 0$.

Thus, if $\operatorname{sp} A$ contains no root of unity, then $a_n = 0$ for all $n \neq 0$ and so $f = \text{const}$. If there is a $\lambda \in \operatorname{sp} A$ such that $\lambda^k = 1$, then the matrix A^k with integer entries has an eigenvector w ($A^k w = w$) with rational components; multiplying by a common denominator we find an integer vector $n \in Z^m$ such that $A^k n = n$. The function

$$f(x) = \sum_{j=0}^{k-1} e^{2\pi i (A^{*j} n, x)}$$

satisfies equation (5.2.3) and is non-constant; the system is not ergodic.

The next theorem shows how the concept of ergodicity of metric dynamical systems is related to the ergodic theorem.

Theorem 5.2.4 *A system* $(X, \mathfrak{M}, \mu, \varphi)$ *is ergodic if and only if for every* $f \in \mathscr{L}^1(X, \mathfrak{M}, \mu)$ *the function* f^* *given by* (5.2.1) *is constant a.e.:* $f^*(x) = \int f \, d\mu$.

The proof is a direct consequence of Theorem 5.2.1 and Exercise 5.2.1.

The condition expressed in Theorem 5.2.4 has a definite physical significance. It means that the time-average of $f \circ \varphi^k$ approaches a limit value equal to the mean value of f over the space X. In the kinetic theory of gas this property seems evident from the viewpoint of physics. On the basis of this observation, it was already Boltzman who posed the hypothesis that the motion of gas particles in a closed vessel is ergodic. This fact was first proved by Sinai (1963).

Exercises

5.2.2.B. Show that the Bernoulli shift (see Example 5.1.2, p. 274) is ergodic.

Hint. Apply the relation

$$\lim_{n \to \infty} \mu(\varphi^{-n}(A) \cap B) = \mu(A)\mu(B),$$

holding for any sets $A, B \in \mathfrak{M}$. (First show that it holds for cylindric sets; then use the fact that such sets generate the entire σ-field.)

5.2.3.B. Prove that each of the following two conditions is equivalent to the ergodicity of a given metric dynamical system $(X, \mathfrak{M}, \mu, \varphi)$:

(a) for any $f, g \in \mathscr{L}^2(X, \mathfrak{M}, \mu)$

$$\lim_{n \to \infty} \frac{1}{n} \sum_{k=0}^{n-1} (U^k f, g) = (f, 1)(1, g),$$

1 denoting the constant function equal identically to 1;

(b) for any sets $A, B \in \mathfrak{M}$

$$\lim_{n \to \infty} \frac{1}{n} \sum_{k=0}^{n-1} \mu(\varphi^{-k}(A) \cap B) = \mu(A)\mu(B).$$

Hint. Prove the condition in (a) for simple functions and then apply the fact that simple functions are dense in $\mathscr{L}^2$; appeal to Birkhoff's theorem.

The concept of topological conjugacy has an analogue in the theory of metric dynamical systems. We only give the definition.

Definition 5.2.2 Two metric dynamical systems $(X, \mathfrak{M}, \mu, \varphi)$ and $(Y, \mathfrak{N}, \nu, \psi)$ are called *isomorphic* iff:

(1) there exist sets $X_0 \in \mathfrak{M}$, $Y_0 \in \mathfrak{N}$ such that

$$\varphi^{-1}(X_0) \subset X_0, \qquad \psi^{-1}(Y_0) \subset Y_0, \qquad \mu(X - X_0) = \nu(Y - Y_0) = 0;$$

(2) there exists a measurable bijection h: $X_0 \to Y_0$ sending $\mathfrak{M}$-measurable sets into $\mathfrak{N}$-measurable sets (and vice versa, with regard to h^{-1}) and such that

$$h \circ \varphi = \psi \circ h;$$

(3) for every $A \in \mathfrak{M}$ we have

$$\mu(A) = \nu(h(A)).$$

Exercises

5.2.4*.B. Prove that any two algebraic automorphisms of the torus T^2 with equal spectra are isomorphic (Adler and Weiss, 1967).

Hint. Apply the results of Section 1.6.

5.2.5.B. Show that the Baker's Transformation (see Exercise 5.1.1) is isomorphic to the Bernoulli shift (Example 5.1.2 with $A = \{0, 1\}$, $p(0) = p(1) = \frac{1}{2}$; see Halmos, 1956).

Hint. Let $(x, y) \in Q$. Combining the binary developments of x and y, produce a single $\{0, 1\}$-valued sequence ξ indexed by n ranging from $-\infty$ to $+\infty$. The mapping h: $(x, y) \mapsto \xi$ carries rectangles into cylinders, preserving measure.

5.2.6.B. Let φ be an algebraic automorphism of the torus T^2. Show that the system $(T^2, \mathfrak{L}, |\cdot|, \varphi)$ is ergodic iff the spectrum of φ has no points in common with the circle $|\lambda| = 1$; if the system is ergodic, then $\operatorname{sp}\varphi \subset \boldsymbol{R}$.

Hint. The characteristic equation for φ is quadratic with coefficients: 1, an integer, 1. Show that if $\lambda \in \operatorname{sp}\varphi$ and $|\lambda| = 1$, then λ is a root of unity; then use Exercise 5.2.1.

5.2.7.A. Prove that if the dynamical systems $(X, \mathfrak{M}, \mu, \varphi)$ and $(X, \mathfrak{M}, \nu, \varphi)$ are ergodic and $\mu \neq \nu$, then the measures μ and ν are orthogonal [37].

Hint. Assume the contrary. Then $\nu = \nu_0 + \nu_1$, $\nu_0 \ll \mu$, $\nu_1 \perp \mu$. Take a set A such that $\nu_0(A) > 0$ and $\nu_1(A) = 0$ and show (applying Theorem 5.2.4) that χ_A^* is not equal ν-a.e. to a constant function.

5.3. Invariant Measures for Topological Dynamical Systems

In Examples 5.1.1, 5.1.2, and 5.1.3, we encountered topological dynamical systems, each of them preserving a certain natural measure on the underlying space: the Lebesgue measure on S^1 and on T^m (in Examples 5.1.1 and 5.1.3) and the product measure on X (in Example 5.1.2). The question arises whether every topological dynamical system preserves some measure, say, a Borel measure; in the case of a smooth system the problem concerns the existence of an invariant measure which is absolutely continuous with respect to the Riemannian measure on the manifold considered. We will examine this question in the present section.

Let $\varphi: X \to X$ be a continuous transformation of a compact metric space X; φ induces an operator U_φ on the space $C(X)$:

$$(U_\varphi f)(x) = f \circ \varphi(x), \qquad f \in C(X).$$

U_φ maps the space $C(X)$ into itself; if φ is a homeomorphism, U_φ is an isometry.

We shall be concerned with Borel measures on X. Thus the σ-field of measurable sets will remain unaltered: the Borel field $\mathfrak{B}(X)$. Speaking of metric dynamical systems we shall just write (X, μ, φ) instead of $(X, \mathfrak{B}(X), \mu, \varphi)$.

The symbol W will denote the class of normalized (= probability) Borel measures on X.

Theorem 5.3.1 (*Bogolyubov–Krylov, 1936*) *Let X be compact metric and let φ: $X \to X$ be a continuous map. Then there exists a φ-invariant probability measure on X.*

Proof Let U_φ: $C(X) \to C(X)$ be the operator defined above and let U_φ^*: $C(X)^* \to C(X)^*$ be adjoint to U_φ. Then U_φ^* maps W into itself. Indeed, taking any $\mu \in W$ and writing $\nu = U_\varphi^*(\mu)$, we have, by the definition of U_φ^*,

$$\int U_\varphi f \mathrm{d}\mu = \int f \mathrm{d}\nu,$$

and by the definition of U_φ,

$$\int U_\varphi f \, d\mu = \int f \circ \varphi \, d\mu;$$

hence $\nu(A) = \mu(\varphi^{-1}(A))$ for every $A \in \mathfrak{B}(X)$ and thus $\nu \in W$. Further, the set W is compact in the weak* topology[31] and is convex: if $\mu_1, \mu_2 \in W$, then $\nu = t\mu_1 + (1-t)\mu_2 \in W$ for $t \in \langle 0, 1 \rangle$. Hence by the Schauder–Tikhonov Theorem (see Dunford and Schwartz, 1958, p. 456, Theorem 5) there exists $\mu \in W$ such that

$$U_\varphi^*(\mu) = \mu,$$

i.e.,

$$\mu(A) = \mu(\varphi^{-1}(A)) \qquad \text{for all } A \in \mathfrak{B}(X).$$

This ends the proof.

The Bogolyubov–Krylov theorem asserts the mere existence of an invariant measure; it gives no information regarding its properties. Such a measure may, e.g., be concentrated only on closed orbits or on closed nowhere dense sets and so on. Thus it is natural to ask: what type of invariant measures does a given dynamical system admit? Of course, one can hardly expect a precise answer to a question of this kind. However, partial answers can be given in certain specific cases.

In the proof of Theorem 5.3.1 we have found an invariant measure viewing it as an invariant functional for the operator U_φ^*. We are now going to obtain some more information on the correspondence between invariant measures and functionals.

Let F be a continuous linear functional on the real Banach space $C(X)$. According to the Riesz representation theorem (see note [29] or Dunford and Schwartz, 1958, Ch. IV, Section 6, Th. 3), F is written in the form

$$F(f) = \int f \, d\mu, \qquad f \in C(X),$$

where μ is a signed Borel measure on X. By the Hahn decomposition theorem, every signed measure μ is written as

$$\mu = \mu^+ - \mu^-,$$

where μ^+, μ^- are mutually singular non-negative Borel measures.

Theorem 5.3.2 If the functional F induced by a signed measure μ is U_φ^*-invariant, then the measures μ^+ and μ^- are φ-invariant.

Proof Let μ^+ and μ^- be concentrated on A_0 and B_0, respectively:

$$\mu^+(X-A_0)=\mu^-(X-B_0)=0, \qquad A_0\cap B_0=\emptyset.$$

By our assumption, $U_\varphi^*(F)=F$; hence

$$\int f\circ\varphi\,\mathrm{d}\mu=\int f\mathrm{d}\mu \qquad \text{for } f\in C(X).$$

Iterating U_φ^*, we get

$$\int f\circ\varphi^n\mathrm{d}\mu=\int f\mathrm{d}\mu \qquad \text{for } f\in C(X),\ n=0,1,\ldots$$

This equality extends to the space $\mathscr{L}^1(X,\mu)$, i.e., to all integrable functions. In particular, we may take $f=\chi_{A_0}$. We obtain

$$\begin{aligned}\mu^+(A_0)=\mu(A_0)&=\int\chi_{A_0}\mathrm{d}\mu=\int\chi_{A_0}\circ\varphi^n\mathrm{d}\mu=\int\chi_{\varphi^{-n}(A_0)}\mathrm{d}\mu\\&=\mu^+\big(\varphi^{-n}(A_0)\cap A_0\big)-\mu^-\big(\varphi^{-n}(A_0)\cap B_0\big).\end{aligned}$$

Since

$$\mu^+(A_0)=\sup_{A\in\mathfrak{B}(X)}\mu(A),$$

we conclude that

$$\mu^-\big(\varphi^{-n}(A_0)\cap B_0\big)=0 \qquad \text{for } n=0,1,2,\ldots$$

In just the same way we show that

$$\mu^+\big(\varphi^{-n}(B_0)\cap A_0\big)=0 \qquad \text{for } n=0,1,2,\ldots$$

Now write

$$A=\bigcup_{n=0}^{\infty}\varphi^{-n}(A_0), \qquad B=\bigcup_{n=0}^{\infty}\varphi^{-n}(B_0).$$

We easily show that

$$\mu^+(X-A)=\mu^-(X-B)=0.$$

Moreover, the sets A and B are φ-invariant,

$$\varphi^{-1}(A)\subset A, \qquad \varphi^{-1}(B)\subset B,$$

and we have

$$\mu|_A=\mu^+, \qquad \mu|_B=\mu^-.$$

Since μ is φ-invariant, so are μ^+ and μ^-. This completes the proof.

In our further considerations we shall often need Yoshida's Ergodic Theorem (see Dunford and Schwartz, 1958, p. 661, Th. 1):

Theorem 5.3.3 *Let $T: E \to E$ be a continuous linear operator on a Banach space E. Assume that the sequence*

$$A_n x := \frac{1}{n}\sum_{k=0}^{n-1} T^k x$$

is bounded for every $x \in E$; then the sequence $(A_n x)_{n=1}^{\infty}$ is convergent at a point x if and only if x belongs to the subspace

$$\left\{y \in E: \frac{1}{n}\|T^n y\| \to 0 \text{ as } n \to +\infty, \text{ the sequence } (A_n y) \text{ being relatively weakly compact in } E\right\}.$$

Exercises

5.3.1.A. Let μ, ν be probability measures on X and let φ: $X \to X$ be a continuous map. Suppose that there exists a constant $c > 0$ such that

$$\frac{1}{n}\sum_{k=0}^{n-1} \mu(\varphi^{-k}(A)) \leqslant c\mu(A) \qquad \text{for every } A \in \mathfrak{B}(X).$$

Show that then the limit

$$\mu_0(A) := \lim_{n\to\infty} \frac{1}{n}\sum_{k=0}^{n-1} \mu(\varphi^{-k}(A))$$

exists and defines a measure μ_0 which is φ-invariant.

Hint. Apply Theorems 5.2.4 and 5.3.3.

5.3.2.A. Let W_φ denote the set of all probability Borel measures on X invariant with respect to a continuous transformation φ: $X \to X$. Clearly, this set is convex and weak* compact. In virtue of the Krein–Milman Theorem[35] (Dunford and Schwartz, 1958) W_φ has extremal points. Show that a measure $\mu \in W_\varphi$ is ergodic if and only if μ is an extremal point of W_φ.

Hint. Apply Theorem 5.2.4 and Exercise 5.2.7. The "if" part is trivial. The "only if" part: observe that any two equivalent measures, one of which is ergodic, are necessarily equal.

5.3.3.A. Let μ be an invariant probability measure for a dynamical system (X, φ). Show that $\mu(X - \Omega(\varphi)) = 0$, $\Omega(\varphi)$ denoting the set of non-wandering points (cf. p. 181).

Hint. Supposing that there exists a set $A \subset X - \Omega(\varphi)$ with $\mu(A) > 0$, find a set $A_1 \subset A$ such that $\mu(A_1) > 0$ and $A_1 \cap \varphi^{-n}(A_1) = \emptyset$ for $n = 1, 2, \ldots$

5.4. Uniquely Ergodic Dynamical Systems

In general, a dynamical system (X, φ) admits more than one invariant probability measure. E.g., if the system has periodic points, we may associate such a measure with each periodic point x_0, namely, the measure concentrated on the orbit of x_0:

$$\mu(\{\varphi^i(x_0)\}) = \frac{1}{p},$$

where p is such that $\varphi^p(x_0) = x_0$, $\varphi^i(x_0) \neq x_0$, $i = 1, \ldots, p-1$.

The class of systems which admit exactly one invariant measure deserves special attention.

Definition 5.4.1 A dynamical system (X, φ) is said to be *uniquely ergodic* iff there exists exactly one probability Borel measure on X which is φ-invariant. Similarly, a flow (X, φ_t) is *uniquely ergodic* iff there is exactly one probability Borel measure on X which is φ_t-invariant, for all t.

For instance, every system (S^1, φ), $\varphi\colon S^1 \to S^1$, such that φ has exactly one periodic trajectory, is uniquely ergodic: any invariant measure is concentrated on the set of non-wandering points (see Exercise 5.3.3), i.e. (in this case) on the periodic trajectory. There is only one such measure.

Observe that in the definition of uniquely ergodic systems we might just as well postulate the existence of a unique *ergodic* invariant probability measure; this is a consequence of the fact that ergodic measures are precisely the extremal points of the set of all invariant probability measures (Exercise 5.3.2) and the Krein–Milman Theorem.

We shall now briefly discuss some properties of uniquely ergodic systems.

Definition 5.4.2 A point $x_0 \in X$ is said to be *generic* for a dynamical system (X, μ, φ) iff the limit condition

$$\lim_{n\to\infty} \frac{1}{n} \sum_{k=0}^{n-1} f \circ \varphi^k(x_0) = \int f \, d\mu$$

is satisfied for every function $f \in C(X)$. A point $x_0 \in X$ is *generic* for a flow (X, μ, φ_t) iff the limit condition

$$\lim_{T \to \infty} \frac{1}{T} \int_0^T f(\varphi_t(x_0))\, dt = \int f\, d\mu$$

is satisfied for every function $f \in C(X)$.

In the following we shall often write $\mu(f)$ in place of $\int f\, d\mu$.

Theorem 5.4.1 (*Oxtoby, 1952*; see also *Furstenberg, 1961*) *Given a dynamical system* (X, φ), *any of the following statements implies the others*:

(1) *the system* (X, φ) *is uniquely ergodic*;

(2) *for every function* $f \in C(X)$ *the sequence*

$$\left(\frac{1}{n} \sum_{k=0}^{n-1} f \circ \varphi^k\right)$$

converges uniformly on X *to a constant function*;

(3) *there is an invariant measure* μ *such that every point* $x \in X$ *is generic for the system* (X, μ, φ).

Proof (1) implies (2). Let μ be the unique φ-invariant probability measure on X. As before, consider the operator U_φ: $C(X) \to C(X)$ given by: $U_\varphi f = f \circ \varphi$. Write

$$E_0 = \{g \in C(X) \colon g = U_\varphi f - f,\ f \in C(X)\}.$$

E_0 is a linear subspace of $C(X)$ and we have $\bar{E}_0 \neq C(X)$, since the constant function $f(x) = 1$, for instance, does not belong to $\bar{E}_0$; indeed, the μ-mean value of any $g \in \bar{E}_0$ is zero.

We now show that $\operatorname{codim} \bar{E}_0 = 1$. Assume the contrary: $\operatorname{codim} \bar{E}_0 > 1$; then there are $g_1, g_2 \in C(X)$ such that

$$\bar{E}_0 \oplus \operatorname{span}\{g_1\} \neq \bar{E}_0 \oplus \operatorname{span}\{g_2\}.$$

By the familiar theorems of functional analysis (see Dunford and Schwartz, 1958) there exist continuous linear functionals ν_1, ν_2 such that

$$\begin{aligned} &\nu_1 = \nu_2 = 0 \qquad \text{on } E_0, \\ &\nu_2(g_1) = \nu_1(g_2) = 0, \qquad \nu_1(g_1) = \nu_2(g_2) = 1. \end{aligned} \tag{5.4.1}$$

Hence

$$\nu_i(U_\varphi f) = \nu_i(f) \qquad \text{for all } f \in C(X), \quad i = 1, 2,$$

i.e., ν_1, ν_2 are φ-invariant. By virtue of Theorem 5.3.2 the non-negative measures $\nu_i^{\pm}$, $i = 1, 2$, occurring in the Hahn decomposition of ν_i are also φ-invariant. By assumption, (X, φ) is uniquely ergodic, and so every invariant finite measure is proportional to μ. Thus $\nu_1 = a\mu$, $\nu_2 = b\mu$ and ν_1, ν_2 are linearly dependent, contrary to (5.4.1).

This shows that

$$C(X) = \bar{E}_0 \oplus \operatorname{span}\{1\}$$

(1 denotes the constant function $f(x) = 1$; $1 \notin \bar{E}_0$).

Now let g be any continuous function and let $\varepsilon > 0$ be arbitrary. According to the above, g is the sum of a constant function and a member of $\bar{E}_0$, i.e., there are $f \in C(X)$ and a constant a such that

$$\|g - a - (U_\varphi f - f)\| < \varepsilon,$$

$\|\cdot\|$ denoting the sup norm. Since the mean value of any function in $\bar{E}_0$ is zero, a is the mean value of g. Applying the operator U_φ^k, $0 \leqslant k \leqslant n-1$, to the function occuring in the last inequality and summing the resulting inequalities we obtain

$$\Big\| \sum_{k=0}^{n-1} U_\varphi^k g - n\mu(g) - (U_\varphi^n f - f) \Big\| < n\varepsilon.$$

Hence

$$\left\| \frac{1}{n} \sum_{k=0}^{n-1} U_\varphi^k g - \mu(g) \right\| < \varepsilon + \frac{2\|f\|}{n} < 2\varepsilon$$

for large n, and this proves condition (2).

(2) implies (3). For any $f \in C(X)$ denote by c_f the constant value defined in condition (2):

$$c_f = \lim_{n\to\infty} \frac{1}{n} \sum_{k=0}^{n-1} f \circ \varphi^k(x).$$

Clearly, the assignment $f \mapsto c_f$ is a continuous linear functional on $C(X)$, and thus it is a measure:

$$c_f = \mu(f).$$

The invariance of μ is evident, as is the condition: $\mu(X) = 1$. Hence every point $x \in X$ is generic.

(3) implies (1). The sequence

$$S_n(f)(x) = \frac{1}{n}\sum_{k=0}^{n-1} U_\varphi^k f(x)$$

converges to $\mu(f)$, for every point $x \in X$ and every function $f \in C(X)$, and we have $|S_n(f)(x)| \leqslant \|f\|$. Thus, taking any probability measure ν we get

$$\nu(S_n(f)) \to \nu(\mu(f)) = \int \mu(f)\,d\nu = \mu(f).$$

If, additionally, ν is φ-invariant, then

$$\nu(S_n(f)) = \nu(f)$$

and consequently

$$\nu(f) = \mu(f) \qquad \text{for all } f \in C(X).$$

It follows that $\mu = \nu$, i.e., μ is the unique invariant probability measure for the system (X, φ).

The foregoing theorem admits the following version for the case of a flow:

Theorem 5.4.1′ Given a flow (X, φ_t), any one of the following statements implies the others:

(1′) *the flow (X, φ_t) is uniquely ergodic;*

(2′) *for every function $f \in C(X)$ we have*

$$\frac{1}{T}\int_0^T f(\varphi_t(x))\,dt \to \text{const} \qquad \text{as } T \to +\infty,$$

the convergence being uniform in $x \in X$;

(3′) *there is an invariant measure μ such that every point $x \in X$ is generic for the flow (X, φ_t).*

We now present certain conditions which are sufficient for a system to be uniquely ergodic.

Theorem 5.4.2 If the system (X, φ) is transitive (i.e., *if there is a trajectory dense in X*) *and if the φ^n are equicontinuous, then (X, φ) is uniquely ergodic.*

Proof Let $f \in C(X)$. The functions $U_\varphi^n f = f \circ \varphi^n$ are equicontinuous; consequently, so are the functions

$$S_n(f) = \frac{1}{n} \sum_{k=0}^{n-1} U_\varphi^k f,$$

and thus they constitute a (relatively) compact set in $C(X)$. By Yoshida's Theorem (Theorem 5.3.3) the sequence $(S_n(f))$ converges in $C(X)$ to a function g. It is easy to see that $U_\varphi g = g$. Since, by assumption, the trajectory $\varphi^n(x_0)$ is dense in X for some x_0, the function $g(x)$ must be constant. According to the preceding theorem, the system (X, φ) is then uniquely ergodic.

Corollary 5.4.1 *If φ: $X \to X$ is an isometry and if the system (X, φ) is transitive, then (X, φ) is uniquely ergodic.*

In particular, we see that an irrational rotation of the circle is uniquely ergodic (cf. Exercise 1.2.1); similarly, the translation on T^2 by a vector (α, β) where α and β are independent over the integers (i.e. such that $n\alpha + m\beta \in Z$ yields $n = m = 0$) is uniquely ergodic (cf. Exercise 1.2.2).

Theorem 5.4.3 (Furstenberg, 1961) *If φ: $S^1 \to S^1$ is a homeomorphism without periodic points, then the system (S^1, φ) is uniquely ergodic.*

Proof Since φ has no periodic points, any measure which is φ-invariant necessarily vanishes on singletons. For $z_1, z_2 \in S^1$ we shall denote by (z_1, z_2) the open arc connecting z_1 with z_2, oriented anticlockwise from z_1 to z_2.

Let $z_0 \in S^1$ be some point. Suppose that μ and ν are two invariant probability measures on S^1. Write

$$\lambda = \tfrac{1}{2}(\mu + \nu).$$

Consider the transformation F: $S^1 \to S^1$ given by:

$$F(z) = e^{2\pi i \lambda((z_0, z))}.$$

Since λ vanishes on singletons, F is continuous. For any three points $z_1, z_2, z_3 \in S^1$ we have

$$\lambda\big((z_1, z_2)\big) + \lambda\big((z_2, z_3)\big) = \lambda\big((z_1, z_3)\big) (\text{mod } 1). \tag{5.4.2}$$

Thus, writing $w_0 = \varphi(z_0)$, $w = \varphi(z)$, we have:

$$F(\varphi(z)) = F(w) = e^{2\pi i \lambda((z_0, w))}$$

$$= e^{2\pi i[\lambda((z_0, w)) + \lambda(w_0, w))]} = e^{2\pi i \lambda((z_0, w_0))} e^{2\pi i \lambda((w_0, w))}$$

$$= e^{2\pi i \alpha} e^{2\pi i \lambda((z_0, z))} = e^{2\pi i \alpha} F(z),$$

where $\alpha = \lambda((z_0 w_0))$.

Suppose that α is rational. Then we get

$$F(\varphi^n(z)) = F(z) \qquad \text{for some } n,$$

i.e.,

$$\lambda((z_0, w_n)) = \lambda((z_0, z)) \qquad \text{for some } n,$$

where $w_n = \varphi^n(z)$. Setting in (5.4.2) $z_1 = z_0$, $z_2 = w_n$, $z_3 = z$, we obtain

$$\lambda((z, w_n)) = 0 (\text{mod } 1).$$

Thus either $\lambda((z, w_n)) = 0$ or $\lambda((z, w_n)) = 1$. The function $z \mapsto \lambda((z, \varphi^n(z))) = \lambda((z, w_n))$ is continuous, and so, if it is zero [one] for some z, then it is zero [one] permanently on S^1. Hence every arc on S^1 is of measure zero [one]. Either possibility is absurd.

Thus α is irrational. The system (S^1, φ) is transformed by F into the system (S^1, r), r denoting the rotation by $2\pi\alpha$:

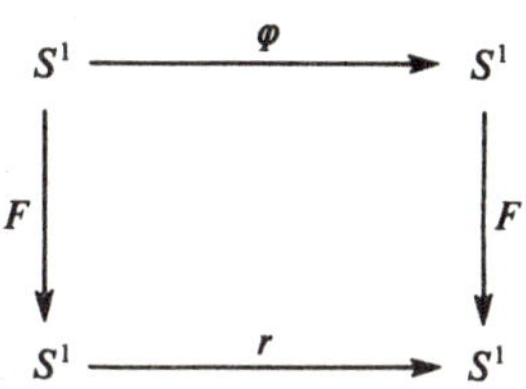

Let $F(\mu)$, $F(\nu)$, $F(\lambda)$ be the measures induced by F:

$$F(\mu)(A) = \mu(F^{-1}(A)),$$
$$F(\nu)(A) = \nu(F^{-1}(A)),$$
$$F(\lambda)(A) = \lambda(F^{-1}(A)).$$

All these measures are invariant under r. Since α is irrational, r is uniquely ergodic, and thus

$$F(\mu) = F(\nu) = F(\lambda) = \frac{1}{2\pi} (\text{Lebesgue measure on } S^1).$$

Let I be any arc on S^1. Write $J = F(I)$, $I_1 = F^{-1}(J)$. Then

$$\mu(I_1) = \nu(I_1) = \lambda(I_1).$$

It follows from the definition of F that $\lambda(I_1 - I) = 0$. Since $\lambda = \frac{1}{2}(\mu+\nu)$, we get

$$\mu(I_1 - I) = \nu(I_1 - I) = \lambda(I_1 - I) = 0$$

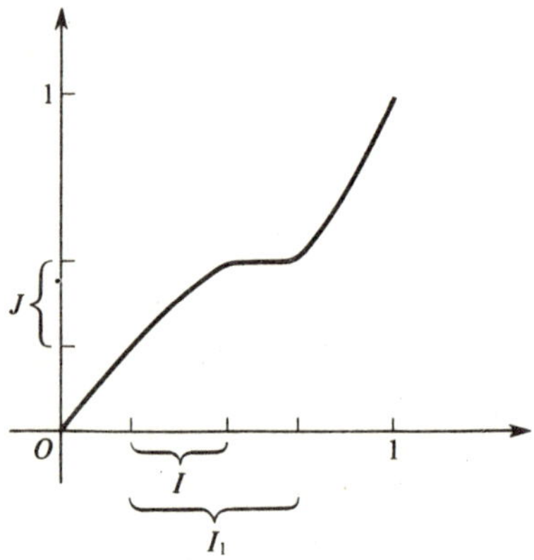

FIG. 5.4.1

(see Fig. 5.4.1). Hence $\mu(I) = \nu(I)$ for every arc I; consequently $\mu(A) = \nu(A)$ for every Borel set $A \subset S^1$, and thus $\mu = \nu$, concluding the proof.

Given a cascade (X, φ), one can consider the special flow (X^f, φ_t) over (X, φ) relative to a given continuous positive function f (cf. Section 1.10). Recall that

$$X^f = X \times \mathbf{R}/\sim, \qquad (x, s) \sim (\varphi(x), s - f(x)).$$

If one of the two systems is uniquely ergodic, need the other be so? Here is the answer:

Theorem 5.4.4 (Strelcyn, 1970a) The cascade (X, φ) is uniquely ergodic if and only if the special flow (X^f, φ_t) is uniquely ergodic.

Proof The "only if" part. Assume that (X, φ) is uniquely ergodic. Let $G(x, s)$ be any continuous function on X^f, let $T > 0$ be any number and (x_0, s) any point of X^f. Write $x_k = \varphi^k x_0$, $k = 1, 2, \ldots$, and take n such that

$$T_n := f(x_0) - s + f(x_1) + \ldots + f(x_n) \leq T$$
$$< f(x_0) - s + f(x_1) + \ldots + f(x_{n+1}).$$

Then

$$T=f(x_0)+\ \dots\ +f(x_n)-s+(T-T_n)$$

and

$$0\leqslant T-T_n<\max_x f(x)<+\infty .$$

The function

$$g(x)=\int_0^{f(x)} G(x,s)\,\mathrm{d}s$$

is continuous on X. By unique ergodicity, we have for every $x\in X$

$$\lim_{n\to\infty}\frac{1}{n}\sum_{k=0}^{n-1} g(\varphi^k(x))=\int g\,\mathrm{d}\mu,$$

where μ denotes the unique φ-invariant probability measure on X.

We intend to show that the limit

$$\lim_{T\to\infty}\frac{1}{T}\int_0^T G(\varphi_t(x_0,s))\,\mathrm{d}t$$

exists and does not depend on the choice of (x_0, s). Indeed,

$$\frac{1}{T}\int_0^T G(\varphi_t(x_0,s))\,\mathrm{d}t$$

$$=\frac{1}{T}\int_0^{f(x_0)-s} G(\varphi_t(x_0,s))\,\mathrm{d}t+\frac{1}{T}\sum_{i=1}^{n} g(x_i)+\frac{1}{T}\int_{T_n}^{T} G(\varphi_t(x_0,s))\,\mathrm{d}t;$$

the outer terms tend to 0 as $T\to+\infty$; the middle term can be rewritten as follows:

$$\frac{\sum_{i=1}^{n} g(x_i)}{f(x_0)+\ \dots\ +f(x_n)-s+(T-T_n)}$$

$$=\frac{\frac{1}{n}\sum_{i=1}^{n} g(\varphi^i(x_0))}{\frac{1}{n+1}\sum_{i=0}^{n} f(\varphi^i(x_0))\cdot\frac{n+1}{n}-\frac{s-(T-T_n)}{n}}.$$

Passing with T(hence also with n) to infinity, we see that the numerator and the denominator tend to $\int g\,d\mu$, $\int f\,d\mu$, respectively. Thus the limit in question exists and is equal to the ratio of the two integrals. Applying Theorem 5.4.1′ we infer that the flow (X^f, φ_t) is uniquely ergodic.

The "if" part. Let μ be an invariant measure for the cascade (X, φ). For any set $B \subset X^f$ of the form

$$B = \{(x, t) \colon t \in I_x,\ x \in A\},$$

where I_x is an interval and A is μ-measurable, we define

$$\mu^f(B) = \int_A \frac{|I_x|}{f(x)}\,d\mu(x).$$

It is not difficult to see that μ^f extends to a Borel measure on X^f and that the extended measure is invariant under φ_t. Thus if (X, φ) admits two distinct ergodic invariant measures μ and ν, then by Exercise 5.2.6 they are singular; hence the corresponding measures μ^f and ν^f are also singular, showing that the flow (X^f, φ_t) admits more than one invariant probability measure. The proof is complete.

Corollary 5.4.2 On the torus, every C^2-flow which has neither critical points nor closed orbits is uniquely ergodic.

Indeed, in view of Theorem 2.5.4 a coordinate system on T^2 can be chosen so that the flow trajectories are straight lines in $\mathbf{R}^2$ (the covering space). Thus the flow under consideration is the special flow over an irrational rotation of the circle. Hence by Theorems 5.4.1 and 5.4.2 the flow is uniquely ergodic.

Exercises

5.4.1.B. Show that for every homeomorphism φ of the circle S^1 the sequence

$$S_n(f)(x) = \frac{1}{n}\sum_{k=0}^{n-1} f \circ \varphi^k(x), \qquad f \in C(X),$$

converges pointwise.

Hint. If φ has no periodic points, apply Theorems 5.4.3 and 5.4.1. If

$\mathrm{Per}\,\varphi \neq \emptyset$, use the fact that $\omega(x) \subset \mathrm{Per}\,\varphi$ for any x (see Exercise 1.11.2).

5.4.2*.B. Show that for every ergodic algebraic automorphism φ of the torus T^2 there exists a continuous function $f \in C(T^2)$ such that the sequence

$$\frac{1}{n}\sum_{k=0}^{n-1} f \circ \varphi^k(x)$$

is divergent at some point $x \in T^2$.

Hint. We consider the situation in the covering space $\boldsymbol{R}^2$, with regard to functions periodic in x^1, x^2, with period 1. Let l_1 and l_2 be the eigenvectors of φ with eigenvalues λ and $\frac{1}{\lambda}$; by ergodicity, we may assume that $|\lambda| > 1$, $\left|\frac{1}{\lambda}\right| < 1$. Let Q_1 and Q_2, $Q_1 \subset Q_2$, be two rhombs with centres at 0 and sides of length ε and 2ε, parallel to l_1 and l_2. Show that for some ε there exists a point $\bar{x}$ such that

$$\limsup_{N\to\infty} \frac{1}{N} \mathrm{Card}\,\{n\colon\ \varphi^n(\bar{x}) \in Q_1 + Z^2,\ 0 \leqslant n < N\}$$

is greater than

$$\limsup_{N\to\infty} \frac{1}{N} \mathrm{Card}\,\{n\colon\ \varphi^n(\bar{x}) \in Q_2 + Z^2,\ 0 \leqslant n < N\}.$$

Now, if f is a function equal to 1 inside Q_1 and 0 outside Q_2, the sequence $\frac{1}{N}\sum_{k=0}^{N-1} f \circ \varphi^k(\bar{x})$ is divergent (see Furstenberg, 1961, Th. 3.2, p. 586).

5.4.3.B. Let (X, φ) be a dynamical system. Write

$$E_0 = \{f \in C(X)\colon\ S_n(f) \to 0 \text{ uniformly}\}.$$

Show that (X, φ) admits exactly m distinct ergodic invariant probability measures if and only if $\mathrm{codim}\, E_0 = m$.

Hint. If μ and ν are distinct ergodic measures for (X, φ), then μ and ν regarded as linear functionals on $C(X)$ are linearly independent (see Exercise 5.2.7).

If $f_1, \ldots, f_m \in E^*$ (E a Banach space) are linearly independent, then the subspace $\{x \in E\colon f_i(x) = 0,\ i = 1, \ldots, m\}$ is of codimension m in E.

The examples of translations on the circle and on the torus suggest the conjecture that minimal systems are uniquely ergodic. However, this is not the case.

Let us consider the following dynamical system:

$$\varphi\colon T^2 \to T^2, \qquad \varphi(x^1, x^2) = (x^1+\alpha,\ x^2+f(x^1)), \tag{5.4.3}$$

where α is irrational and $f\colon S^1 \to S^1$. A. Furstenberg (1961) has shown that if $\deg f \neq 0$, the system is uniquely ergodic. Nevertheless, if $\deg f = 0$, the system can admit more than one invariant measure (clearly, the system preserves the Lebesgue measure). We shall now show an example of such an f.

Let ν_n be a sequence of integers increasing rapidly; say, let $\nu_{n+1} = 2^{\nu_n}$, $\nu_0 = 1$. Write

$$\alpha = \sum_{n=0}^{\infty} \frac{1}{\nu_n}$$

and define

$$f(t) = \sum_{n=0}^{\infty} a_n(e^{2\pi i \nu_n \alpha} - 1)e^{2\pi i \nu_n t}, \tag{5.4.4}$$

where (a_n) is any sequence with the properties

$$\sum a_n^2 < +\infty, \qquad \sum |a_n| = +\infty$$

(e.g. $a_n = 1/n$). Then there exists a function $H(t)$ satisfying the equation

$$H(t+\alpha) - H(t) = f(t),$$

namely

$$H(t) = \sum_{n=-\infty}^{+\infty} a_n e^{2\pi i \nu_n t}, \qquad a_{-n} = \overline{a_n}.$$

Since the sequence (a_n) is square-summable, the function H is square-integrable: $H \in \mathscr{L}^2(S^1, \mathfrak{L}, |\cdot|)$.

The curves

$$B_C = \{(x^1, x^2)\colon x^2 = H(x^1) + C\}$$

are φ-invariant:

$$(x^1, H(x^1)+C) \overset{\varphi}{\mapsto} (x^1+\alpha, H(x^1)+C+f(x^1))$$
$$= (x^1+\alpha, H(x^1+\alpha)+C).$$

The function H is equivalent to a Borel function; we may thus assume that it is Borel itself. The curves under discussion are then Borel subsets of the torus.

For each C transport the Lebesgue measure from S^1 onto B_C by the mapping

$$S^1 \ni x^1 \mapsto (x^1, H(x^1)+C) \in B_C$$

and denote the resulting measure on B_C by μ_C; then extend μ_C to a measure on T^2 by putting

$$\mu_C(A) = \mu_C(A \cap B_C) \qquad \text{for } A \in \mathfrak{B}(T^2).$$

Clearly μ_C is φ-invariant.

It follows that the system admits uncountably many invariant ergodic probability measures.

Minimality is left to the reader as an exercise:

Exercise

5.4.4*.B. Prove that the transformation φ of the torus defined by (5.4.3) with f given by (5.4.4) is minimal (i.e. the system (T^2, φ) is minimal).

Hint. For any $m \in N$ and $k = 0, 1, \ldots, \nu_m$ consider the vector

$$w_{m,k} = (e^{2\pi i \nu_1 \alpha k}, \ldots, e^{2\pi i \nu_m \alpha k}) \in T^m.$$

Take $\varepsilon > 0$. Show that there exists an m such that for every $w \in T^m$ and every n, $1 \leqslant n \leqslant m$,

$$\min_{k \leqslant \nu_m} |(w - w_{m,k})_n| \leqslant \frac{\varepsilon}{2^n};$$

here $(\cdot)_n$ denotes the n-th entry of the vector in brackets.

As A. B. Katok has informed the present author, the example just discussed was already known to A. N. Kolmogorov in the thirties. It was first published in Furstenberg's paper (1961) with a proof of unique ergodicity. I. Kornfeld (1972) was the first to state "in writing" that the system is minimal.

Problem Given a natural number k, does there exist a diffeomorphism φ of a compact manifold M such that the dynamical system (M, φ) is minimal and admits exactly k ergodic invariant probability measures?

It seems likely that the answer is positive. A. B. Katok (1973) and

E. A. Sataev (1975) constructed flows on compact two-dimensional orientable manifolds M of genus p with the following properties:

1° the flow has $2p-2$ critical points,

2° the trajectory of every non-critical point is dense in M,

3° given $k \leqslant p$, the flow (M, φ_t) admits exactly k non-trivial ergodic invariant measures.

It would be reasonable to call such systems quasi-minimal.

M. S. Štilman (1971) constructed an example of a transitive dynamical system (X, φ) admitting exactly k ergodic invariant measures (k prescribed). In his example X is a space consisting of sequences with zero-one terms (yet X is not a manifold).

5.5. Invariant Measures for Smooth Dynamical Systems

The starting point is the problem of existence of an invariant measure for flows in $\boldsymbol{R}^m$. Consider a system of ordinary differential equations

$$\frac{\mathrm{d}x}{\mathrm{d}t} = f(x), \tag{5.5.1}$$

$$x = (x^1, \ldots, x^m) \in D, \qquad f = (f_1, \ldots, f_m);$$

D is a connected open set in $\boldsymbol{R}^m$. Assume, for simplicity, that for every initial value $x \in D$ system (5.5.1) has a solution $\varphi_t(x)$ defined for $t \in (-\infty, +\infty)$ and such that $\varphi_0(x) = x$.

Definition 5.5.1 A real-valued function $\varrho(x) \geqslant 0$, $x \in D$, is called an *integral invariant* of equation (5.5.1) iff the measure

$$\mu(A) := \int_A \varrho(x)\,\mathrm{d}x, \qquad A \subset D,$$

is invariant for the flow φ_t.

Theorem 5.5.1 (Liouville) Let f and ϱ be of class C^1. Then ϱ is an integral invariant of equation (5.5.1) *if and only if*

$$\operatorname{div}(\varrho f) = 0 \text{ on } D. \tag{5.5.2}$$

Proof Let $A \subset D$ be a compact set and let t be any real number. Then we have, for $h > 0$,

$$\int_{\varphi_{t+h}(A)} \varrho(x)\,dx - \int_{\varphi_t(A)} \varrho(x)\,dx$$

$$= \int_{\varphi_t(A)} \varrho(\varphi_h(x'))\left|\det\frac{\partial\varphi_h(x')}{\partial x}\right| dx' - \int_{\varphi_t(A)} \varrho(x')\,dx'$$

$$= \int_{\varphi_t(A)} \left(\varrho(\varphi_h(x))\left|\det\frac{\partial\varphi_h(x)}{\partial x}\right| - \varrho(x)\right) dx. \tag{5.5.3}$$

Applying the Taylor formula to the function $h \mapsto \varrho(\varphi_h(x))$, we may write

$$\varrho(\varphi_h(x)) = \varrho(x) + \operatorname{grad}\varrho(x) f(x) h + R(h), \tag{5.5.4}$$

where $R(h)/h \to 0$ as $h \to 0$.

Calculate $\partial\varphi_h(x)/\partial x$: since $\varphi_h(x)$ is of class C^2 in h, we have

$$\varphi_h(x) = x + f(x)h + \frac{1}{2}\frac{\partial^2}{\partial t^2}\varphi_\xi(x)h^2,$$

where ξ is a number between 0 and h. Hence

$$\frac{\partial\varphi_h(x)}{\partial x} = I + df(x)h + F(x, h)h^2,$$

$F(x, h)$ denoting a function continuous in the region $D \times \boldsymbol{R}$. Thus

$$\det\frac{\partial\varphi_h(x)}{\partial x} = \det\left(I + df(x)h + F(x, h)h^2\right).$$

It follows from the definition of the determinant that the Jacobian under consideration is of the form

$$\begin{aligned}\det\frac{\partial\varphi_h(x)}{\partial x} &= 1 + \sum_{i=1}^{m}\frac{\partial f^i(x)}{\partial x^i}h + G(x, h)h^2 \\ &= 1 + \operatorname{div} f(x)h + G(x, h)h^2,\end{aligned} \tag{5.5.5}$$

$G(x, h)$ being a continuous function. Inserting (5.5.5) and (5.5.4) into (5.5.3) we obtain

$$\int_{\varphi_{t+h}(A)} \varrho(x)\,\mathrm{d}x - \int_{\varphi_t(A)} \varrho(x)\,\mathrm{d}x$$

$$= \int_{\varphi_t(A)} \Big(\big(\varrho(x)+\operatorname{grad}\varrho(x) f(x)h + R(h)\big)\big(1+\operatorname{div} f(x)h +$$

$$+G(x, h)h^2\big) - \varrho(x)\Big)\,\mathrm{d}x$$

$$= \int_{\varphi_t(A)} \Big(\big(\operatorname{grad}\varrho(x) f(x) + \varrho(x)\operatorname{div} f(x)\big)h + L(x, h)h^2\Big)\,\mathrm{d}x,$$

where $L(x, h)$ is a continuous function. Dividing by h and passing with h to zero, we get in the limit

$$\frac{\mathrm{d}}{\mathrm{d}t}\int_{\varphi_t(A)} \varrho(x)\,\mathrm{d}x = \int_{\varphi_t(A)} \big(\operatorname{grad}\varrho(x) f(x) + \varrho(x)\operatorname{div} f(x)\big)\,\mathrm{d}x$$

$$= \int_{\varphi_t(A)} \operatorname{div}\big(\varrho(x) f(x)\big)\,\mathrm{d}x.$$

We immediately conclude, by the continuity of the last integrand, that the value of $\int_{\varphi_t(A)} \varrho(x)\,\mathrm{d}x$ is independent of t (for any $A \subset D$) if and only if

$$\operatorname{div}\big(\varrho(x) f(x)\big) = 0 \qquad \text{for all } x \in D.$$

This is exactly the assertion of the theorem.

Corollary 5.5.1 If (5.5.1) *is a Hamiltonian system, then the Lebesgue measure is invariant for the flow induced by the system.*

Proof Indeed, if (5.5.1) is of the form

$$\frac{\mathrm{d}q}{\mathrm{d}t} = \frac{\partial H}{\partial p}, \qquad \frac{\mathrm{d}p}{\mathrm{d}t} = -\frac{\partial H}{\partial q}, \tag{5.5.6}$$

where $H(q, p)$ is the energy function, then

$$\operatorname{div}\left(\frac{\partial H}{\partial p}, -\frac{\partial H}{\partial q}\right) = \frac{\partial^2 H}{\partial q \partial p} - \frac{\partial^2 H}{\partial p \partial q} = 0 \qquad \text{for all } (p, q).$$

Thus the function $\varrho(x) = 1$ is an integral invariant, i.e., the flow preserves the measure $\int_A \mathrm{d}q\mathrm{d}p = |A|$.

As is well known, in a Hamiltonian system motion occurs on surfaces of constant energy $\{(q, p)\colon H(q, p) = h = \text{const}\}$. On every such surface there is a measure invariant for the flow generated by (5.5.6).

Theorem 5.5.2 In a Hamiltonian system (5.5.6), *the measure* μ *defined below is an invariant measure on any surface of constant energy* $\{H = \text{const}\}$ *such that* $\operatorname{grad} H \neq 0$ *at each of its points*:

$$\mu(A) = \int_A \frac{\mathrm{d}\sigma}{\|\operatorname{grad} H\|}, \qquad A \subset \{H = \text{const}\};$$

$\mathrm{d}\sigma$ *denotes the non-oriented surface measure* (*on* $\{H = \text{const}\}$) *induced from the phase space* $\boldsymbol{R}^{2m}$.

Proof Assume that $\operatorname{grad} H(q_0, p_0) \neq 0$. Then the equation $H(q, p) = h$ can be locally solved with respect to some variable, say p^m (we write, as usual, $q = (q^1, \ldots, q^m)$, $p = (p^1, \ldots, p^m)$). Thus let $p^m = F(q, p')$, where $p' = (p^1, \ldots, p^{m-1})$, be the function satisfying the identity

$$H(q, p', F(q, p')) = h$$

for (q, p') close to (q_0, p'_0). As it is known from the course of advanced calculus,

$$\sigma(A) = \int_{\pi(A)} \sqrt{\|\operatorname{grad} F\|^2 + 1}\, \mathrm{d}q\mathrm{d}p' = \int_{\pi(A)} \frac{\|\operatorname{grad} H\|}{\left|\dfrac{\partial H}{\partial p^m}\right|} \mathrm{d}q\mathrm{d}p'.$$

for any Borel set

$$A \subset \{H = h\} \cap \text{domain of } F;$$

$\pi(A)$ denotes the projection of A onto the hyperplane (q, p'). In short formal notation we rewrite the last formula as follows:

$$\frac{\mathrm{d}\sigma}{\|\operatorname{grad} H\|} = \frac{\mathrm{d}l_{2m-1}}{\left|\dfrac{\partial H}{\partial p^m}\right|},$$

where l_{2m-1} is the Lebesgue measure in $\mathbf{R}^{2m-1}$, the space of variables (q, p'). Setting in (5.5.6) $p^m = F(q, p')$ and leaving out the last equation of (5.5.6) (the one involving $\mathrm{d}p^m/\mathrm{d}t$), we obtain

$$\frac{\mathrm{d}q^i}{\mathrm{d}t} = \frac{\partial H}{\partial p^i}(q, p', F(q, p')),$$
$$\frac{\mathrm{d}q^m}{\mathrm{d}t} = \frac{\partial H}{\partial p^m}(q, p', F(q, p')), \qquad i = 1, \ldots, m-1.$$
$$\frac{\mathrm{d}p^i}{\mathrm{d}t} = -\frac{\partial H}{\partial q^i}(q, p', F(q, p')),$$

This is a system of equations in $\mathbf{R}^{2m-1}$. Let us now apply Liouville's theorem to the function

$$\varrho = \frac{1}{\left|\dfrac{\partial H}{\partial p^m}\right|}.$$

Since

$$\frac{\dfrac{\partial H}{\partial p'}}{\dfrac{\partial H}{\partial p^m}} = -\frac{\partial F}{\partial p'}, \qquad \frac{\dfrac{\partial H}{\partial q}}{\dfrac{\partial H}{\partial p^m}} = -\frac{\partial F}{\partial q},$$

we have

$$\operatorname{div} \ (\partial H/\partial p^m)^{-1}\left[\frac{\partial H}{\partial p'}, \frac{\partial H}{\partial p^m}, -\frac{\partial H}{\partial q}\right] = \operatorname{div}\left[-\frac{\partial F}{\partial p'}, 1, \frac{\partial F}{\partial q}\right]$$
$$= -\frac{\partial^2 F}{\partial q \partial p'} + \frac{\partial^2 F}{\partial p' \partial q} = 0,$$

and this together with the previous conclusions yields our assertion, in virtue of Theorem 5.5.1.

As a further application of Liouville's theorem we now show that, on a Riemannian manifold, the Riemannian measure is invariant for the geodesic flow.

Let M be a complete Riemannian manifold and let (M, T, V) be a holonomic system, i.e., a system described locally by equations (5.5.6),

where

$$H = T + V, \qquad T = \tfrac{1}{2} \sum_{i,j=1}^{m} a_{ij}(q)\dot{q}^i\dot{q}^j = \tfrac{1}{2}\big(A(q)\dot{q}, \dot{q}\big),$$

$$a_{ij}(\cdot) = a_{ji}(\cdot) \qquad \text{positive definite}$$

and $V\colon M \to \boldsymbol{R}$ is some smooth function. Let $\big(a^{ij}(\cdot)\big)$ be the matrix inverse to $\big((a_{ij}(\cdot)\big)$. The interplay between impulse p and velocity $\dot{q}$ is expressed by

$$p = A(q)\dot{q}, \qquad \text{i.e.,} \qquad p_i = \sum_{j=1}^{m} a_{ij}(q)\dot{q}^j$$

(passage from Lagrange equations to Hamilton equations). Hence

$$T = \tfrac{1}{2} \sum_{i,j=1}^{m} a^{ij}(q)\, p_i\, p_j = \tfrac{1}{2}\big(p, A(q)^{-1} p\big).$$

System (5.5.6) preserves the Lebesgue measure $\mathrm{d}q\mathrm{d}p$. Let $C = C(q)$ be a matrix such that $C^T A^{-1} C = I$ (C exist since A^{-1} is symmetric and positive definite). By the change of variables

$$q = q', \qquad p = C(q)\,p'$$

we get

$$\mathrm{d}q\mathrm{d}p = \mathrm{d}q'\mathrm{d}p'\,|\det C(q)| = \mathrm{d}q'\mathrm{d}p'\sqrt{\det A(q)}. \tag{5.5.7}$$

In the new variables the surface of constant energy $\{H = h = \text{const}\}$ is described by

$$\tfrac{1}{2} \sum_{i=1}^{m} p_i'^2 + V(q') = h.$$

Now we pass to the spherical coordinates $r, \vartheta_1, \ldots, \vartheta_{m-1}$ for p'. Let $\mathrm{d}\vartheta$ be the surface measure on the $(m-1)$-sphere S^{m-1}. Then

$$\mathrm{d}q'\mathrm{d}p'\sqrt{\det A(q')} = \sqrt{\det A(q')}\, r^{m-1}\mathrm{d}q'\mathrm{d}r\mathrm{d}\vartheta. \tag{5.5.8}$$

Since $r = \|p'\| = \sqrt{2\big(h - V(q')\big)}$, we obtain

$$\sqrt{\det A(q')}\, r^{m-1}\mathrm{d}q'\mathrm{d}r\mathrm{d}\vartheta = 2^{\frac{1}{2}m-1}\sqrt{\det A(q')}\,\big(h - V(q')\big)^{\frac{1}{2}m-1}\mathrm{d}q'\mathrm{d}h\mathrm{d}\vartheta. \tag{5.5.9}$$

In the coordinates q', h, $\vartheta_1, \ldots, \vartheta_{m-1}$ the surface of constant energy has the equation $h = \text{const}$. In view of (5.5.7), (5.5.8), (5.5.9) the measure

$$\sqrt{\det A(q')}\,(h - V(q'))^{\frac{1}{2}m-1}\mathrm{d}q'\mathrm{d}h\mathrm{d}\vartheta$$

is invariant for the system. Since also the surface $\{h = \text{const}\}$ is invariant, we see that the following measure is invariant for the system on that surface:

$$\mathrm{d}\mu = \sqrt{\det A(q')}\,(h - V(q'))^{\frac{1}{2}m-1}\mathrm{d}q'\mathrm{d}\vartheta. \tag{5.5.10}$$

In the configuration manifold $M = \{(q^1, \ldots, q^m)\}$ we introduce the Riemannian metric

$$\mathrm{d}s^2 = (h - V(q')) \sum_{i,j=1}^{m} a_{ij}(q')\,\mathrm{d}q^i\,\mathrm{d}q^j$$

(assuming that $h - V(q') \neq 0$). Hence the Riemannian measure $\mathrm{d}\sigma$ on M is of the form

$$\begin{aligned} \mathrm{d}\sigma &= \sqrt{\det\big((h - V(q'))a_{ij}(q')\big)}\mathrm{d}q' \\ &= (h - V(q'))^{\frac{1}{2}m}\sqrt{\det A(q')}\,\mathrm{d}q'. \end{aligned} \tag{5.5.11}$$

Comparing (5.5.10) with (5.5.11) we get

$$\mathrm{d}\mu = \frac{\mathrm{d}\sigma\mathrm{d}\vartheta}{h - V(q')}.$$

In particular, if $h - V(q') = 1$, i.e., if the flow generated by (5.5.6) is the geodesic flow in the metric, the measure $\mathrm{d}\sigma\mathrm{d}\vartheta$ is invariant. Clearly, $\mathrm{d}\sigma\mathrm{d}\vartheta$ is the Riemannian measure on the sphere bundle over M. In this fashion we have proved the following theorem.

Theorem 5.5.3 The geodesic flow of the sphere bundle W_M over the Riemannian manifold M preserves the Riemannian measure on W_M, i.e., the measure $\mathrm{d}\mu = \mathrm{d}\sigma\mathrm{d}\vartheta$, *where* $\mathrm{d}\sigma$ *is the Riemannian measure on M and* $\mathrm{d}\vartheta$ *is the usual measure on the sphere* S^{m-1}, $m = \dim M$.

Anosov (1967) has shown (see also Anosov and Sinai, 1967) that if M is a manifold of negative curvature, then the geodesic flow is ergodic.

Now we shall prove that expanding mappings, sufficiently smooth, of compact manifolds preserve a measure absolutely continuous with respect to the Riemannian measure (see Krzyżewski and Szlenk, 1969).

Theorem 5.5.4 *If φ: $M \to M$ is an expanding C^2-map of a compact Riemannian manifold M, then:*

1° φ preserves a certain measure μ, which is equivalent to the Riemannian measure,

2° the system (M, μ, φ) is ergodic.

Before passing to the proof we introduce some notation. Let $\|\cdot\|$ be the Riemannian metric such that

$$\|\mathrm{d}\varphi(x)(v)\| \geqslant c\,\|v\| \qquad \text{for } x \in M,\ v \in T_x(M),$$

where $c > 1$; let μ_0 and $d(\cdot, \cdot)$ denote the Riemannian measure and the distance (metric) induced by the Riemannian metric $\|\cdot\|$. Let μ_x be the Lebesgue measure in $T_x(M)$.

We introduce the function

$$(D\varphi)(x) = \mu_{\varphi(x)}(\mathrm{d}\varphi(x)(B_x)), \qquad x \in M,$$

where B_x is any set in $T_x(M)$ of Lebesgue measure 1: $\mu_x(B_x) = 1$. The function $D\varphi$ is called the *scalar derivative* of φ.

We shall need the following lemma:

Lemma 5.5.1 *Under the assumptions of Theorem 5.5.4 there exists a constant $a > 0$ such that for every Borel set $A \subset M$ we have*

$$\mu_0(\varphi^{-n}(A)) \leqslant a\mu_0(A), \qquad n = 0, 1, 2, \dots$$

Proof of the lemma *Step* I. We know that expanding mappings cover M (Exercise 4.6.3). Thus there exists an open cover (U_i), $i = 1, \dots, p$, such that:

(1) φ is a diffeomorphism between U_i and $\varphi(U_i)$, for each $i = 1, \dots, p$,

(2) for any points $x, y \in U_i$ there exists a curve γ connecting x with y, $\hat{\gamma} \subset U_i$, of length equal to $d(x, y)$.

Let δ be the Lebesgue number of the cover (U_i). We claim that for some k and for every $n = 0, 1, 2, \dots$ there are families of open sets $A_{i_0 \dots i_n}$, $1 \leqslant i_0 \leqslant k$, $1 \leqslant i_1 \leqslant N, \dots, 1 \leqslant i_n \leqslant N$, where $N = \deg \varphi$, with the following properties:

(3) $\mu_0\left(M - \bigcup_{i_0=1}^{k} A_{i_0}\right) = 0$ and $A_{i_0} \neq \emptyset$ for $i_0 = 1, \dots, k$;

(4) $A_{i_0 \dots i_n} \cap A_{i'_0 \dots i'_n} = \emptyset$ whenever $(i_0, \dots, i_n) \neq (i'_0, \dots, i'_n)$;

(5) $\varphi^{-1}(A_{i_0 \dots i_n}) = \bigcup_{i_{n+1}=1}^{N} A_{i_0 \dots i_n i_{n+1}}$;

(6) φ maps $A_{i_0\ldots i_{n+1}}$ diffeomorphically onto $A_{i_0\ldots i_n}$;

(7) diam $A_{i_0\ldots i_n} < \frac{\delta}{c^n}$;

(8) $d(\varphi(x), \varphi(y)) \geqslant cd(x, y)$ for any pair of points x, y belonging to a common $A_{i_0\ldots i_n}$.

The sets $A_{i_0\ldots i_n}$ are constructed by induction. Let (B_s), $s = 1, \ldots, k'$, be a covering family of open balls of radii $\leqslant \frac{1}{2}\delta$, each contained in some U_i. Define

$$A_1' = B_1, \qquad A_s' = B_s - \bigcup_{j=1}^{s-1} \bar{B}_j, \qquad s = 2, \ldots, k'.$$

Disregarding those A_s' which are empty and re-indexing, we obtain the family of sets A_{i_0}, $i_0 = 1, \ldots, k$:

$$\{A_s' \colon s = 1, \ldots, k'\} = \{A_{i_0} \colon i_0 = 1, \ldots, k\} \cup (k' - k \text{ empty sets}).$$

Each B_s is contained in some U_i; hence the same regards the sets A_{i_0}.

Now assume that $A_{i_0\ldots i_j}$ are defined for $j \leqslant n$. By (7), for any multi-index $(i_0, \ldots, i_n)$ there is $r_{i_0\ldots i_n}$ such that $A_{i_0\ldots i_n} \subset U_{r_{i_0\ldots i_n}}$. By (1), we have

$$\varphi^{-1}(A_{i_0\ldots i_n}) = \bigcup_{i_{n+1}=1}^{N} A_{i_0\ldots i_n i_{n+1}},$$

where the members of the union on the right denote the intersections of $\varphi^{-1}(A_{i_0\ldots i_n})$ with the components of $\varphi^{-1}(U_{r_{i_0\ldots i_n}})$.

It is not difficult to verify that conditions (3)–(8) are satisfied. This proves our claim.

Step II. We now show that there is a constant $b > 0$ such that

$$\frac{(D\varphi^n)(x)}{(D\varphi^n)(y)} \leqslant b \tag{5.5.12}$$

whenever x and y belong to a common $A_{i_0\ldots i_n}$, $n = 1, 2, \ldots$

Thus let $x, y \in A_{i_0\ldots i_n}$. By the chain rule for differentiation we have

$$\begin{aligned}\frac{(D\varphi^n)(x)}{(D\varphi^n)(y)} &= \prod_{i=0}^{n-1} \frac{(D\varphi)(\varphi^i(x))}{(D\varphi)(\varphi^i(y))} \leqslant \prod_{i=0}^{n-1}\left(1 + \frac{|(D\varphi)(\varphi^i(x)) - (D\varphi)(\varphi^i(y))|}{(D\varphi)(\varphi^i(y))}\right) \\ &\leqslant \exp \sum_{i=0}^{n-1} \frac{|(D\varphi)(\varphi^i(x)) - (D\varphi)(\varphi^i(y))|}{(D\varphi)(\varphi^i(y))}.\end{aligned} \tag{5.5.13}$$

By assumption, φ is of class C^2. Hence $D\varphi$ is of class C^1 and thus fulfils the Lipschitz condition (we omit the details):

$$|(D\varphi)(x)-(D\varphi)(y)| \leqslant Ld(x, y).$$

Using this and writing $\alpha = \inf_{x \in M} (D\varphi)(x)$, we get from (5.5.13)

$$\frac{(D\varphi^n)(x)}{(D\varphi^n)(y)} \leqslant \exp\left(\frac{L}{\alpha}\sum_{i=0}^{n-1} d(\varphi^i(x), \varphi^i(y))\right). \tag{5.5.14}$$

On account of property (8), we have

$$d(\varphi^i(x), \varphi^i(y)) \geqslant cd(\varphi^{i-1}(x), \varphi^{i-1}(y)),$$

whence

$$d(\varphi^{n-1}(x), \varphi^{n-1}(y)) \geqslant c^{n-1-i} d(\varphi^i(x), \varphi^i(y)), \qquad i = 0, \ldots, n-1.$$

Estimate (5.5.14) together with the last inequality results in

$$\begin{aligned}\frac{(D\varphi^n)(x)}{(D\varphi^n)(y)} &\leqslant \exp\left(\frac{L}{\alpha}\sum_{i=0}^{n-1} \frac{d(\varphi^{n-1}(x), \varphi^{n-1}(y))}{c^{n-1-i}}\right)\\ &\leqslant \exp\left(\frac{L}{\alpha}\,\frac{c}{c-1}\operatorname{diam} M\right).\end{aligned}$$

This number is the desired constant b.

Step III. We are now in a position to prove the lemma. On account of (4), (5) and (6) we may write

$$\mu_0(\varphi^{-n}(A)) = \sum_{\substack{1 \leqslant i_1 \leqslant N\\ \cdots\cdots\\ 1 \leqslant i_n \leqslant N}} \int_A (D\varphi^{-n}_{i_0 \ldots i_n})(y)\, \mathrm{d}\mu_0(y), \tag{5.5.15}$$

where A is any Borel set contained in some A_{i_0} and $(D\varphi^{-n}_{i_0 \ldots i_n})$ denotes the scalar derivative of the mapping inverse to φ^n restricted to $A_{i_0 \ldots i_n}$.

It follows from (5.5.12) that

$$\frac{\sup_{y \in A_{i_0}} (D\varphi^{-n}_{i_0 \ldots i_n})(y)}{\inf_{x \in A_{i_0}} (D\varphi^{-n}_{i_0 \ldots i_n})(x)} \leqslant b. \tag{5.5.16}$$

From (5.5.15) and (5.5.16) we obtain

$$\mu_0(\varphi^{-n}(A)) \leqslant \mu_0(A)\, b \sum_{\substack{1 \leqslant i_1 \leqslant N\\ \cdots\cdots\\ 1 \leqslant i_n \leqslant N}} \inf_{x \in A_{i_0}} (D\varphi^{-n}_{i_0 \ldots i_n})(x).$$

Setting $A = A_{i_0}$ in (5.5.15), we also get

$$\mu_0(\varphi^{-n}(A_{i_0})) \geqslant \mu_0(A_{i_0}) \sum_{\substack{1\leqslant i_1 \leqslant N\\ \cdots\cdots\\ 1\leqslant i_n\leqslant N}} \inf_{x\in A_{i_0}} (D\varphi^{-n}_{i_0\ldots i_n})(x).$$

Division yields

$$\mu_0(\varphi^{-n}(A)) \leqslant b\frac{\mu_0(A)}{\mu_0(A_{i_0})}\mu_0(\varphi^{-n}(A_{i_0})) \leqslant \frac{b\mu_0(M)}{\mu_0(A_{i_0})}\mu_0(A).$$

Writing

$$a = \frac{b\mu_0(M)}{\min\limits_{1\leqslant i_0\leqslant k} \mu_0(A_{i_0})}$$

we hence obtain

$$\mu_0(\varphi^{-n}(A)) \leqslant a\mu_0(A)$$

for any Borel set which is contained in some A_{i_0}. In view of property (3), this estimate extends to arbitrary Borel sets $A \subset M$. The lemma is proved.

Proof of Theorem 5.5.4, assertion 1° Lemma 5.5.1 shows that

$$\frac{1}{n}(\mu_0(A)+\mu_0(\varphi^{-1}(A))+ \ \ldots \ +\mu_0(\varphi^{-n+1}(A))) \leqslant a\mu_0(A).$$

Thus, in view of Exercise 5.3.1 on p. 289, the formula

$$\mu(A) := \lim_{n\to\infty}\frac{1}{n}(\mu_0(A)+ \ \ldots \ +\mu_0(\varphi^{-n+1}(A)))$$

defines an invariant measure μ. Clearly, this measure is μ_0-continuous: $\mu \ll \mu_0$.

We shall show that also $\mu_0 \ll \mu$. We first show that if, for some index i_0, $\mu_0(\varphi^{-n}(A_{i_0})) \not\to 0$, then

$$\mu(A) = 0 \Rightarrow \mu_0(A) = 0 \qquad \text{for } A \subset A_{i_0}.$$

In Lemma 5.5.1 the measure μ_0 may be replaced by any measure $\mu_s = \mu_0 \circ \varphi^{-s}$; by the conclusion of the lemma we then have

$$\inf_n \mu_0(\varphi^{-n}(A_{i_0})) = a_{i_0} > 0,$$

whenever $\mu_0(\varphi^{-n}(A_{i_0})) \not\to 0$. From (5.5.15) and (5.5.16) we get, for $A \subset A_{i_0}$,

$$\mu_0(\varphi^{-n}(A)) \geqslant \mu_0(A) \sum_{\substack{1 \leqslant i_1 \leqslant N \\ \cdots\cdots \\ 1 \leqslant i_n \leqslant N}} \inf_{x \in A_{i_0}} (D\varphi_{i_0 \ldots i_n}^{-n})(x)$$

$$\geqslant \frac{1}{b} \mu_0(A) \sum_{\substack{1 \leqslant i_1 \leqslant N \\ \cdots\cdots \\ 1 \leqslant i_n \leqslant N}} \sup_{x \in A_{i_0}} (D\varphi_{i_0 \ldots i_n}^{-n})(x)$$

and

$$\mu_0(\varphi^{-n}(A_{i_0})) \leqslant \mu_0(A_{i_0}) \sum_{\substack{1 \leqslant i_1 \leqslant N \\ \cdots\cdots \\ 1 \leqslant i_n \leqslant N}} \sup_{x \in A_{i_0}} (D\varphi_{i_0 \ldots i_n}^{-n})(x).$$

Just as before, division yields:

$$\mu_0(\varphi^{-n}(A)) \geqslant \frac{1}{b} \frac{\mu_0(A)}{\mu_0(A_{i_0})} \mu_0(\varphi^{-n}(A_{i_0})) \geqslant \frac{1}{b} \frac{a_{i_0}}{\mu_0(A_{i_0})} \mu_0(A). \tag{5.5.17}$$

Hence

$$\mu(A) \geqslant \frac{1}{b} \frac{a_{i_0}}{\mu_0(A_{i_0})} \mu_0(A) \qquad \text{for } A \subset A_{i_0},$$

and so

$$\mu(A) = 0 \Rightarrow \mu_0(A) \qquad \text{for } A \subset A_{i_0},$$

as claimed.

Thus, in order to prove that $\mu_0 \ll \mu$, it suffices to show that actually $\mu_0(\varphi^{-n}(A_{i_0})) \not\to 0$ for every i_0. Assume the contrary:

$$\mu_0(\varphi^{-n}(A_{i_0'})) \to 0 \qquad \text{for some } i_0'.$$

Since

$$\mu_0(M) = \sum_{i_0=1}^{k} \mu_0(\varphi^{-n}(A_{i_0})) \qquad \text{for all } n,$$

we have

$$\mu_0(\varphi^{-n}(A_{i_0''})) \not\to 0 \qquad \text{for some } i_0''.$$

According to Theorem 4.11.1 on p. 258, there is an n_0 such that $\varphi^{n_0}(A_{i_0''}) \supset A_{i_0'}$. Consider the set $B = \varphi^{-n_0}(A_{i_0'}) \cap A_{i_0''}$. B is open and non-empty and is contained in $A_{i_0''}$. Using inequality (5.5.17) with $i_0 = i_0''$, we conclude that

$$\mu_0(\varphi^{-n}(B)) \not\to 0.$$

On the other hand, we have

$$\varphi^{-n}(B) \subset \varphi^{-n-n_0}(A_{i_0'}) \qquad \text{for } n = 1, 2, \ldots,$$

whence

$$\mu_0(\varphi^{-n}(B)) \to 0.$$

We arrive at a contradiction, which ends the proof of assertion 1°.

Proof of assertion 2° Let $A \subset M$ be a Borel set of positive measure, and suppose that A is φ-invariant, i.e., $\varphi^{-1}(A) = A$. It is known from the theory of functions of a real variable (see e.g. Natanson, 1954, Ch. XIII, Section 1) that the following limit relation holds for almost every x:

$$\lim_{h \to 0} \frac{1}{h^m} \int_{Q(x,h)} \chi_A = \chi_A(x),$$

where $Q(x, h)$ is the cube (in a chart) with centre x and with sides of length h parallel to the coordinate axes, $m = \dim M$ and χ_A is the characteristic function of the set A. Hence the inequality

$$\frac{|Q(x, h) \cap A|}{|Q(x, h)|} \geqslant 1 - \varepsilon$$

holds for almost every $x \in A$, every $\varepsilon > 0$ and h small enough. Thus, according to condition (7) on p. 310, for any n sufficiently large there is at least one set $A_{i_0 \ldots i_n}$ such that

$$\frac{|A_{i_0 \ldots i_n} \cap A|}{|A_{i_0 \ldots i_n}|} \geqslant 1 - \varepsilon;$$

or, equivalently,

$$\frac{|A_{i_0 \ldots i_n} - A|}{|A_{i_0 \ldots i_n}|} \leqslant \varepsilon.$$

Applying Lemma 5.5.1 (properties (3)–(8) and estimate (5.5.12)), we obtain

$$\frac{|A_{i_0}-A|}{|A_{i_0}|}=\Big(\int\limits_{A_{i_0\ldots i_n}-A} D\varphi^n\Big)\Big(\int\limits_{A_{i_0\ldots i_n}} D\varphi^n\Big)^{-1}$$

$$\leqslant \big(|A_{i_0\ldots i_n}-A| \sup_{A_{i_0\ldots i_n}} (D\varphi^n)\big)\big(|A_{i_0\ldots i_n}| \inf_{A_{i_0\ldots i_n}} (D\varphi^n)\big)^{-1}\leqslant b\varepsilon.$$

Since i_0 can assume only finitely many values and since ε is arbitrary, there is an i_0 for which

$$\frac{|A_{i_0}-A|}{|A_{i_0}|}=0,$$

that means,

$$\frac{|A_{i_0}\cap A|}{|A_{i_0}|}=1.$$

By virtue of Theorem 4.11.1 (3) there exists an n such that $\varphi^n(A_{i_0}) \supset M$. Hence, omitting easy details, we infer that $\mu_0(M-A)=0$. We have already shown (assertion 1°) that the measures μ_0 and μ are equivalent. Thus $\mu(M-A)=0$ and so φ is ergodic. The proof is complete.

Remark 5.5.1 K. Krzyżewski (1971a – announcement) has proved that if φ is an expanding C^r-map, $r \geqslant 2$, of a compact Riemannian manifold M, then the Radon–Nikodym derivative $d\mu/d\mu_0$ (i.e., the function unique up to equivalence, satisfying

$$\mu(A)=\int\limits_A \frac{d\mu}{d\mu_0}(x)\, d\mu_0(x)$$

for any Borel set $A \subset M$) is of class C^{r-1} (Independently R. Sachsteder (1973) proved the same results.)

To conclude this section let us mention a very interesting result of Lifschitz and Sinai (1972):

Theorem 5.5.5 Let $\varphi \in \mathrm{Diff}^2(M)$ be an Anosov diffeomorphism of a compact Riemannian manifold M. Then φ preserves a measure equivalent to the Riemannian measure on M if and only if for every periodic point

$x_0 \in M$ we have

$$\det |d\varphi^p(x_0)| = 1,$$

p denoting the prime period of x_0.

The proof is based on certain facts of statistical physics and therefore it is omitted here.

5.6. Flows on the Torus T^2(II). Linearization

We already encountered flows on the torus in Section 2.5. We have proved that every vector field $\Phi \in \mathscr{F}^r(T^2)$, $r \geqslant 2$, without critical points and closed orbits has a phase portrait equivalent to the constant vector field on T^2 (i.e., constant in the covering plane $\boldsymbol{R}^2$). Now we shall discuss the problem of topological conjugacy: given a vector field Φ on T^2, find conditions under which the flow generated by Φ is topologically conjugate to the flow of a constant field. The flow under discussion is then said to be *linearizable*; the problem stated is called: the problem of *linearization* of a system of differential equations. No general solution is known and, with the problem posed in this way it is doubtful whether a solution can be given at all (by saying "general solution", we mean here something corresponding to a classification of C^2-diffeomorphisms of the circle without periodic points). Up to the present, certain sufficient conditions on Φ have been found, under very strong additional assumptions. The results which we present below have been obtained by Kolmogorov (1953) (under the assumption of analyticity of flows) and strengthened by Katok (1967). Thus it is reasonable to call this series of results: the Katok–Kolmogorov theory.

Consider the following system of equations on T^2:

$$\frac{dx^1}{dt} = f^1(x^1, x^2), \qquad \frac{dx^2}{dt} = f^2(x^1, x^2), \qquad (x^1, x^2) \in \boldsymbol{R}^2; \tag{5.6.1}$$

i.e., f^1 and f^2 are assumed to be periodic in each variable, with period 1. Moreover, we assume that the vector field $\Phi = (f^1, f^2)$ is of class C^r, $r \geqslant 2$, and that Φ has neither critical points nor closed orbits in T^2. Further, we assume that there exists an integral invariant $g(x^1, x^2) > 0$ for system (5.6.1) and that g is of class C^{r-1}.

By Theorem 2.5.3 there exists a Siegel curve for the vector field Φ. Suppose that $x^1 = 0$ is that curve (i.e., the lines $x^1 = n \in Z$ are copies of that curve in $\boldsymbol{R}^2$). In view of Lemma 2.5.1 we may assume that $f^1(x^1, x^2) > 0$ for $(x^1, x^2) \in \boldsymbol{R}^2$. For $x = (0, v) \in \{(x^1, x^2): x^1 = 0\}$ denote by $\psi(v)$ the abscissa of the earliest point at which the trajectory $\varphi(t, x)$ of (5.6.1) meets the line $x^1 = 1$. Then the function $\psi(\text{mod } 1)$ is a C^r-diffeomorphism of the circle S^1 (see Theorem 1.5.1); since Φ has no critical elements, ψ has no periodic points. By Corollary 2.4.2 there is a homeomorphism $H: S^1 \to S^1$ such that $H \circ \psi = r \circ H$, where r is the

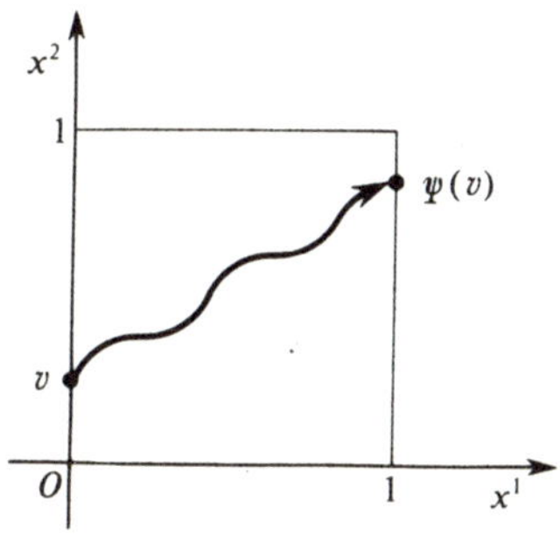

FIG. 5.6.1

rotation of S^1 by $2\pi\varrho$, ϱ denoting the rotation number of ψ.

We now prove:

Lemma 5.6.1 Under the above assumptions on the system (5.6.1), *H is a C^r-diffeomorphism of the circle.*

Proof Instead of (5.6.1) we consider the system

$$\frac{\mathrm{d}x^1}{\mathrm{d}t} = 1, \qquad \frac{\mathrm{d}x^2}{\mathrm{d}t} = \frac{f^2(x^1, x^2)}{f^1(x^1, x^2)}. \tag{5.6.2}$$

(5.6.2) is equivalent to (5.6.1) in the sense that the trajectories of the two systems are identical and thus they induce the same diffeomorphism ψ of the circle. The first equation of (5.6.2) yields $x^1(t) = x_0^1 + t$. The function

$$g_1(x^1, x^2) = g(x^1, x^2) f^1(x^1, x^2) > 0, \qquad (x^1, x^2) \in \boldsymbol{R}^2,$$

is an integral invariant for the system (5.6.2); this can be easily verified, for instance with the use of Liouville's theorem. Thus

$$\int_A g_1(x^1, x^2)\, \mathrm{d}x^1 \mathrm{d}x^2 = \int_{A_t} g_1(x^1, x^2)\, \mathrm{d}x^1 \mathrm{d}x^2$$

for any measurable set A; of course, $A_t = \varphi_t(A)$, φ_t now standing for the flow generated by the vector field $(1, f^2/f^1)$. Take, in particular,

$t = 1$ and
$$A = \{(x^1, x^2)\colon\ a < x^1 < b,\ x^2 \in I\},$$

where I is some fixed interval of the circle S^1, i.e., of the x^2-axis. We obtain

$$\int_a^b \int_{I_u} g_1(u, x^2)\, dx^2\, du = \int_{a+1}^{b+1} \int_{J_u} g_1(u, x^2)\, dx^2\, du$$

where

$$J_u = \varphi_1(I_u),$$
$$I_u = \{(x^1, x^2)\colon\ x^1 = u,\ x^2 \in I\}, \qquad a < u < b.$$

Since g_1 is periodic in x^1 with period 1, we have

$$\int_a^b \int_{I_u} g_1(u, x^2)\, dx^2\, du = \int_a^b \int_{J_u} g_1(u, x^2)\, dx^2\, du,$$

whence

$$\int_a^b \Big(\int_{I_u} g_1(u, x^2)\, dx^2 - \int_{J_u} g_1(u, x^2)\, dx^2\Big) du = 0.$$

This holds for any interval (a, b). The function g_1 is continuous; thus the last equality implies

$$\int_{I_u} g_1(u, x^2)\, dx^2 = \int_{J_u} g_1(u, x^2)\, dx^2 \qquad \text{for all } u.$$

It follows that ν given by

$$\nu(I) := \int_I g_1(0, x^2)\, dx^2, \qquad I \subset \langle 0, 1),$$

is a measure invariant for the diffeomorphism $\psi = \varphi_1$.

Since $H \circ \psi = r \circ H$, we have

$$|H \circ \psi(I)| = |r \circ H(I)| = |H(I)|$$

(clearly, r preserves the Lebesgue measure). Define

$$\sigma(I) := |H(I)|, \qquad I \subset \langle 0, 1).$$

This equality shows that the measure σ is ψ-invariant. Now, ψ is uniquely ergodic by Theorem 5.4.3. Hence $\sigma = \nu$. Thus we have

$$H(b) - H(a) = \int_a^b g_1(0, x^2)\, dx^2$$

for any interval (a, b); consequently, H is of class C^r, since g_1 is of class C^{r-1}. Moreover, we get

$$\frac{\mathrm{d}H}{\mathrm{d}x^2} = g_1(0, x^2) > 0,$$

which shows that H is a diffeomorphism.

Theorem 5.6.1 *Under the assumptions imposed on the system* (5.6.1), *this system admits reduction by a suitable C^r-change of variables to a system of the form*

$$\frac{\mathrm{d}x^1}{\mathrm{d}t} = G(x^1, x^2), \qquad \frac{\mathrm{d}x^2}{\mathrm{d}t} = \varrho G(x^1, x^2), \tag{5.6.3}$$

where $G > 0$ is of class C^r and ϱ denotes the rotation number of ψ.

Proof This is an immediate consequence of Lemma 5.6.1 and Corollary 2.5.2.

The number ϱ is also termed the *rotation number* of the system considered.

It is easily observed that system (5.6.3) has an integral invariant: namely, the function $1/G$. This follows directly from Liouville's theorem:

$$\operatorname{div}\left(\frac{1}{G(x^1, x^2)}[G(x^1, x^2), \varrho G(x^1, x^2)]\right) = 0 \qquad \text{for } (x^1, x^2) \in \boldsymbol{R}^2,$$

proving our claim.

Definition 5.6.1 We say that a number ϱ fulfils *condition* (An), where $n = 1, 2, \ldots,$ if there is a constant $d > 0$ such that for any $p, q \in Z$, $q \neq 0$, we have

$$\left|\varrho - \frac{p}{q}\right| > \frac{d}{q^n}.$$

Exercise

5.6.1.A. Show that the set of all real numbers fulfilling condition (A3) is of full measure.

Hint. The complement of the set in question is a countable union of sets of type $B_d := \{\varrho$: the inequality $|\varrho - pq^{-1}| \leqslant dq^{-3}$ has infinitely many solutions in integer pairs $(p, q)\}$. For every q the set $B_d \cap \langle a, b\rangle$ can be covered by intervals of joint length not greater than $2dq^{-2}$; here $\langle a, b\rangle$ denotes any interval. Hence easily follows $|B_d| = 0$.

It is well known (see Chandrasekharan, 1968, Ch. III, Theorem 4) that no number satisfies condition (A2); a proof of this fact is also suggested to the reader as an exercise.

We now pass to the final result of this section:

Theorem 5.6.2 Consider system (5.6.1) *on the torus. Suppose that* $\Phi = (f^1, f^2)$ *is of class* C^4, *without critical points or periodic orbits. Further, suppose that there exists an integral invariant of class* C^4. *Finally, suppose that the rotation number* ϱ *of the system satisfies condition* (A3). *Then the system can be reduced, by a diffeomorphic change of variables* $(x^1, x^2) \mapsto (u, v)$, *to the form*

$$\frac{\mathrm{d}u}{\mathrm{d}t} = \alpha, \qquad \frac{\mathrm{d}v}{\mathrm{d}t} = \beta, \tag{5.6.4}$$

where α *and* β *are constants (i.e., the system is linearizable).*

In the proof we shall need the following lemma;

Lemma 5.6.2 If $F \in C^4(\boldsymbol{R})$ *is periodic with period* 1, *then the equation*

$$F(s) - K(s) + K(s+\varrho) = c = \int_0^1 F(s)\,\mathrm{d}s \tag{5.6.5}$$

has a solution $K \in C^1(\boldsymbol{R})$ *which is also periodic with period* 1.

Proof We use Fourier expansions. Given

$$F(s) = \sum_{n=-\infty}^{+\infty} a_n \mathrm{e}^{2\pi i n s},$$

we postulate

$$K(s) = \sum_{n=-\infty}^{+\infty} w_n \mathrm{e}^{2\pi i n s}.$$

Equation (5.6.5) takes the form of the system of equations

$$a_n - w_n + w_n \mathrm{e}^{2\pi i n \varrho} = 0 \qquad \text{for } n \neq 0,$$

$$a_0 = \int_0^1 F(s)\,\mathrm{d}s = c.$$

Hence we get

$$w_n = \frac{a_n}{1 - e^{2\pi i n \varrho}}.$$

Condition (A3) ensures that $|q\varrho - p| > dq^{-2}$ for $p, q \in Z$. Consequently

$$|e^{2\pi i n \varrho} - 1| > \frac{d_1}{n^2},$$

where d_1 is some constant. This yields the following estimate for w_n:

$$n^2 |w_n| \leqslant \frac{n^2 |a_n|}{|e^{2\pi i n \varrho} - 1|} < \frac{n^2 |a_n|}{d_1 n^{-2}} = \frac{1}{d_1} n^4 |a_n|.$$

Observe that $(2\pi i)^4 \sum_{-\infty}^{+\infty} n^4 a_n e^{2\pi i n s}$ is the Fourier series of the continuous function $\frac{d^4}{ds^4} F(s)$. Hence the series of its coefficients is square-summable,

$$\sum_{n=-\infty}^{+\infty} (n^4 |a_n|)^2 < +\infty,$$

and thus also

$$\sum_{n=-\infty}^{+\infty} (n^2 |w_n|)^2 < +\infty,$$

showing that $K'' \in L^2(0, 1)$. It follows that $K \in C^1(\boldsymbol{R})$, ending the proof of the lemma.

Proof of Theorem 5.6.2 According to Theorem 5.6.1, we may assume that the system under consideration is of the form (5.6.3). Denote by $F(s)$ the time needed for getting from the point $(0, s)$ to the point $(1, s_1)$ along the trajectory; of course, $s_1 = s + \varrho$. The first equation of (5.6.3) gives

$$F(s) = \int_0^1 \frac{dx^1}{G(x^1, \varrho x^1 + s)};$$

thus clearly $F \in C^4(\boldsymbol{R})$. By Lemma 5.6.2 equation (5.6.5) has a solution $K \in C^1(\boldsymbol{R})$.

Let us consider the curve $\gamma = \gamma(s)$ defined as follows:

$$\gamma(s) = \varphi_{K(s)}((0, s));$$

γ is a closed C^1-curve on the torus. We claim that γ has the following property: for any point $x \in \hat{\gamma}$, the earliest moment $t > 0$ at which the trajectory $\varphi_t(x)$ again intersects γ is

$$t = c = \int_0^1 F(s)\,\mathrm{d}s.$$

We prove this. Choose $x \in \hat{\gamma}$. Considering the situation in the covering plane, we may assume that x belongs to the strip $\{0 \leqslant x^1 < 1\}$. The trajectory $\varphi_t(x)$ intersects the lines $\{x^1 = 0\}$ and $\{x^1 = 1\}$ at the points which we denote by $(0, s)$ and x_1, respectively (see Fig. 5.6.2).

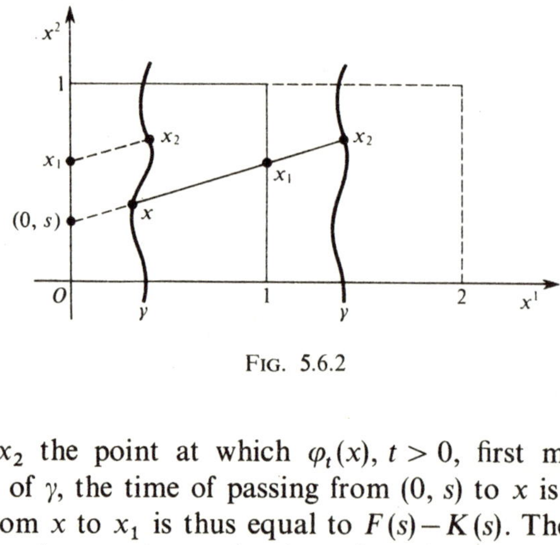

FIG. 5.6.2

Denote by x_2 the point at which $\varphi_t(x)$, $t > 0$, first meets $\hat{\gamma}$. By the construction of γ, the time of passing from $(0, s)$ to x is $K(s)$; the time of passing from x to x_1 is thus equal to $F(s) - K(s)$. The x^2-coordinate of the point x_1 is $\varrho + s$; hence the time of passing from x_1 to x_2 is equal to $K(s+\varrho)$. Summing up, we see that the time period t needed for getting from x to x_2 is

$$t = F(s) - K(s) + K(s+\varrho),$$

and, since K satisfies equation (5.6.5), we obtain $t = c$, as claimed.

Now we introduce the new coordinate system

$$u(x) = \frac{1}{c}\tau(x), \qquad v(x) = x^2 - \varrho x^1 + \varrho u(x),$$

where $\tau(x)$ denotes the time of getting from $\hat{\gamma}$ to $x = (x^1, x^2)$.

In the original coordinates (x^1, x^2) the torus is considered as the strip $\{0 \leqslant x^1 < 1, x^2 \in \boldsymbol{R}\}$ with x^2 taken modulo 1. In the new coordinates (u, v) we view the torus as a curvilinear strip limited by two copies of the curve γ and taking v modulo 1. Obviously, the mapping $(x^1, x^2) \mapsto (u, v)$ is bijective. The function $u(x)$ is differentiable and the derivative $\mathrm{d}u(x)$ is continuous (this easily follows from the Implicit Function Theorem). Also the mapping $(u, v) \mapsto (x^1, x^2)$ is differentiable: using equations (5.6.3) we obtain

$$\frac{\partial(u, v)}{\partial(x^1, x^2)} = \det \begin{bmatrix} \dfrac{\partial u}{\partial x^1} & \dfrac{\partial u}{\partial x^2} \\ \dfrac{\partial v}{\partial x^1} & \dfrac{\partial v}{\partial x^2} \end{bmatrix}$$

$$= \det \begin{bmatrix} \dfrac{\partial u}{\partial x^1} & \dfrac{\partial u}{\partial x^2} \\ \varrho\left(\dfrac{\partial u}{\partial x^1} - 1\right) & 1 + \varrho \dfrac{\partial u}{\partial x^2} \end{bmatrix}$$

$$= \frac{\partial u}{\partial x^1} + \varrho \frac{\partial u}{\partial x^2} = \frac{1}{G}\left(\frac{\partial u}{\partial x^1} G + \frac{\partial u}{\partial x^2} \varrho G\right)$$

$$= \frac{1}{G}\left(\frac{\partial u}{\partial x^1}\frac{\mathrm{d}x^1}{\mathrm{d}t} + \frac{\partial u}{\partial x^2}\frac{\mathrm{d}x^2}{\mathrm{d}t}\right)$$

$$= \frac{1}{G}\frac{\mathrm{d}}{\mathrm{d}t} u(x^1(t), x^2(t)) = \frac{1}{cG} > 0,$$

and this proves that the mapping $(x^1, x^2) \mapsto (u, v)$ is a diffeomorphism.

In the coordinates (u, v) system (5.6.3) assumes the form of the system of equations

$$\frac{\mathrm{d}u}{\mathrm{d}t} = \frac{1}{c}$$

and (since $x^2 = \varrho x^1 - \text{const}$)

$$\frac{\mathrm{d}v}{\mathrm{d}t} = \frac{\mathrm{d}x^2}{\mathrm{d}t} - \varrho \frac{\mathrm{d}x^1}{\mathrm{d}t} + \varrho \frac{\mathrm{d}u}{\mathrm{d}t} = \frac{\varrho}{c}.$$

Thus $\alpha = 1/c$ and $\beta = \varrho/c$ are the desired constants; the theorem is proved.

Remark 5.6.1 The assumption that ϱ satisfies condition (A3) is essential: M. Šklover (1967) proved that if ϱ is very well approximated by rationals (i.e., if ϱ does not fulfil any of the conditions (An)), then system (5.6.3) is not linearizable.

Exercise

5.6.2.B. Consider system (5.6.1) supposing that f^1, f^2 are analytic functions, that there exists an integral invariant which is also analytic, and that ϱ satisfies condition (An) for some n. Prove that the system is linearizable.

Hint. If a function $\sum_{n=-\infty}^{+\infty} a_n e^{2\pi i n t}$ is analytic, then $\sum_{n=-\infty}^{+\infty} n^k |a_n| < +\infty$ for every k. Copy the proof of Theorem 5.6.2 (of Lemma 5.6.2, in fact).

5.7. Metric Entropy

We present the concept of metric entropy and list its properties very briefly in the form of isolated definitions and theorems, omitting proofs. Readers who want to study this subject in more detail should resort to Billingsley (1965), Sinai (1973, 1970) and Totoki (1970).

Let $(X, \mathfrak{M}, \mu)$ be a given probability measure space. A family of subsets of X: $\xi = \{A_i : i = 1, \ldots, n\}$ is called a *finite partition* iff

$$A_i \in \mathfrak{M} \qquad \text{for } i = 1, \ldots, n,$$

$$A_i \cap A_j = \emptyset \qquad \text{for } i \neq j,$$

$$\bigcup_{i=1}^{n} A_i = X \text{ (modulo a } \mu\text{-nullset).}$$

Let $\xi = \{A_i\}_1^n$ and $\eta = \{B_j\}_1^m$ be two partitions. Define

$$\xi \vee \eta := \{A_i \cap B_j : A_i \in \xi, B_j \in \eta\}. \tag{5.7.1}$$

Clearly, $\xi \vee \eta$ is also a partition.

Given partitions $\xi = \{A_i\}_{i=1}^n$ and $\eta = \{B_j\}_{j=1}^m$, we say that η is *finer* than ξ iff each member of ξ is the union of some members of η (possibly, modulo nullsets):

$$A_i = \bigcup_{j \in J_i} B_j \text{ (mod } \mu = 0\text{)}, \qquad J_i \subset \{1, \ldots, m\}.$$

This is written symbolically as

$$\xi \prec \eta. \tag{5.7.2}$$

Given a finite sequence of partitions $\xi_1, \ldots, \xi_k$,

$$\xi_j = \{A_{j,i}\}_{i=1}^{n_j}, \qquad j = 1, \ldots, k,$$

we write

$$\bigvee_{j=1}^{k} \xi_j := \xi_1 \vee \ldots \vee \xi_k$$
$$= \{A_{1,i_1} \cap \ldots \cap A_{k,i_k} \colon A_{1,i_1} \in \xi_1, \ldots, A_{k,i_k} \in \xi_k\}.$$

For any finite partition $\xi = \{A_i\}_1^n$ we introduce the quantity

$$H(\xi) := \sum_{i=1}^{n} \mu(A_i) \log \mu(A_i)$$

called the *entropy* of ξ (if for some i $\mu(A_i) = 0$, we set $\mu(A_i)\log\mu(A_i) = 0$). If $\xi \prec \eta$, we have

$$H(\xi) \leqslant H(\eta). \tag{5.7.3}$$

Exercise

5.7.1.A. Prove (5.7.3).
Hint. If $c = a+b$, $a, b \geqslant 0$, then $c\log c \geqslant a\log a + b\log b$.

Further, let φ be a measure preserving transformation:

$$\varphi\colon X \to X, \qquad \mu(\varphi^{-1}(A)) = \mu(A) \qquad \text{for } A \in \mathfrak{M}.$$

If $\xi = \{A_i\}_1^m$ is a partition, $\varphi^{-1}(\xi) := \{\varphi^{-1}(A_i)\}_1^m$ is a partition as well. It is readily seen that

$$H(\xi) = H(\varphi^{-1}(\xi)). \tag{5.7.4}$$

Theorem 5.7.1 For any partition $\xi = \{A_i\}_1^m$ *the following limit exists:*

$$\lim_{n\to\infty} \frac{1}{n} H(\xi \vee \varphi^{-1}(\xi) \vee \ldots \vee \varphi^{-n+1}(\xi)).$$

For the proof see e.g. Billingsley (1965), Ch. 2, Section 6.
Denote this limit by $h_\mu(\varphi, \xi)$.

Definition 5.7.1 The *metric entropy* of a metric dynamical system $(X, \mathfrak{M}, \mu, \varphi)$ is the quantity $h_\mu(\varphi)$ defined by the formula

$$h_\mu(\varphi) = \sup_\xi h_\mu(\varphi, \xi),$$

the supremum being taken over all finite partitions.

Now we formulate certain theorems concerning the concept just introduced.

Theorem 5.7.2 If the systems $(X, \mathfrak{M}, \mu, \varphi)$ *and* $(Y, \mathfrak{N}, \nu, \psi)$ *are isomorphic, then*

$$h_\mu(\varphi) = h_\nu(\psi).$$

Theorem 5.7.3 For every metric dynamical system $(X, \mathfrak{M}, \mu, \varphi)$

$$h_\mu(\varphi^n) = nh_\mu(\varphi) \qquad \text{for } n \in Z^+.$$

If φ *is invertible, then also*

$$h_\mu(\varphi^n) = |n| h_\mu(\varphi) \qquad \text{for } n \in Z.$$

Theorem 5.7.4 Given a metric dynamical system $(X, \mathfrak{M}, \mu, \varphi)$, φ *invertible, and a sequence of partitions*

$$\xi_1 \prec \xi_2 \prec \ldots \prec \xi_n \prec \ldots$$

Suppose that the σ-field generated by all sets which are members of the partitions

$$\bigvee_{k=-N}^{N} \varphi^{-k}(\xi_n), \qquad n, N = 1, 2, \ldots,$$

(i.e., the smallest σ-field containing all such sets) is the entire $\mathfrak{M}$. *Then*

$$h_\mu(\varphi) = \lim_{n \to \infty} h_\mu(\varphi, \xi_n).$$

A partition ξ_0 is termed a *generator* iff the σ-field generated by all the sets which enter the partitions

$$\bigvee_{k=-N}^{N} \varphi^{-k}(\xi_0), \qquad N = 1, 2, \ldots,$$

coincides with $\mathfrak{M}$. Theorem 5.7.4 implies

Theorem 5.7.5 *If ξ_0 is a generator for a system $(X, \mathfrak{M}, \mu, \varphi)$ with φ invertible, then*

$$h_\mu(\varphi) = h_\mu(\varphi, \xi_0).$$

Krieger (1970) has proved that every ergodic dynamical system with an invertible φ on a Lebesgue space (i.e., on a measure space isomorphic [39] to $(\langle 0, 1\rangle, \mathfrak{L}, |\cdot|)$) with finite entropy has a generator.

Applying Theorem 5.7.5 it is easy e.g. to calculate the entropy of the Bernoulli automorphism; we leave this to the reader.

Exercises

5.7.2.A. Show that if φ is the Bernoulli shift with $\mathscr{A} = \{a_1, \ldots, a_m\}$, $p(a_i) = p_i$, $i = 1, \ldots, m$, then

$$h_\mu(\varphi) = -\sum_{i=1}^{m} p_i \log p_i.$$

Hint. Take $A_i = \{(x_n): x_0 = a_i\}$, $i = 1, \ldots, m$. The partition $\xi = (A_i)_1^m$ is generating and we have

$$H\Big(\bigvee_{k=0}^{n-1} \varphi^{-k}(\xi)\Big) = nH(\xi).$$

5.7.3.A. Show that if $p_i > 0$, $\sum_{i=1}^{m} p_i = 1$, then $-\sum_{i=1}^{m} p_i \log p_i \leqslant \log m$, equality holding if and only if $p_1 = \ldots = p_m = 1/m$.

Hint. Apply Jensen's inequality [40] to the function $-\log x$ at the points $x_i = 1/p_i$ with the coefficients $t_i = p_i$, $i = 1, \ldots, m$.

5.7.4.A. Show that

$$H(\xi \vee \eta) \leqslant H(\xi) + H(\eta)$$

holds for any finite partitions ξ, η.

Hint. Let $\xi = (A_i)_1^n$, $\eta = (B_j)_1^m$, $\mu(A_i) \neq 0$. For each i consider the conditional measure $\mu_i = \mu(\cdot | A_i)$ given by $\mu_i(B) = \mu(B \cap A_i)/\mu(A_i)$ and define

$$H(\eta|\xi) := -\sum_{i=1}^{n} \mu(A_i) \sum_{j=1}^{m} \mu_i(B_j) \log \mu_i(B_j).$$

Show that $H(\xi \vee \eta) = H(\xi) + H(\eta|\xi)$ and that $H(\eta|\xi) \leqslant H(\eta)$; this follows from the convexity of the function $x\log x$; see Billingsley (1965).

5.7.5.A. Let λ and v be probability measures on $(X, \mathfrak{M})$. Show that the measure $\mu = t\lambda + (1-t)v$, $0 \leqslant t \leqslant 1$, satisfies the inequality

$$H_\mu(\xi) \geqslant tH_\lambda(\xi) + (1-t)H_v(\xi),$$

for any partition ξ.

Hint. Apply Jensen's inequality [40] to the function $x\log x$.

It has been conjectured for quite a long time that any two Bernoulli automorphisms are isomorphic. The problem was solved in the negative by A. N. Kolmogorov (1959), who introduced the concept of entropy as an isomorphy invariant and used it to produce counterexamples. This notion has been later modified by Ya. G. Sinai (1959) to its present practical form.

D. Ornstein (1970) proved

Theorem 5.7.6 Any two Bernoulli isomorphisms with the same entropy are isomorphic.

Thus e.g., two systems with the measure distributions $\{\frac{1}{4}, \frac{1}{4}, \frac{1}{4}, \frac{1}{4}\}$ and $\{\frac{1}{2}, \frac{1}{8}, \frac{1}{8}, \frac{1}{8}, \frac{1}{8}\}$, respectively, are isomorphic (since the entropy of either of them is 2log2), despite the fact that the corresponding sets $\mathscr{A}$ are of different cardinalities. This example is due to Meshalkin; it was known earlier than Ornstein's theorem.

It thus turns out that metric entropy is a complete invariant of Bernoulli automorphisms.

5.8. Topological Entropy

The notion of topological entropy was introduced about 20 years ago by Adler, Konheim, and McAndrew (1965).

Let X be a compact metric space and let α denote an open cover:

$\alpha = \{A_i\}$, $\bigcup_i A_i = X$, A_i open.

Definition 5.8.1 Given two open covers $\alpha = \{A_i\}$, $\beta = \{B_j\}$, we define a new cover as

$$\alpha \vee \beta := \{A_i \cap B_j\colon\ A_i \in \alpha,\ B_j \in \beta\},$$

and we call $\alpha \vee \beta$ the *common refinement* of α and β.

Definition 5.8.2 A cover β is said to be *finer* than α, in symbols: $\alpha \prec \beta$, iff each member of β is a subset of some member of α.

It is readily noted that the relation just introduced is transitive and defines a partial order in the family of all open covers of X.

Definition 5.8.3 Given an open cover α, one can find a subcover of minimal cardinality; any such subcover is called *minimal*. The minimal cardinality (of a subcover of α) is denoted by $N(\alpha)$.

By compactness, $N(\alpha)$ is finite.

The number $H_0(\alpha) := \log N(\alpha)$ is called the *entropy* of the cover α.

Let us write down certain trivial properties of the symbols $N(\alpha)$ and $H_0(\alpha)$:

1° if $\alpha \prec \beta$ then $N(\alpha) \leqslant N(\beta)$;
2° if $\alpha \prec \beta$ then $N(\alpha \vee \beta) = N(\beta)$;
3° $N(\alpha \vee \beta) \leqslant N(\alpha)N(\beta)$,
$H_0(\alpha \vee \beta) \leqslant H_0(\alpha) + H_0(\beta)$.

For a continuous map $\varphi\colon X \to X$ write $\varphi^{-1}(\alpha) = \{\varphi^{-1}(A_i)\}$ if $\alpha = \{A_i\}$; we have:

4° if $\alpha \prec \beta$ then $\varphi^{-1}(\alpha) \prec \varphi^{-1}(\beta)$;
5° $\varphi^{-1}(\alpha \vee \beta) = \varphi^{-1}(\alpha) \vee \varphi^{-1}(\beta)$;
6° $N(\alpha) \geqslant N(\varphi^{-1}(\alpha))$, the equality holding if φ is surjective.

Theorem 5.8.1 For every continuous map $\varphi\colon X \to X$ and every open cover α the following limit exists and is finite:

$$\lim_{n\to\infty} \frac{1}{n} H_0\Big(\bigvee_{k=0}^{n-1} \varphi^{-k}(\alpha)\Big)$$

$$= \lim_{n\to\infty} \frac{1}{n} H_0\big(\alpha \vee \varphi^{-1}(\alpha) \vee \ldots \vee \varphi^{-n+1}(\alpha)\big).$$

Proof Let m, n be positive integers. Using properties 3° and 6° we see that

$$\begin{aligned} &H_0(\alpha \vee \dots \vee \varphi^{-m-n+1}(\alpha)) \\ &= H_0((\alpha \vee \dots \vee \varphi^{-m+1}(\alpha)) \vee \varphi^{-m}(\alpha \vee \dots \vee \varphi^{-n+1}(\alpha))) \\ &\leqslant H_0(\alpha \vee \dots \vee \varphi^{-m+1}(\alpha)) + H_0(\alpha \vee \dots \vee \varphi^{-n+1}(\alpha)). \end{aligned}$$

Denoting $H_0(\alpha \vee \dots \vee \varphi^{-n+1}(\alpha))$ by a_n we rewrite the last inequality as $a_{m+n} \leqslant a_m + a_n$; and we have $a_n \geqslant 0$. As is known from elementary calculus, the limit $\lim a_n/n$ then exists and is finite.

Definition 5.8.4 The limit

$$h(\varphi, \alpha) := \lim_{n \to \infty} \frac{1}{n} H_0\Big(\bigvee_{k=0}^{n-1} \varphi^{-k}(\alpha)\Big)$$

is called the *topological entropy* of φ with respect to the cover α.

Entropy can also be defined relative to closed covers; the characteristic that results is of an entirely different nature.

Theorem 5.8.2 The symbol $h(\varphi, \alpha)$ has the following properties:

(1) $h(\varphi, \alpha) \leqslant H_0(\alpha)$;

(2) *if* $\alpha \prec \beta$ *then* $h(\varphi, \alpha) \leqslant h(\varphi, \beta)$;

(3) *if* φ *is a homeomorphism then* $h(\varphi, \alpha) = h(\varphi^{-1}, \alpha)$.

Proof (1) and (2) are obvious: (1) follows from 3° and (2) from the definition. (3) is also clear:

$$\begin{aligned} &H_0(\alpha \vee \dots \vee \varphi^{-n+1}(\alpha)) = H_0(\varphi^{n-1}(\alpha \vee \dots \vee \varphi^{-n+1}(\alpha))) \\ &= H_0(\alpha \vee \dots \vee \varphi^{n-1}(\alpha)) \\ &= H_0(\alpha \vee (\varphi^{-1})^{-1}(\alpha) \vee \dots \vee (\varphi^{-1})^{-n+1}(\alpha)) \end{aligned}$$

and (3) follows.

Definition 5.8.5 The quantity

$$h(\varphi) := \sup_{\alpha} h(\varphi, \alpha)$$

is called the *topological entropy* of the map $\varphi: X \to X$; the supremum is taken over all open covers α.

The entropy $h(\varphi)$ is a certain numerical parameter of a given dynamical system (X, φ). It indicates how quickly the points of X are "mixed" by φ. As we shall see in a moment, isometries have zero entropy (they preserve distance, hence do not mix points), whereas algebraic automorphisms of the torus, for instance, have in general positive entropy.

Let (α_n), $n = 1, 2, \ldots$, be a sequence of open covers such that:

7° $\alpha_n \prec \alpha_{n+1}$;

8° for every open cover β there is an α_n with $\beta \prec \alpha_n$.

Then, according to Definition 5.8.5, we have

$$h(\varphi) = \lim_{n\to\infty} h(\varphi, \alpha_n). \tag{5.8.1}$$

For any cover $\alpha = \{A_i\}$ of X write

$$d(\alpha) = \sup_i \operatorname{diam} A_i$$

and let $L(\alpha)$ be the Lebesgue number of α. It is easy to see that if for some two covers α and β we have $d(\beta) < L(\alpha)$, then $\alpha \prec \beta$. Hence in view of (5.8.1) we get

Corollary 5.8.1 *If (α_n) is a sequence of open covers such that $d(\alpha_n) \to 0$, then*

$$h(\varphi) = \lim_{n\to\infty} h(\varphi, \alpha_n).$$

The premise is satisfied, for instance, for the sequence (α_n) where α_n is the cover of X by balls with arbitrary centres and with radii less than $1/n$.

Theorem 5.8.3 *Topological entropy is an invariant of topological conjugacy. In other words: if the systems (X, φ) and (Y, ψ) are topologically conjugate, then $h(\varphi) = h(\psi)$.*

Proof Let $g\colon X \to Y$ be the conjugating homeomorphism: $g \circ \varphi = \psi \circ g$, and let α be any open cover of X. Then, writing ψ as $g \circ \varphi \circ g^{-1}$, we obtain

$$\begin{aligned}
h(\psi, g(\alpha)) &= \lim_{n\to\infty} \frac{1}{n} H_0\big(g(\alpha) \vee \psi^{-1}(g(\alpha)) \vee \ldots \vee \psi^{-n+1}(g(\alpha))\big) \\
&= \lim_{n\to\infty} \frac{1}{n} H_0\big(g(\alpha) \vee (g \circ \varphi^{-1})(\alpha) \vee \ldots \vee (g \circ \varphi^{-n+1})(\alpha)\big) \\
&= \lim_{n\to\infty} \frac{1}{n} H_0\big(g(\alpha \vee \varphi^{-1}(\alpha) \vee \ldots \vee \varphi^{-n+1}(\alpha))\big) \\
&= \lim_{n\to\infty} \frac{1}{n} H_0\big(\alpha \vee \varphi^{-1}(\alpha) \vee \ldots \vee \varphi^{-n+1}(\alpha)\big) = h(\varphi, \alpha).
\end{aligned}$$

Hence we immediately see that $h(\psi) \geqslant h(\varphi)$. By symmetry we have also $h(\varphi) \geqslant h(\psi)$.

Theorem 5.8.4 *Let (X, φ) be any dynamical system. Then*

$$h(\varphi^k) = kh(\varphi) \qquad \textit{for every } k \in Z^+.$$

If, moreover, φ is a homeomorphism, then

$$h(\varphi^k) = |k| h(\varphi) \qquad \textit{for every } k \in Z.$$

Proof Let α be an open cover and let $k \in Z^+$. Then

$$\begin{aligned} h(\varphi^k) &\geqslant h\big(\varphi^k, \alpha \vee \varphi^{-1}(\alpha) \vee \ldots \vee \varphi^{-k+1}(\alpha)\big) \\ &= \lim_{n \to \infty} \frac{k}{nk} H_0\big(\alpha \vee \ldots \vee \varphi^{-k+1}(\alpha) \vee \varphi^{-k}(\alpha) \vee \ldots \vee \varphi^{-nk+1}(\alpha)\big) \\ &= kh(\varphi, \alpha). \end{aligned}$$

Hence

$$h(\varphi^k) \geqslant kh(\varphi).$$

On the other hand,

$$\begin{aligned} h(\varphi, \alpha) &= \lim_{n \to \infty} \frac{1}{nk} H_0\big(\alpha \vee \ldots \vee \varphi^{-nk+1}(\alpha)\big) \\ &\geqslant \lim_{n \to \infty} \frac{1}{nk} H_0\big(\alpha \vee (\varphi^k)^{-1}(\alpha) \vee \ldots \vee (\varphi^k)^{-n+1}(\alpha)\big) \\ &= \frac{1}{k} h(\varphi^k, \alpha), \end{aligned}$$

and so,

$$h(\varphi) \geqslant \frac{1}{k} h(\varphi^k).$$

The two inequalities result in the required equality. If φ is a homeomorphism, then $h(\varphi) = h(\varphi^{-1})$, as can easily be verified; consequently $h(\varphi^k) = |k| h(\varphi)$.

Examples

5.8.1. Let $\varphi\colon X \to X$ be an isometry. Then $h(\varphi) = 0$. To see this, take α_k to consist of all balls with radii $< 1/k$. Since φ is isometric, we have

$\alpha_k \vee \varphi^{-1}(\alpha_k) = \alpha_k$ for each k. Hence

$$H_0(\alpha_k \vee \varphi^{-1}(\alpha_k) \vee \ldots \vee \varphi^{-n+1}(\alpha_k)) = H_0(\alpha_k),$$

and so

$$h(\varphi, \alpha_k) = \lim_{n\to\infty} \frac{1}{n} H_0(\alpha_k) = 0.$$

Then by Corollary 5.8.1, $h(\varphi) = 0$.

Thus, in particular, every rotation of the circle and every translation on the torus have zero entropy.

5.8.2. Let (X, φ) be a dynamical system and assume that the iterates φ^n, $n \in Z$, constitute an equicontinuous family. Then $h(\varphi) = 0$. To show this, let us replace the original metric ϱ in X by the new metric ϱ^* defined by

$$\varrho^*(x, y) = \sup_{n\in Z} \varrho(\varphi^n(x), \varphi^n(y)).$$

The metric ϱ^* is equivalent to ϱ because φ^n are equicontinuous. In the new metric φ is an isometry. Hence $h(\varphi) = 0$.

5.8.3. Let G be a compact metric group and let $\varphi\colon G \to G$ be given by

$$\varphi(x) = axb, \qquad \text{where } a, b \in G.$$

Then $h(\varphi) = 0$: defining

$$\varrho^*(x, y) = \sup_{u,v\in G} \varrho(uxv, uyv)$$

we obtain a new metric in G equivalent to the original metric ϱ, as is not difficult to check, and φ is an isometry with respect to ϱ^*.

5.8.4. Every homeomorphism of the circle has zero entropy. To prove this, let α_k be the cover of S^1 consisting of intervals of lengths $< 1/k$ and let $\varphi\colon S^1 \to S^1$ be a homeomorphism. Now, for any covers α and β which consist of intervals we have $N(\alpha \vee \beta) \leqslant N(\alpha) + N(\beta)$ (cf. Exercise 5.8.2). Consequently,

$$\begin{aligned} &N(\alpha_k \vee \varphi^{-1}(\alpha_k) \vee \ldots \vee \varphi^{-n+1}(\alpha_k)) \\ &\quad \leqslant N(\alpha_k) + N(\varphi^{-1}(\alpha_k)) + \ldots + N(\varphi^{-n+1}(\alpha_k)) = nN(\alpha_k), \end{aligned}$$

whence by Corollary 5.8.1

$$h(\varphi) = \lim_{k\to\infty} h(\varphi, \alpha_k) = \lim_{k\to\infty} \lim_{n\to\infty} \frac{1}{n} \log nN(\alpha_k) = 0.$$

5.8.5. Let $X = \prod_{n=-\infty}^{+\infty} \mathscr{A}_n$, where $\mathscr{A}_n = \mathscr{A} = \{a_1, \ldots, a_m\}$. Let ϱ_0 be the discrete metric in $\mathscr{A}$: $\varrho_0(a_i, a_j) = \delta^i_j$. Consider the following metric in X:

$$\varrho(x, y) = \sum_{n=-\infty}^{+\infty} 2^{-|n|}\varrho_0(x_n, y_n)$$

if

$$x = (x_n), \quad y = (y_n), \quad x_n, y_n \in \mathscr{A}.$$

X is a compact metric space. Let φ be the shift to the left:

$$\text{for } x = (x_n) \qquad \varphi(x) = (y_n),\ y_n = x_{n+1}.$$

We shall calculate $h(\varphi)$.

Let $\alpha_0 = \{A_i\}_1^m$, $A_i = \{x = (x_n)\colon x_0 = a_i\}$. Write

$$\alpha_k = \bigvee_{j=-k}^{k} \varphi^j(\alpha_0).$$

Then

$$d(\alpha_k) = \sup_{A_i^{(k)} \in \alpha_k} \operatorname{diam} A_i^{(k)} \leqslant \sum_{|n|>k} 2^{-|n|} = 2^{-k+1} \to 0 \qquad \text{as} \quad k \to +\infty.$$

Thus $h(\varphi) = \lim_{k\to\infty} h(\varphi, \alpha_k)$. Now, we have by definition

$$\begin{aligned} h(\varphi, \alpha_k) &= \lim_{n\to\infty} \frac{1}{n} H_0\Big(\bigvee_{j=0}^{n-1} \varphi^{-j}(\alpha_k)\Big) \\ &= \lim_{n\to\infty} \frac{1}{n} H_0\Big(\bigvee_{j=-k}^{k} \varphi^j(\alpha_0) \vee \bigvee_{j=-k-1}^{k-1} \varphi^j(\alpha_0) \vee \ldots \vee \\ &\qquad\qquad \vee \bigvee_{j=-k-(n-1)}^{k-(n-1)} \varphi^j(\alpha_0)\Big) \\ &= \lim_{n\to\infty} \frac{1}{n} H_0\Big(\bigvee_{j=-k-(n-1)}^{k} \varphi^j(\alpha_0)\Big) \\ &= \lim_{n\to\infty} \frac{1}{n} H_0\Big(\varphi^k\Big(\bigvee_{j=-k-(n-1)}^{k} \varphi^j(\alpha_0)\Big)\Big) \\ &= \lim_{n\to\infty} \frac{1}{n} H_0\Big(\bigvee_{j=1}^{2k} \varphi^j(\alpha_0) \vee \bigvee_{j=-(n-1)}^{0} \varphi^j(\alpha_0)\Big) \\ &\leqslant \lim_{n\to\infty} \frac{1}{n}\Big(H_0\Big(\bigvee_{j=1}^{2k} \varphi^j(\alpha_0)\Big) + H_0\Big(\bigvee_{j=0}^{n-1} \varphi^{-j}(\alpha_0)\Big)\Big) = h(\varphi, \alpha_0). \end{aligned}$$

It follows that $h(\varphi) = h(\varphi, \alpha_0)$.

Since $\varphi^{-j}(A_s) \cap \varphi^{-j}(A_r) = \emptyset$ for $s \neq r$, we have

$$N(\alpha_0 \vee \varphi^{-1}(\alpha_0) \vee \ldots \vee \varphi^{-n+1}(\alpha_0)) = m^n,$$

whence

$$H_0(\alpha_0 \vee \varphi^{-1}(\alpha_0) \vee \ldots \vee \varphi^{-n+1}(\alpha_0)) = n \log m$$

and we arrive at the result

$$h(\varphi) = h(\varphi, \alpha_0) = \log m.$$

If we took X to be the product of countably many copies of an infinite set $\mathscr{A}$, the entropy of the shift defined analogously would be infinite.

5.8.6. Let φ be an algebraic automorphism of the torus T^2 and suppose that φ is ergodic. Let λ and $\varepsilon\lambda^{-1}$ denote the eigenvalues of φ; $\varepsilon = \operatorname{sign}\det\varphi$. According to the conclusion of Exercise 5.2.6 λ is a real number and $|\lambda| \neq 1$; we may assume that $|\lambda| > 1$. We shall show that $h(\varphi) = \log|\lambda|$. Let α_k be the cover of T^2 consisting of open parallelograms P with sides of lengths $< 1/k$ parallel to the eigenvectors of φ. Denote these vectors by a and b: $\|a\| = \|b\| = 1$, $\varphi(a) = \lambda a$, $\varphi(b) = \varepsilon\lambda^{-1}b$.

For every $P \in \alpha_k$ the set $\varphi^{-1}(P)$ is also a parallelogram with sides parallel to a and b, of length less than $\frac{1}{k}|\lambda|^{-1}$ and $\frac{1}{k}|\lambda|$, respectively. Consequently the cover $\alpha_k \vee \varphi^{-1}(\alpha_k)$ consists of parallelograms with sides parallel to a and b, of length less than $\frac{1}{k}|\lambda|^{-1}$ and $\frac{1}{k}$. We show by induction that the members of the cover $\alpha_k \vee \varphi^{-1}(\alpha_k) \vee \ldots \vee \varphi^{-n+1}(\alpha_k)$ are also such parallelograms, with sides of length less than $\frac{1}{k}|\lambda|^{-(n-1)}$ and $\frac{1}{k}$. The area of each of them is less than $k^{-2}|\lambda|^{-(n-1)}$. Thus

$$N(\alpha_k \vee \varphi^{-1}(\alpha_k) \vee \ldots \vee \varphi^{-n+1}(\alpha_k))k^{-2}|\lambda|^{-(n-1)} > 1$$

(T^2 being of unit surface), whence

$$N(\alpha_k \vee \varphi^{-1}(\alpha_k) \vee \ldots \vee \varphi^{-n+1}(\alpha_k)) > k^2|\lambda|^{n-1}.$$

This leads to the estimate

$$h(\varphi, \alpha_k) = \lim_{n\to\infty} \frac{1}{n} H_0\big(\alpha_k \vee \varphi^{-1}(\alpha_k) \vee \ldots \vee \varphi^{-n+1}(\alpha_k)\big)$$

$$\geqslant \lim_{n\to\infty} \frac{1}{n}(\log k^2 + (n-1)\log|\lambda|) = \log|\lambda|.$$

On the other hand, each parallelogram $P \in \alpha_k$ can be covered by $|\lambda|^{n-1}+1$ parallelograms which enter the cover $\alpha_k \vee \varphi^{-1}(\alpha_k) \vee \ldots \vee \varphi^{-n+1}(\alpha_k)$. It follows that

$$N\big(\alpha_k \vee \varphi^{-1}(\alpha_k) \vee \ldots \vee \varphi^{-n+1}(\alpha_k)\big) \leqslant (|\lambda|^{n-1}+1) N(\alpha_k),$$

yielding the estimate

$$h(\varphi, \alpha_k) \leqslant \log|\lambda|.$$

The two estimates result in

$$h(\varphi, \alpha_k) = \log|\lambda|.$$

By Corollary 5.8.1 we hence obtain

$$h(\varphi) = \lim_{k\to\infty} h(\varphi, \alpha_k) = \log|\lambda|.$$

Exercises

5.8.1.B. We say that a dynamical system (Y, ψ) is a *continuous image* of the system (X, φ) iff there exists a continuous mapping f of X onto Y such that $f \circ \varphi = \psi \circ f$. Show that, if (Y, ψ) is a continuous image of (X, φ), then $h(\varphi) \geqslant h(\psi)$.

Hint. If α is an open cover of Y, then $f^{-1}(\alpha)$ is an open cover of X.

5.8.2.A. Let α and β be covers of S^1 consisting of intervals. Show that

$$N(\alpha \vee \beta) \leqslant N(\alpha) + N(\beta).$$

Hint. Assume that α and β are minimal (hence finite). Split the members $I_i \cap J_k$ of $\alpha \vee \beta$ into three groups; 1° $I_i \supset J_k$, 2° $I_i \subset J_k$, 3° neither inclusion holds. The conclusion then follows easily.

5.8.3.B. Let φ be an algebraic automorphism of the m-torus T^m. Prove that

$$h(\varphi) = \sum_{\lambda_i \in \mathrm{sp}\,\varphi} \log^+ |\lambda_i|,$$

where $\log^+ x = \max(0, \log x)$; eigenvalues are counted with multiplicities.

Hint. Repeat the argument of Example 5.8.6.

5.9. Properties of Entropy. Interrelation between Topological and Metric Entropy

The concepts of topological entropy and metric entropy are related in a significant way. Namely, the topological entropy of a dynamical system is equal to the least upper bound of the values of metric entropy with respect to all invariant probability measures. This fact was already conjectured by Adler, Konheim, and McAndrew (1965).

For the case of a finite-dimensional space X and a homeomorphism φ the theorem was proved by Dinaburg (1971). Goodwyn (1969) showed that the topological entropy is greater than or equal to the metric entropy. Eventually, Goodman (1971) proved the theorem in the formulation given above. The original proofs are complicated. They have been simplified by Denker (1972). Quite recently Misiurewicz (1976a, 1976b) found surprisingly simple proofs, which we now present.

As usual, X denotes a compact metric space, $\varphi: X \to X$ is a continuous map and μ is an invariant probability measure on X. We prove

Theorem 5.9.1 (Goodwyn, 1969) For the dynamical system (X, φ) the following inequality holds true:

$$h_\mu(\varphi) \leqslant h(\varphi).$$

In the proof we shall need two lemmas. Let $\alpha = \{A_i\}$ be a finite open cover and let $\xi = \{C_i\}$ be a finite partition. Write

$$\xi^n = \bigvee_{i=0}^{n-1} \varphi^{-i}(\xi), \qquad \alpha^n = \bigvee_{i=0}^{n-1} \varphi^{-i}(\alpha),$$

$$p(\alpha, \xi) = \max_{A \in \alpha} \operatorname{card} \{C \in \xi : A \cap C \neq \emptyset\}.$$

Lemma 5.9.1 $h_\mu(\varphi, \xi) \leqslant h(\varphi, \alpha) + \log p(\alpha, \xi)$.

Proof According to Exercise 5.7.4 and to the definition of $p(\alpha, \xi)$ we have

$$H(\xi^n) \leqslant \log N(\xi^n) \leqslant \log\left(p^n(\alpha, \xi) N(\alpha^n)\right).$$

Dividing by n and passing with n to infinity, we obtain our inequality.

Lemma 5.9.2 For every partition ξ there exist a partition $\tilde{\xi}$ and a cover α such that

(1) $h_\mu(\varphi, \tilde{\xi}) \geqslant h_\mu(\varphi, \xi) - 1$,
(2) $p(\alpha, \tilde{\xi}) \leqslant 2$.

Proof Let $\xi = \{C_1, \ldots, C_s\}$. Choose compact sets $K_i \subset C_i$, $i = 1, \ldots, s$, such that $\mu(C_i - K_i)$ are small enough for (1) to hold for the partition

$$\tilde{\xi} = \{K_1, \ldots, K_s, X - \bigcup_{i=1}^{s} K_i\}.$$

Put $U_{s+1} = X - \bigcup_{i=1}^{s} K_i$ and $U_i = K_i \cup U_{s+1}$ and define α:

$$\alpha = \{U_1, \ldots, U_s, U_{s+1}\}.$$

Condition (2) is satisfied, as is easy to see.

Proof of Theorem 5.9.1 Let ξ be any finite partition and let $\tilde{\xi}$ and α be as in Lemma 5.9.2 with φ^n in place of φ (n being fixed for the moment). In virtue of Lemmas 5.9.1 and 5.9.2 we get

$$\begin{aligned} h_\mu(\varphi^n, \xi) &\leqslant h_\mu(\varphi^n, \tilde{\xi}) + 1 \\ &\leqslant h(\varphi^n, \alpha) + \log p(\alpha, \tilde{\xi}) + 1 \\ &\leqslant h(\varphi^n) + \log 2 + 1. \end{aligned}$$

Hence

$$h_\mu(\varphi^n) \leqslant h(\varphi^n) + \log 2 + 1.$$

Applying Theorems 5.7.3 and 5.8.4 we obtain

$$nh_\mu(\varphi) \leqslant nh(\varphi) + \log 2 + 1.$$

Division by n yields in the limit ($n \to +\infty$)

$$h_\mu(\varphi) \leqslant h(\varphi).$$

Before further considerations we produce an equivalent definition of topological entropy. As usual, X is a compact metric space, ϱ denoting the metric, and $\varphi: X \to X$ is a continuous map.

Definition 5.9.1 (1) A set $E \subset X$ is said to be (n, ε)-*separated* iff, for any two distinct points $x, y \in E$, there is a j, $0 \leqslant j < n$, such that

$$\varrho(\varphi^j(x), \varphi^j(y)) > \varepsilon.$$

(2) A set $F \subset X$ is said to be (n, ε)-*spanning* iff for every $x \in X$ there exists an $y \in F$ such that

$$\varrho(\varphi^j(x), \varphi^j(y)) \leqslant \varepsilon \qquad \text{for } 0 \leqslant j < n.$$

Introduce the notation:

$$r_n(\varepsilon) = \min\{\operatorname{card} F\colon F \text{ is } (n, \varepsilon)\text{-spanning}\},$$

$$s_n(\varepsilon) = \max\{\operatorname{card} E\colon E \text{ is } (n, \varepsilon)\text{-separated}\},$$

$$\bar{r}(\varepsilon) = \limsup_{n\to\infty} \frac{1}{n} \log r_n(\varepsilon),$$

$$\bar{s}(\varepsilon) = \limsup_{n\to\infty} \frac{1}{n} \log s_n(\varepsilon).$$

Lemma 5.9.3 (1) $r_n(\varepsilon) \leqslant s_n(\varepsilon) \leqslant r_n(\tfrac{1}{2}\varepsilon) < +\infty$;
(2) *if* $\varepsilon_1 < \varepsilon_2$ *then* $\bar{r}(\varepsilon_1) \geqslant \bar{r}(\varepsilon_2)$, $\bar{s}(\varepsilon_1) \geqslant \bar{s}(\varepsilon_2)$.

Proof It is readily noted that any maximal (n, ε)-separated set is automatically (n, ε)-spanning; otherwise it might be enlarged. Hence $r_n(\varepsilon) \leqslant s_n(\varepsilon)$. Now suppose that E is (n, ε)-separated and F is $(n, \tfrac{1}{2}\varepsilon)$-spanning. Thus, for every $x \in X$, there is a $g(x) \in F$ such that $\varrho(\varphi^j(x), \varphi^j(g(x))) \leqslant \tfrac{1}{2}\varepsilon$ for $0 \leqslant j < n$. If $g(x) = g(y)$, then $\varrho(\varphi^j(x), \varphi^j(y)) \leqslant \varepsilon$ for $0 \leqslant j < n$. Hence, for $x, y \in E$ we have $g(x) \neq g(y)$ and so $\operatorname{card} F \geqslant \operatorname{card} E$. Consequently $r_n(\tfrac{1}{2}\varepsilon) \geqslant s_n(\varepsilon)$. The finiteness of $r_n(\cdot)$, $s_n(\cdot)$ is evident, and so are inequalities (2).

Lemma 5.9.3 implies the existence of the limit

$$d(\varphi) := \lim_{\varepsilon\to 0} \bar{r}(\varepsilon) = \lim_{\varepsilon\to 0} \bar{s}(\varepsilon). \tag{5.9.1}$$

Theorem 5.9.2 (*Bowen, 1971*) *For every dynamical system* (X, φ) *the following equality holds true*:

$$h(\varphi) = d(\varphi).$$

Proof Let α_ε be a finite cover consisting of balls with radii less than ε and let $\tilde{\alpha}_\varepsilon^n$ be a minimal subcover in the cover $\alpha_\varepsilon^n = \bigvee_{k=0}^{n-1} \varphi^{-k}(\alpha_\varepsilon)$. From each member of $\tilde{\alpha}_\varepsilon^n$ select one point; denote the resulting set by F. It is easy to see that F is (n, ε)-spanning. Therefore

$$N(\alpha_\varepsilon^n) = N(\tilde{\alpha}_\varepsilon^n) = \operatorname{card} F \geqslant r_n(\varepsilon).$$

Taking logarithms, dividing by n and passing with n to infinity we obtain

$$h(\varphi) \geqslant h(\varphi, \alpha_\varepsilon) \geqslant \bar{r}(\varepsilon),$$

whence, passing with ε to 0,

$$h(\varphi) \geqslant d(\varphi). \tag{5.9.2}$$

Now let $\alpha = \{A_i\}$ be an arbitrary open cover of mesh less than ε. Denote by δ the Lebesgue number of α. Let F_n be minimal $(n, \frac{1}{2}\varepsilon)$-spanning. For $x \in F_n$ put

$$B_x = \bigcap_{k=0}^{n-1} \{y \in X;\ \varrho(\varphi^k(x), \varphi^k(y)) < \tfrac{1}{2}\delta\}.$$

Since δ is the Lebesgue number for α and since $\operatorname{diam} B_x < \delta$, we see that each B_x is contained in some member of α^n: for $x \in \bigcap_{k=0}^{n-1} \varphi^{-k}(A_{i_k})$ we have

$$\varphi^k(B_x) \subset \{y\colon \varrho(\varphi^k(x), y) < \tfrac{1}{2}\delta\}, \qquad 0 \leqslant k < n,$$

and so

$$\varphi^k(B_x) \subset A_{i_k}, \qquad 0 \leqslant k < n,$$

which means that

$$B_x \subset \bigcap_{k=0}^{n-1} \varphi^{-k}(A_{i_k}).$$

Consider the cover

$$\beta_n = \{B_x\colon x \in F_n\}.$$

As we have shown, it is finer than α^n. Hence

$$N(\alpha^n) \leqslant N(\beta_n) = \operatorname{card} F_n = r_n(\tfrac{1}{2}\delta),$$

and thus

$$h(\varphi, \alpha) = \lim_{n\to\infty} \frac{1}{n} \log N(\alpha^n) \leqslant \limsup_{n\to\infty} \frac{1}{n} \log r_n(\tfrac{1}{2}\delta) = \bar{r}(\tfrac{1}{2}\delta).$$

Letting $\varepsilon \to 0$ (which yields $\delta \to 0$), we get

$$h(\varphi) \leqslant d(\varphi),$$

which together with (5.9.2) gives the assertion of the theorem.

Now we prove

Theorem 5.9.3 (Goodman, 1971) For every dynamical system (X, φ) (where, as before, X is compact and φ is continuous) the following equality holds true:

$$h(\varphi) = \sup_{\mu} h_\mu(\varphi),$$

the supremum being taken over all invariant probability measures.

Proof (*Misiurewicz, 1976b*) For any probability measure μ write $S_\mu^k(A) = \mu(\varphi^{-k}(A))$. Fix $\varepsilon > 0$. For any positive integer n let E_n be an (n, ε)-separated set such that $\operatorname{card} E_n = s_n(\varepsilon)$. Denote by σ_n the measure concentrated on E_n:

$$\sigma_n(\{x\}) = \frac{1}{s_n(\varepsilon)}, \qquad x \in E_n.$$

Define

$$\mu_n = \frac{1}{n} \sum_{k=0}^{n-1} S_{\sigma_n}^k.$$

The measures μ_n are normalized; hence the set $\{\mu_n\colon n = 1, 2, \ldots\}$ is weak *-compact in $C(X)^*$. Let μ be a cluster point of the sequence (μ_n). We may require that μ be the limit of a certain subsequence (μ_{n_i}) such that

$$\lim_{i\to\infty} \frac{1}{n_i} s_{n_i}(\varepsilon) = \bar{s}(\varepsilon).$$

Clearly, μ is an invariant measure, which depends on ε.

We now construct a partition α with the following properties:

$$\operatorname{diam}(\alpha) < \varepsilon,$$

$$\mu\Big(\bigcup_{A \in \alpha^n} \operatorname{Fr} A\Big) = 0 \qquad \text{for all } n$$

(Fr denotes boundary). This is done as follows:

Let $E = \{x_1, \ldots, x_m\}$ be a $\frac{1}{4}\varepsilon$-net. The sets $\{y\colon \varrho(x_i, y) = r\}$, $\frac{1}{4}\varepsilon < r < \frac{1}{2}\varepsilon$, are closed and disjoint for different r. Thus there are at most countably many values of r such that

$$\mu\Big(\bigcup_{i=1}^{m} \operatorname{Fr}\{y\colon \varrho(x_i, y) \leqslant r\}\Big) \neq 0.$$

Take an r for which this is not the case and define

$$A_1 = \{y\colon \varrho(x_1, y) \leqslant r\},$$
$$A_2 = \{y\colon \varrho(x_2, y) \leqslant r\} - A_1,$$
$$\cdots\cdots\cdots\cdots\cdots\cdots$$
$$A_m = \{y\colon \varrho(x_m, y) \leqslant r\} - \bigcup_{i=1}^{m-1} A_i$$

and

$$\alpha = \{A_1, \ldots, A_m\}.$$

Clearly,

$$\operatorname{diam}(\alpha) \leqslant 2r < \varepsilon$$

and

$$\mu\Big(\bigcup_{i=1}^{m} \operatorname{Fr} A_i\Big) \leqslant \mu\Big(\bigcup_{i=1}^{m} \operatorname{Fr}\{y\colon \varrho(x_i, y) \leqslant r\}\Big) = 0.$$

Using the elementary inclusions

$$\operatorname{Fr}(\varphi^{-k}(B)) \subset \varphi^{-k}(\operatorname{Fr} B), \quad \operatorname{Fr}\Big(\bigcup_{i=1}^{k} B_i\Big) \subset \bigcup_{i=1}^{k} \operatorname{Fr} B_i,$$

we hence obtain

$$\mu\Big(\bigcup_{A \in \alpha^n} \operatorname{Fr} A\Big) = 0.$$

Now, let n, m be integers, $n \geqslant 2m > 0$, and let $0 \leqslant j < m$. Write

$$s(j) = E\left(\frac{n-j}{m}\right) - 1.$$

We have

$$\alpha^n = \bigvee_{k=0}^{s(j)} \varphi^{-km-j}(\alpha^m) \vee \bigvee_{k\in I} \varphi^{-k}(\alpha),$$

where

$$I = \{0, 1, \ldots, j-1\} \cup \{ms(j)+j+m, \ldots, n-1\}.$$

Since

$$I \subset \{0, 1, \ldots, m-1\} \cup \{n-m, \ldots, n-1\},$$

we have

$$\operatorname{card} I \leqslant 2m.$$

By the conclusions of Exercises 5.7.3 and 5.7.4 we obtain the estimate

$$\begin{aligned} H_{\sigma_n}(\alpha^n) &= H_{\sigma_n}\Big(\bigvee_{k=0}^{s(j)} \varphi^{-km-j}(\alpha^m) \vee \bigvee_{k\in I} \varphi^{-k}(\alpha)\Big) \\ &\leqslant \sum_{k=0}^{s(j)} H_{\sigma_n}(\varphi^{-km-j}(\alpha^m)) + \sum_{k\in I} H_{\sigma_n}(\alpha) \\ &= \sum_{k=0}^{s(j)} H_{\sigma_n}(\varphi^{-km-j}(\alpha^m)) + (\operatorname{card} I)(\log \operatorname{card} \alpha) \\ &\leqslant \sum_{k=0}^{s(j)} H_{\sigma_n}(\varphi^{-km-j}(\alpha^m)) + 2m \log \operatorname{card} \alpha. \end{aligned}$$

This inequality holds for $j = 0, 1, \ldots, m-1$. Summing these m inequalities together, we get

$$mH_{\sigma_n}(\alpha^n) \leqslant \sum_{j=0}^{m-1} \sum_{k=0}^{s(j)} H_{\sigma_n}(\varphi^{-km-j}(\alpha^m)) + 2m^2 \log \operatorname{card} \alpha.$$

Rearrange the double sum increasingly with respect to iteration exponents. We obtain (possibly enlarging the value of the sum by introducing some non-negative terms)

$$mH_{\sigma_n}(\alpha^n) \leqslant \sum_{k=0}^{n-1} H_{\sigma_n}(\varphi^{-k}(\alpha^m)) + 2m^2 \log \operatorname{card} \alpha. \tag{5.9.3}$$

The set E_n is (n, ε)-separated. Thus each member of α^n contains at most one point of E_n. Since $\sigma_n(\{x\}) = 1/s_n(\varepsilon)$ for $x \in E_n$, we have

$$H_{\sigma_n}(\alpha^n) = -\sum_{x\in E_n} \frac{1}{s_n(\varepsilon)} \log \frac{1}{s_n(\varepsilon)} = \log s_n(\varepsilon).$$

Further, observe that

$$H_{\sigma_n}(\varphi^{-k}(\alpha^m)) = H_{S^k_{\sigma_n}}(\alpha^m).$$

Inserting these values into (5.9.3) and dividing by n we obtain

$$\frac{m}{n}\log s_n(\varepsilon) \leqslant \frac{1}{n}\sum_{k=0}^{n-1} H_{S^k_{\sigma_n}}(\alpha^m) + \frac{2m^2}{n}\log\operatorname{card}\alpha.$$

Hence, in view of Exercise 5.7.5 (applied to an arbitrary finite convex combination of measures), we get

$$\frac{m}{n}\log s_n(\varepsilon) \leqslant H_{\mu_n}(\alpha^m) + \frac{2m^2}{n}\log\operatorname{card}\alpha \tag{5.9.4}$$

(recall that $\mu_n = \frac{1}{n}\sum_{k=0}^{n-1} S^k_{\sigma_n}$). Taking $n = n_i$, $i = 1, 2, \ldots$, and using the fact that the members of α^m have boundaries of measure zero, we see that

$$\lim_{i\to\infty} H_{\mu_{n_i}}(\alpha^m) = H_\mu(\alpha^m).$$

Thus passing to the limit in (5.9.4) as $n_i \to \infty$, we obtain

$$\bar{s}(\varepsilon) \leqslant \frac{1}{m} H_\mu(\alpha^m).$$

This yields, by letting m tend to infinity,

$$\bar{s}(\varepsilon) \leqslant h_\mu(\varphi, \alpha) \leqslant h_\mu(\varphi).$$

The measure μ depends on ε: $\mu = \mu(\varepsilon)$. Taking on both sides the supremum over $\varepsilon > 0$ we get, on account of Theorem 5.9.2,

$$h(\varphi) = \sup_{\varepsilon>0} \bar{s}(\varepsilon) \leqslant \sup_{\varepsilon>0} h_{\mu(\varepsilon)}(\varphi) \leqslant \sup_{\mu} h_\mu(\varphi).$$

The opposite inequality is asserted by Theorem 5.9.1. The proof is thus complete.

Remark 5.9.1 B. Gurevich (1969) has constructed an example of a transitive dynamical system (X, φ) in which the strict inequality

$$h_\mu(\varphi) < h(\varphi)$$

holds for every φ-invariant probability measure μ; in other words, the supremum $\sup_\mu h_\mu(\varphi)$ is not attained.

Invariant measures μ which enjoy the property

$$h_\mu(\varphi) = h(\varphi)$$

are called *maximal.*

In Gurevich's example (X, φ) is not a smooth dynamical system (i.e., X is not a manifold). M. Misiurewicz (1973) proved that every manifold M of dimension $\geqslant 4$ admits a diffeomorphism $\varphi\colon M \to M$ without any maximal invariant measure.

As before let $\Omega = \Omega(\varphi)$ denote the set of all nonwandering points of the system (X, φ). The set Ω is compact and φ-invariant; thus it makes sense to consider the dynamical system (Ω, φ).

Theorem 5.9.4 The topological entropy of the system (X, φ) *is attained on* Ω, *i.e.,*

$$h(\varphi) = h(\varphi|_\Omega).$$

This is a direct consequence of Exercise 5.3.3 on p. 289 and of Theorem 5.9.3.

Now we turn our attention to the entropy of smooth dynamical systems. Let M be a Riemannian manifold, $\dim M = m$, and let $\varphi\colon M \to M$ satisfy the Lipschitz condition

$$\varrho(\varphi(x), \varphi(y)) \leqslant L\varrho(x, y) \qquad \text{for } x, y \in M, \tag{5.9.5}$$

where L is some constant and ϱ denotes the metric on M induced by the Riemannian metric. It is not difficult to see that, by the compactness of M, condition (5.9.5) does not depend on the choice of the metric; the constant L itself may depend on it.

Theorem 5.9.5 (Krzyżewski, 1968) If φ *satisfies condition* (5.9.5), *then the topological entropy* $h(\varphi)$ *is finite.*

The proof is based on the following lemma, which is sufficiently obvious to justify the omission of a proof.

Lemma 5.9.4 There exist a system of charts (U_i, ψ_i) *on* M, $i = 1, \ldots, k$, *an open set* $G \subset \mathbf{R}^m$, *and constants* $a, b > 0$ *such that*

(1) $I^m \subset G = \psi_i(U_i), \qquad i = 1, \ldots, k,$

(2) $\bigcup_{i=1}^{k} \psi_i^{-1}(I^m) = M,$

(3) $ad(x_1, x_2) \leqslant \varrho(\psi_i^{-1}(x_1), \psi_i^{-1}(x_2)) \leqslant bd(x_1, x_2)$ *for* $x_1, x_2 \in G$, $i = 1, \ldots, k$;

here $d(\cdot, \cdot)$ *denotes the usual metric in* $\mathbf{R}^m$ *and* I^m *the unit cube in* $\mathbf{R}^m$.

Proof of Theorem 5.9.5 Applying Lemma 5.9.4, for any p large enough (say, $p \geqslant p_0$) one can find a cover α_p of M such that

$$\operatorname{diam}(\alpha_p) \leqslant 2^{1-p} b\sqrt{m},$$

$$\operatorname{card} \alpha_p \leqslant k2^{pm}.$$

Let n be any positive integer. Select one point from each member of α_p. According to (5.9.5) the resulting set of cardinality $\leqslant k2^{pm}$ is $(n, 2^{1-p} b\sqrt{m}\, L^n)$-spanning.

Fix $\varepsilon > 0$ and take

$$p = E\left(\frac{\log \dfrac{b\sqrt{m}\, L^n}{\varepsilon}}{\log 2} + 2\right).$$

Then $2^{1-p} b\sqrt{m}\, L^n < \varepsilon$, and thus the set just constructed is (n, ε)-spanning. Therefore

$$r_n(\varepsilon) \leqslant k2^{pm},$$

whence after an easy calculation we get

$$\frac{1}{n} \log r_n(\varepsilon) \leqslant \frac{1}{n} \log (kb^m m^{\frac{1}{2}m} 2^{2m} \varepsilon^{-m}) + m \log L.$$

Thus

$$\bar{r}(\varepsilon) = \limsup_{n \to \infty} \frac{1}{n} \log r_n(\varepsilon) \leqslant m \log L.$$

In virtue of Theorem 5.9.2 with formula (5.9.1), this yields the estimate

$$h(\varphi) \leqslant m \log L,$$

ending the proof.

Corollary 5.9.1 If $\varphi\colon M \to M$ *is of class* C^1, *then* $h(\varphi) < +\infty$.

Corollary 5.9.2 If $\varphi\colon M \to M$ *is of class* C^1, *then for every* φ*-invariant probability measure* μ *the entropy* $h_\mu(\varphi)$ *is finite.*

Proof Theorems 5.9.1 and 5.9.5 show that

$$h_\mu(\varphi) \leqslant h(\varphi) < +\infty.$$

This fact was first proved by Kushnirenko (1965) by a quite different method.

Theorem 5.9.6 If φ: $M \to M$ is an expanding map of a compact Riemannian manifold M, then

$$h(\varphi) = \log \deg \varphi.$$

Proof Let $N = \deg \varphi$ and let E be any $(1, \varepsilon)$-separated set, where ε is so small that $\varrho(y_1, y_2) \geqslant \varepsilon$ for any $x \in M$ and any $y_1, y_2 \in \varphi^{-1}(x)$: such an ε exists because φ is a covering. Then the set $\varphi^{-1}(E)$ is $(2, \varepsilon)$-separated, $\varphi^{-2}(E)$ is $(3, \varepsilon)$-separated, and by induction, $\varphi^{-n}(E)$ is $(n+ +1, \varepsilon)$-separated. Since, by assumption, φ is an n-fold covering, we have

$$\operatorname{card} \varphi^{-n}(E) = N^n \operatorname{card} E,$$

whence

$$s_n(\varepsilon) \geqslant N^{n-1} \operatorname{card} E$$

and

$$\bar{s}(\varepsilon) = \limsup_{n \to \infty} \frac{1}{n} \log s_n(\varepsilon) \geqslant \lim_{n \to \infty} \frac{1}{n} \log (N^{n-1} \operatorname{card} E) = \log N.$$

Letting $\varepsilon \to 0$ we obtain, in view of Theorem 5.9.2,

$$h(\varphi) \geqslant \log N. \tag{5.9.6}$$

Now let F be a $(1, \varepsilon)$-spanning set and write

$$U_x = \{y\colon \varrho(x, y) < \varepsilon\} \qquad \text{for } x \in F.$$

The components of $\varphi^{-n}(U_x)$, $x \in F$, constitute a cover of M and the diameter of each component is less than ε, for any n (because φ is expanding). Thus if y belongs to some component of $\varphi^{-n}(U_x)$, then

$$\varrho(\varphi^k(x), \varphi^k(y)) < \varepsilon \qquad \text{for } k = 0, \ldots, n,$$

and so $\varphi^{-n}(F)$ is $(n+1, \varepsilon)$-spanning. Since

$$\operatorname{card} \varphi^{-n}(F) = N^n \operatorname{card} F,$$

we have

$$r_n(\varepsilon) \leqslant N^{n-1} \operatorname{card} F,$$

whence

$$\bar{r}(\varepsilon) = \limsup_{n\to\infty} \frac{1}{n} \log r_n(\varepsilon) \leqslant \lim_{n\to\infty} \frac{1}{n} \log(N^{n-1} \operatorname{card} F) = \log N.$$

Letting $\varepsilon \to 0$ we obtain, again by Theorem 5.9.2,

$$h(\varphi) \leqslant \log N. \tag{5.9.7}$$

Inequalities (5.9.6) and (5.9.7) jointly give the assertion of the theorem.

We conclude the section with a every interesting theorem regarding entropy:

Theorem 5.9.7 (*Bowen, 1970b*) *Let M be a compact manifold and let $\varphi \in \operatorname{Diff}^2(M)$. If φ satisfies Axiom A, then*

$$\lim_{n\to\infty} \frac{1}{n} \log C(\varphi^n) = h(\varphi),$$

where $C(\varphi^n)$ denotes the number of fixed points of φ^n.

We have proved this theorem in the very particular case of the torus T^2 and an algebraic automorphism φ (see Corollary 1.6.1).

In the general case the proof of Theorem 5.9.7 is not easy and is therefore omitted.

North Carolina State University

School of Physical and Mathematical Sciences

Department of Mathematics
Box 8205
North Carolina State University
Raleigh, N. C. 27695-8205
(919) 737-2382

Date: April 28, 1986

To: Dr. Charlton

From: Math Library

The following book is being held for you in the Math Library until May 12, 1986:

QA614.8 An Introduction to the Theory of Smooth
S91413 Dynamical Systems by Wieslaw Szlenk
1984

Notes

Complementary Definitions, Facts and Comments

[1] A family $\{\varphi^\alpha: \alpha \in \mathfrak{A}\}$ of mappings of a metric space X into itself is called a *continuous semigroup* (of mappings) iff the index set $\mathfrak{A}$ has the structure of a commutative metric semigroup and the following conditions are satisfied:

1° $\{\varphi^\alpha: \alpha \in \mathfrak{A}\}$ is a semigroup with respect to the operation of composition:

$$\varphi^{\alpha+\beta} = \varphi^\alpha \circ \varphi^\beta \qquad \text{for all } \alpha, \beta \in \mathfrak{A};$$

2° the operation $(\alpha, x) \mapsto \varphi^\alpha(x)$ is jointly continuous ($\alpha \in \mathfrak{A}$, $x \in X$);
3° φ^0 is the identity map.

[2] Let X and Y be topological spaces. A continuous mapping $\pi: Y \to X$ is said to be *covering* iff every point $x_0 \in X$ has a neighbourhood U such that $\pi^{-1}(U)$ is the disjoint union of open subsets of Y, each of them being mapped by π homeomorphically onto U; then Y is called a *covering space* and X the *underlying space*, or the *base space*.

[3] Suppose that Y_0 is a covering space of X such that, for any other covering space Y, the space Y_0 is covering for Y. Then Y_0 is called a *universal* covering space.

If Y_0 is covering for X, Y_0 is simply connected and X is locally path-connected, then Y_0 is a universal covering space of X.

If X is a locally simply connected space (e.g. a connected differentiable manifold), then every universal covering space of X is simply connected.

Suppose that Y_0 is a universal covering space of X and let $f: X \to X$ be a continuous map. Then there exists a mapping $\tilde{f}: Y_0 \to Y_0$ such that the following diagram commutes:

$$\begin{array}{ccc} Y_0 & \xrightarrow{\tilde{f}} & Y_0 \\ \big\downarrow{\scriptstyle \pi} & & \big\downarrow{\scriptstyle \pi} \\ X & \xrightarrow[f]{} & X \end{array}$$

Such an $\tilde{f}$ is called a *lift* of f to Y_0 (see Spanier, 1966).

[4] Let X be a complete metric space with metric $\varrho(\cdot, \cdot)$. A mapping $f\colon X \to X$ is called a *contraction* iff there is a number α, $0 < \alpha < 1$, such that

$$\varrho(f(x), f(y)) \leqslant \alpha\varrho(x, y) \qquad \text{for all } x, y \in X.$$

The Banach Fixed Point Theorem If f is a contraction, then there exists a unique point $x \in X$ such that $f(x) = x$; x can be obtained as the limit of any sequence (x_n) where $x_{n+1} = f(x_n)$ and $x_0 \in X$ is arbitrary.

[5] *Corollary to Brouwer's theorem on the invariance of domain* Let $U \subset \boldsymbol{R}^m$ be an open set and let f be a one-to-one continuous mapping of U onto a set $V \subset \boldsymbol{R}^m$. Then V is open and f maps U onto V homeomorphically (see Spanier, 1966, p. 224).

[6] Let $\alpha = \{U_i\}$ be a family of open subsets of a metric space X such that $\bigcup U_i = X$ (shortly, an *open cover* of X). A number $\delta > 0$ is called a *Lebesgue number* of α iff every set $A \subset X$ with diameter $\leqslant \delta$ is contained some U_i. If X is compact, then for any open cover there is a Lebesgue number $\delta > 0$.

[7] A *vector bundle* is a quadruple (E, B, F, p), where E and B are topological spaces, E being called the *total space* and B the *underlying space*, or the *base space*, F is a linear topological space called the *fibre*, and $p\colon E \to B$ is a continuous map called the *projection map*; it is assumed that there exist an open cover α of B and homeomorphisms $h_U\colon U \times F \to p^{-1}(U)$, $U \in \alpha$, such that the following diagram commutes:

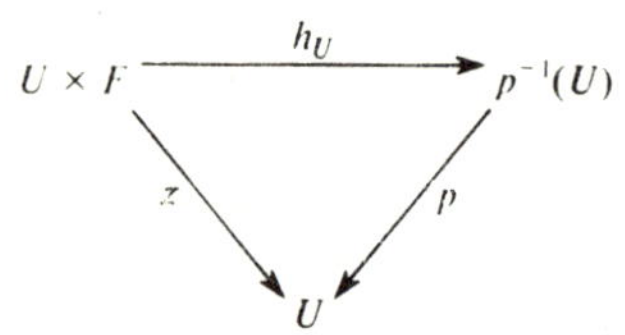

π denoting the projection onto the first factor. Hence, the bundle projection map $p\colon E \to B$ and the natural projection mapping $\pi\colon U \times F \to U$ are locally equivalent. The fibre over $b \in B$ is defined to be equal to $p^{-1}(b)$; it is demanded that for any $b \in B$ the fibre $p^{-1}(b)$ should be a linear space isomorphic to F, the isomorphism being established by the mapping $F \ni x \mapsto h_U(b, x) \in p^{-1}(b)$, where U denotes a suitable neighbourhood of b, $U \in \alpha$ (see Spanier, 1966, Ch. 2, §7).

[8] By a *section* of a vector bundle (E, B, F, p) we mean any mapping $s\colon B \to E$ such that $p \circ s = \mathrm{id}_B$. For any $U \in \alpha$ (notation as in Note 7) we have $s\colon U \to p^{-1}(U)$ and

$$h_U^{-1} \circ s(b) = (b, x(b)), \; b \in U, \; x(b) \in E.$$

If the mapping s is continuous, we say that the section is *continuous*; s is the *zero-section* iff $h_U^{-1} \circ s(b) = (b, 0)$ for all $b \in B$.

[9] Let M, N be topological spaces and let $f, g\colon M \to N$ be continuous mappings. We say that f and g are *homotopic* iff there exists a continuous function $h\colon M \times \langle 0, 1\rangle \to N$

such that

$$h(x, 0) = f(x), \qquad h(x, 1) = g(x) \qquad \text{for all } x \in M.$$

If M, N have differentiable structure, we say that f, g are *smoothly homotopic* whenever there exists an h of class C^1 with properties as above. If, moreover, f, g and $h(\cdot, t)$ (for all t) are diffeomorphisms, then f and g are *isotopic*.

[10] Suppose that M is a compact differentiable manifold and denote by $\mathrm{Diff}^r(M)$ the class of all C^r-diffeomorphisms of M. We define the *C^r-topology* (*C^r-convergence*) in $\mathrm{Diff}^r(M)$ as follows:

$$\varphi_n \to \varphi_0 \text{ in } C^r\text{-topology}, \qquad \varphi_n, \varphi_0 \in \mathrm{Diff}^r(M)$$

iff

$$d^k\varphi_n(x) \to d^k\varphi_0(x) \text{ uniformly for } 0 \leqslant k \leqslant r.$$

In the usual metric induced from $C^r(M, M)$ (the sup metric for all $d^k\varphi$, $0 \leqslant k \leqslant r$) this space is incomplete. It is made into a complete metric space by redefining the metric (introducing the sup metric for all $d^k\varphi$ and $d^k\varphi^{-1}$, $0 \leqslant k \leqslant r$). The two metrics are equivalent, because the assignment $\varphi \mapsto \varphi^{-1}$ is continuous in C^r-topology (this follows from the compactness of M).

[11] A topological space M is called an *m-dimensional differentiable manifold* of class C^r, $0 \leqslant r \leqslant \infty$, $m < \infty$, iff there exist an open cover $\{U_i\}$ of M and a family of homeomorphisms $\alpha_i \colon U_i \to \boldsymbol{R}^m$ such that $\alpha_i(U_i)$ are open subsets of $\boldsymbol{R}^m$ and, for any i, j the composite mapping $\alpha_i \circ \alpha_j^{-1}$ (whenever it is defined) is of class C^r on its domain. Each pair (U, α) (we omit the lower indices) is called a *chart*, and the whole family $\{(U, \alpha)\}$ is called an *atlas*.

A continuous mapping $\gamma\colon J \to M$ ($J \subset \boldsymbol{R}$ being an interval) is called a *curve* of class C^s, $s \leqslant r$, iff the composite mapping $\alpha(\gamma(t))$ is a C^s-curve in $\boldsymbol{R}^m$ for any admissible chart α. The term: curve is also used with regard to the set $\hat{\gamma} := \gamma(J)$.

The class of all differentiable curves $\gamma(t)$, $t \in \langle 0, b)$ such that $\gamma(0) = x_0$ and that $\alpha(\gamma(t))$ have a common tangent half-line in $\boldsymbol{R}^m$ at the point $\alpha(x_0)$ determine (up to a factor) a *tangent vector* to M at x_0. If M is a C^∞-manifold, tangent vectors can be alternatively introduced as follows: denote by $C_0^\infty(M, x_0)$ the space of all real-valued C^∞-functions defined in a neighbourhood of x_0 (depending on the function). By a *tangent vector* to M at x_0 we mean any mapping of $C_0^\infty(M, x_0)$ into itself fulfilling the conditions;

1° $D(af + bg) = aD(f) + bD(g)$,

2° $D(fg) = D(f)g + fD(g)$

for $a, b \in \boldsymbol{R}$, $f, g \in C_0^\infty(M, x_0)$. The two definitions are equivalent.

The set of all tangent vectors to M at a point $x \in M$ is called the *tangent space* to M at x and is denoted by $T_x(M)$. The space of all pairs (x, v) where $x \in M$, $v \in T_x(M)$, is called the *tangent bundle* and is denoted by $T(M)$. This space has a natural structure of a differentiable manifold (see Gromoll, Klingenberg and Meyer, 1968; Sternberg, 1964).

[12] Suppose that a manifold M admits an atlas consisting of charts satisfying the following compatibility condition: if (U, α) and (V, β) are any two of these charts, then the derivative $d(\beta \circ \alpha^{-1})\colon \alpha(U \cap V) \to \beta(U \cap V)$ is orientation preserving (in $\boldsymbol{R}^m$). Then

M is said to be *orientable*. By an *orientation* of M we mean a maximal atlas satisfying the above condition. A connected orientable manifold admits exactly two distinct orientations.

Standard examples of non-orientable manifolds are: the *Möbius band* and the *Klein bottle*.

Every non-orientable manifold has a doubly covering space, which is an orientable manifold (see Gromoll, Klingenberg, and Meyer, 1968; Sternberg, 1964).

[13] Introduce an equivalence relation in $\boldsymbol{R} \times \boldsymbol{R}^m$ as follows:

$$(t, x^1, x^2, \ldots, x^m) \sim (t+k, (-1)^k x^1, x^2, \ldots, x^m), \ k \in Z.$$

The quotient space $(\boldsymbol{R} \times \boldsymbol{R}^m)/\sim$ is an m-dimensional manifold called the *generalized Möbius* band. It is non-orientable.

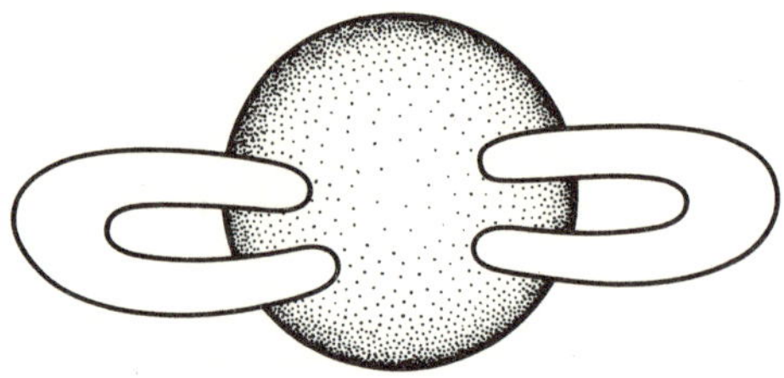

[14] Every orientable two-dimensional compact connected manifold is homeomorphic to a sphere with a number of "handles" (Wallace, 1968; see the figure). The number of handles, denote it by p, is equal to the maximal number of closed curves on M whose union does not disconnect M; this number is called the *genus* of M. It is connected with $\chi(M)$, the *Euler characteristic* of M, by the formula: $\chi(M) = 2-2p$ (see Spanier, 1966).

[15] A *Riemannian manifold* is defined as a differentiable manifold such that, for every $x \in M$, in the tangent space $T_x(M)$ there is defined a scalar product $(\cdot, \cdot)_x$ depending on x continuously.

The *Riemannian metric* on M is the function

$$\|\dot{x}\|_x^2 = (\dot{x}, \dot{x})_x$$

defined for $\dot{x} \in T_x(M)$, $x \in M$. In local coordinate systems (i.e. in charts) the Riemannian metric is a homogeneous quadratic form of the vector $\dot{x} = (\dot{x}^1, \ldots, \dot{x}^m)$:

$$\|\dot{x}\|_x^2 = \sum_{i,j=1}^{m} g_{ij}(x)\dot{x}^i\dot{x}^j,$$

where $(g_{ij}(x))$ is a positive-definite matrix.

For any C^1-curve $\gamma(t)$, $t \in \langle a, b \rangle$, which is a simple arc in M, define its length to be the quantity

$$\int_a^b \|\dot{\gamma}(t)\| \, dt.$$

If M is connected, a metric can be introduced in M by means of lengths of curves:

$$\varrho(x, y) := \inf\{\text{length of } \gamma\colon \langle a, b\rangle \to M,\ \gamma(a) = x,\ \gamma(b) = y\}.$$

If M is compact, this infimum is attained for some curve.

[16] Let M be a Riemannian manifold and suppose that $\gamma = \gamma(t)$, $t \in J$ (J being an interval), is a curve such that every point $x_0 \in \hat{\gamma}$ has a neighbourhood U with the following property: if $x, y \in U \cap \hat{\gamma}$, $x = \gamma(a)$, $y = \gamma(b)$, are any two points such that the piece of $\hat{\gamma}$ corresponding to the values of $t \in \langle a, b\rangle$ is contained in U, then the length of that piece is equal to the distance $\varrho(x, y)$ (see Note 15). Any such curve $\hat{\gamma}$ is called a *geodesic line* in M.

In local coordinates $x = (x^1, \ldots, x^m)$, $m = \dim M$, the geodesics are described by the differential equation:

$$\frac{d^2x^k}{dt^2} = \sum_{i,j=1}^{m} \Gamma^k_{i,j}(x)\frac{dx^i}{dt}\frac{dx^j}{dt},$$

where $\Gamma^k_{i,j}(x)$ are called *Christoffel's symbols*; every solution of this equation is a geodesic line. Hence, for any $x \in M$ and any $v \in T_x(M)$ there is exactly one geodesic line passing through x and having v as its tangent vector at x.

Every point in M has a neighbourhood such that any two points in it can be connected, in a unique way, by a geodesic line of length equal to $\varrho(x, y)$ (see Gromoll, Klingenberg and Meyer, 1968; Sternberg, 1964).

[17] A very specific notion in Riemannian geometry is the so-called *exponential map* $\exp\colon T(M) \to M$ (we assume that M is compact). It is defined as follows: given a pair $(x, v) \in T(M)$, we draw a geodesic line through x in the direction of v and we pick a point x' on that line such that the distance from x to x' is equal to τ – a prescribed constant value; by definition, $x' = \exp_x v$. Usually one takes $\tau = 1$; in general, τ has to be so small that any point y whose distance from x is less than τ could be *uniquely* joined with x by a geodesic path of length $\varrho(x, y)$.

The mapping exp is of class C^1 and we have $d\exp_x(0) = \text{identity}$. Moreover, for a fixed $x \in M$, $\exp_x(\cdot)$ is a diffeomorphism of the ball $B(0, \tau) \subset T_x(M)$ onto a neighbourhood of x in M (see Gromoll, Klingenberg and Meyer, 1968; Sternberg, 1964).

[18] On a Riemannian manifold M we introduce the *Riemannian measure*: if A is a Borel set contained within a single chart (U, α) (i.e. $A \subset U$), we put

$$\lambda(A) := \int_{\alpha(A)} \sqrt{\det(g_{ij}(y))}\,dy, \qquad y = \alpha(x);$$

integration is performed with respect to the Lebesgue measure in $\boldsymbol{R}^m$; if A is not contained in a single chart, $\lambda(A)$ is defined "piecewise". It can be proved that the value of $\lambda(A)$ does not depend on the choice of charts.

[19] By the *Gaussian curvature* (or simply: the *curvature*) of a two-dimensional Riemannian manifold M at a point $x \in M$ we mean the number $K(x)$ defined as follows: let U be a neighbourhood of x such that any two points in U can be joined uniquely by a

geodesic line realizing the distance; for any pair of points x', $x'' \in U$ consider the geodesic triangle with vertices x, x', x'' (i.e. the curvilinear triangle whose sides are the corresponding geodesics) and denote by a, a', a'' the angles of that triangle, i.e., the angles between the respective tangent vectors in $T_x(M)$, $T_{x'}(M)$, $T_{x''}(M)$ (they are well defined, since the tangent spaces are equipped with scalar products); then

$$K(x) := \lim_{x',x'' \to x} \frac{a+a'+a''-\pi}{\lambda(\Delta xx'x'')},$$

where the expression in the denominator stands for the Riemannian measure of the geodesic triangle $xx'x''$.

If $\dim M = m > 2$, one can examine the curvatures of two-dimensional sections: given $x \in M$, choose a two-dimensional subspace P of $T_x(M)$, project it locally (in a neighbourhood of 0) onto M via the mapping exp and compute the curvature $K_P(x)$ of the resulting submanifold. If $K_P(x) = \text{const}$ for all $P \subset T_x(M)$, $\dim P = 2$, and for all $x \in M$, we say that M has *constant curvature*; e.g. the m-sphere S^m_R with radius R has constant curvature equal to R^{-2} (see Cartan, 1951).

[20] Let M be a Riemannian manifold. Denote by $T^*_x(M)$ the space of all linear functionals on the tangent space $T_x(M)$, $x \in M$. The set of all pairs (x, p), where $x \in M$, $p \in T^*_x(M)$, is called the *co-tangent bundle* and is denoted by the symbol $T^*(M)$. In the terminology of mechanics, the pair: position and velocity is an element of the tangent bundle, whereas the pair: position and momentum is an element of the co-tangent bundle.

[21] Let M be a Riemannian manifold of class C^∞ and let $h(x)$ be a real-valued C^1-function on M. Fix $x \in M$. With any tangent vector $v \in T_x(M)$ we associate the directional derivative $D = \partial_v$ (cf. Note 11); $\partial_v h(x)$ depends linearly on v. Every linear functional $f(v)$ on a Euclidean space is of the form $f(v) = (v, v_0)$, where v_0 is some vector. Hence, in particular, $\partial_v h(x) = (v, v_0)$; the vector v_0 is called the *gradient* of h at x.

If M is (a piece of) $\boldsymbol{R}^m$ and $x = (x^1, \ldots, x^m)$ are the natural coordinates, v_0 takes on the form

$$v_0 = \left[\frac{\partial h}{\partial x^1}, \ldots, \frac{\partial h}{\partial x^m}\right].$$

[22] A system of differential equations

$$\frac{\mathrm{d}x}{\mathrm{d}t} = f(x, t), \qquad x \in \boldsymbol{R}^m, \qquad f\colon \boldsymbol{R}^m \times \boldsymbol{R} \to \boldsymbol{R}^m,$$

is called *autonomous* whenever $f(x, t)$ does not depend explicitly on t, i.e., $f(x, t) = f(x)$, $f\colon \boldsymbol{R}^m \to \boldsymbol{R}^m$.

[23] *Whitney's Extension Theorem* Let r be a given positive integer, $D \subset \boldsymbol{R}^m$ an open set and $X \subset D$ a closed subset. Suppose that with any multi-index $\alpha = (\alpha_1, \ldots, \alpha_m)$, α_i being non-negative integers, $|\alpha| = \sum \alpha_i \leqslant r$, there is associated a continuous function f_α on X. Then a necessary and sufficient condition for the existence of a function $f \in C^r(D)$ such

that $D^\alpha f|_X = f_\alpha$, $|\alpha| \leqslant r$, is that the limit relation

$$f_\alpha(x) = \sum_{|\beta| \leqslant r-|\alpha|} \frac{1}{\beta!} f_{\alpha+\beta}(y)(x-y)^\beta + o(\|x-y\|^{r-|\alpha|})$$

should be satisfied uniformly on every compact subset of X as $\|x-y\| \to 0$ (see Malgrange, 1966).

[24] Suppose that E is a vector space over $\boldsymbol{R}$ or $\boldsymbol{C}$ and that a norm is defined in E, i.e. a functional $\|\cdot\|$ with the following properties:

1° $\|x\| \geqslant 0$ for all x; $\|x\| = 0$ iff $x = 0$;

2° $\|tx\| = |t|\,\|x\|$ for $x \in E$, $t \in \boldsymbol{R}(t \in \boldsymbol{C})$;

3° $\|x+y\| \leqslant \|x\| + \|y\|$ for $x, y \in E$.

If E is a complete metric space with the distance $\varrho(x, y) = \|x-y\|$, we call E a *Banach space*.

Standard examples:

1. $E = \boldsymbol{R}^m$.
2. $E = C(X)$, the space of continuous functions on X, X compact metric, $\|f\| = \sup_{x \in X} |f(x)|$ for $f \in C(X)$.
3. $E = L^2\langle 0, 1\rangle$, the space of square-integrable functions on $\langle 0, 1\rangle$, $\|f\| = \left(\int_0^1 |f(x)|^2 \mathrm{d}x\right)^{1/2}$.

[25] Let E be a Banach space with norm $\|\cdot\|$ and suppose that $|\cdot|$ is another norm in E. The two norms are said to be *equivalent* if there exist constants $a, b > 0$ such that

$$a\|x\| \leqslant |x| \leqslant b\|x\| \qquad \text{for all } x \in E.$$

[26] Let E be a Banach space and let E_0 be a closed subspace. We say that E_0 is a *complemented* subspace of E iff there is another closed subspace $E_1 \subset E$ such that $E = E_0 \oplus E_1$.

Non-complemented subspaces exist; e.g., in $C\langle 0, 1\rangle$, any subspace isomorphic to a Hilbert space is non-complemented (Pełczyński, 1968). Lindenstrauss (1973) proved that if every subspace $E_0 \subset E$ is complemented in E then E is isomorphic to a Hilbert space.

[27] Let E and F be Banach spaces. A mapping $A\colon E \to F$ is called a *linear operator* iff

$$A(\alpha x + \beta y) = \alpha Ax + \beta Ay \qquad \text{for } x, y \in E,\ \alpha, \beta \in \boldsymbol{R} \text{ (or } \boldsymbol{C}).$$

A is a *bounded* linear operator iff there exists a constant c such that $\|Ax\| \leqslant c\|x\|$ for all $x \in E$. The least such constant is called the *norm* of A (denoted by $\|A\|$); $\|A\| = \sup_{\|x\| \leqslant 1} \|Ax\|$. A linear operator is continuous iff it is bounded. The space of all bounded linear operators from E to F is itself a Banach space (with norm $\|A\|$) and is denoted by the symbol $\mathscr{L}(E, F)$.

[28] *The Banach–Steinhaus Theorem* If $A_n \in \mathscr{L}(E, F)$, $n = 1, 2, \ldots$, E, F Banach spaces, and if for every $x \in E$ the limit $\lim_{n\to\infty} A_n x$ exists, then denoting this limit by Ax we define an operator A, which belongs to $\mathscr{L}(E, F)$ (see Dunford and Schwartz, 1958).

[29] The space of all continuous linear functionals on a Banach space E is called the *conjugate* space to E and is denoted by E^*.

Riesz' Representation Theorem If $E = C(X)$ is the space of all continuous real-valued functions on a compact metric space X, then $E^* = C(X)^*$ is identified with the space of all signed bounded Borel measures on X; a measure λ, viewed as a functional on $C(X)$, acts on a function $f \in C(X)$ according to the formula

$$\lambda(f) = \int f \, d\lambda$$

and we have $\|\lambda\| = \operatorname{Var} \lambda$ (the total variation of λ) (see Dunford and Schwartz, 1958).

[30] Let E and F be Banach spaces and let $A \in \mathscr{L}(E, F)$. We define the *dual operator* $A^*: F^* \to E^*$ by the formula

$$(A^*y^*)(x) = y^*(Ax) \qquad \text{for } x \in E,\ y \in F,\ y^* \in F^*.$$

[31] Let E be a separable Banach space. A sequence (x_n^*) of functionals $x_n^* \in E^*$ is said to be *weak* convergent* to $x^* \in E^*$ iff for every $x \in E$ the numerical sequence $(x_n^*(x))$ converges to $x^*(x)$. Accordingly, we define weak* closedness, weak* compactness, etc.

Every set $B \subset E^*$ which is bounded (i.e. $B \subset \{x^*: \|x^*\| \leqslant r\}$ for some r) and weak* closed is weak* compact (see Dunford and Schwartz, 1958, Chapter 5).

[32] Let E and F be Banach spaces and let $A \in \mathscr{L}(E, F)$. The *kernel* of A is the set

$$\ker A = \{x \in E: \ Ax = 0\};$$

the *image* of A is the set

$$\operatorname{im} A = \{Ax: \ x \in E\}.$$

[33] Let E and F be Banach spaces and let $B: E \times E \to F$ be a continuous bilinear mapping. Then there exists a constant c such that

$$\|B(x, y)\| \leqslant c\|x\| \cdot \|y\| \qquad \text{for } x, y \in E.$$

The least such constant, i.e. the number $\sup_{\|x\| \leqslant 1, \|y\| \leqslant 1} \|B(x, y)\|$ is called the *norm* of B (denoted by $\|B\|$).

[34] Let W be a convex set in a linear space E. A point $x_0 \in W$ is called an *extremal point* of W iff the relation $x_0 = tx_1 + (1-t)x_2$, $x_1, x_2 \in W$, $0 < t < 1$, implies $x_1 = x_2 = x_0$.

[35] *The Krein–Milman Theorem* Let E be a locally convex topological vector space (i.e. such that the linear algebra operations are continuous in its topology and there exists a base of neighbourhoods of zero consisting of convex sets). If W is a compact subset of E, then W is equal to $\operatorname{conv} G$, the smallest convex set containing the set G of extremal points of W.

In applications, one usually takes E to be the space conjugate to some Banach space, with the weak* topology (Note 31) (see Dunford and Schwartz, 1958).

[36] A topological space M is called a *Banach manifold* modelled on a Banach space E iff for every $x \in M$ there exist a neighbourhood U of x in M and a homeomorphism $\alpha: U \to E$ such that $\alpha(U)$ is an open subset of E, the regularity conditions imposed on the α's being the same as in the case of a finite-dimensional manifold (cf. Note 11). In particular, if the compositions $\beta \circ \alpha^{-1}$ are of class C^r, $0 \leqslant r \leqslant \infty$, we say that M is of class C^r (see Lang, 1962).

[37] A *measure space* is a triple $(X, \mathfrak{M}, \mu)$, where X is a set, $\mathfrak{M}$ is a σ-field of subsets of X and μ is a (non-negative) measure on $\mathfrak{M}$.

Suppose that μ and ν are two measures defined on the same σ-field $\mathfrak{M}$. They are said to be *mutually singular* (or *orthogonal*), in symbols $\mu \perp \nu$, iff there exists a set $A \in \mathfrak{M}$ such that $\nu(A) = \mu(X - A) = 0$.

[38] Let $(X, \mathfrak{M}, \mu)$ and $(Y, \mathfrak{N}, \nu)$ be two measure spaces. Write $W = X \times Y$ and denote by $\mathfrak{S}$ the least σ-field of subsets of W containing all product sets $C = A \times B$, $A \in \mathfrak{M}$, $B \in \mathfrak{N}$. Put

$$\lambda(C) = \mu(A)\nu(B) \qquad \text{for } C = A \times B.$$

Then λ extends uniquely to a measure on $\mathfrak{S}$; it is called the *product measure* of μ and ν. One defines similarly the product of infinitely (countably) many probability measures. Given a sequence of measure spaces $(X_n, \mathfrak{M}_n, \mu_n)$, $\mu_n(X_n) = 1$, write $W = \prod_{n=1}^{\infty} X_n$ and denote by $\mathfrak{S}$ the least σ-field of subsets of W containing all cylindric sets, i.e., sets of the form

$$C = \{(x_n) \in \prod_{n=1}^{\infty} X_n \colon x_{n_i} \in A_i,\ i = 1, \ldots, m\},$$

where $(n_1, \ldots, n_m)$ is any m-tuple of indices, m being arbitrary natural, and $A_i \in \mathfrak{M}_{n_i}$. For any such cylinder C put

$$\lambda(C) = \prod_{i=1}^{m} \mu_{n_i}(A_i).$$

Then λ extends uniquely to a measure on $\mathfrak{S}$; it is called the *product measure* of the μ_n's (see Halmos, 1950).

[39] Two measure spaces $(X, \mathfrak{M}, \mu)$ and $(Y, \mathfrak{N}, \nu)$ are called *isomorphic* whenever there exist sets $X_0 \in \mathfrak{M}$, $Y_0 \in \mathfrak{N}$, $\mu(X - X_0) = \nu(Y - Y_0) = 0$, and a bijective bimeasurable mapping $h\colon X_0 \to Y_0$ such that $\nu(h(A)) = \mu(A)$ for every $A \in \mathfrak{M}$, $A \subset X_0$.

Write $X_a = \{x \in X\colon \mu(\{x\}) > 0\}$; the points in X_a are called the *atoms* of μ. $(X, \mathfrak{M}, \mu)$ is a *Lebesgue space* iff $(X - X_a, \mathfrak{M}, \mu)$ is isomorphic to $(\langle 0, 1\rangle, \mathfrak{L}, |\cdot|)$ (the unit interval with Lebesgue measure) and X_a is at most countable.

[40] *Jensen's Inequality* If f is a convex continuous real-valued function in an interval $J \subset \boldsymbol{R}$:

$$f\left(\frac{x_1 + x_2}{2}\right) \leqslant \tfrac{1}{2}(f(x_1) + f(x_2)) \qquad \text{for } x_1, x_2 \in J,$$

then the inequality

$$f(t_1x_1 + \dots + t_nx_n) \leqslant t_1f(x_1) + \dots + t_nf(x_n)$$

holds for any $x_1, \dots, x_n \in J$ and $t_1, \dots, t_n \geqslant 0, \sum t_i = 1$.

[41] *The Arizonian Meteorite* In 1891 an enormous crater (the so-called Diablo Crater) was discovered in Arizona, USA; it had been bored by a meteorite – a small planetoid, in fact. Its diameter is about 1200 metres.

Lindenstrauss, J. (1973). *Classical Banach Spaces*, Lecture Notes in Mathematics, 338, Springer-Verlag, Berlin.

Malgrange, B. (1966). *Ideals of Differentiable Functions*, Univ. Press, Oxford.

Manning, A. (1974). *Amer. J. Math.*, **96**, 422–9.

Marcinkiewicz, J. and Zygmund, A. (1936). *Fund. Math.*, **26**, 1–43.

Markus, L. (1961). *Ann. of Math.*, **73**, 2, 1–19.

Mather, J. (1968). *Indag. Math.*, **30**, 5, 473–83.

Milnor, J. W. (1965). *Topology from the Differentiable Viewpoint*, Univ. Press of Virginia, Charlottesville.

Misiurewicz, M. (1973). *Bull. Acad. Pol. Sci. sér. Math.*, **21**, 902–10.

Misiurewicz, M. (1976a). *Stud. Math.*, **55**, 175–200.

Misiurewicz, M. (1976b). *Bull. Acad. Pol. Sci. sér. Math.*, **24**, 1069–75.

Moser, J. (1969). *J. Diff. Eq.*, **5**, 411–40.

Moser, J. (1973). *Stable and Random Motions in Dynamical Systems*, Princeton Univ. Press, Princeton.

Natanson, I. P. (1954). *Theorie der Funktionen einer reellen Verändlerlichen*, Akademie-Verlag, Berlin.

Nemytskii, V. V. and Stepanov, V. V. (1960). *Qualitative Theory of Differential Equations*, Princeton Univ. Press, Princeton.

Newhouse, S. (1970). *Global Analysis, Proc. Symp. Pure Math.*, **14**, Amer. Math. Soc., Providence, R. I., 191–202.

Nitecki, Z. (1971). *Differentiable Dynamics*, MIT Press, Cambridge, Mass.

Ornstein, D. (1970). *Adv. in. Math.*, **4**, 3, 337–52.

Oxtoby, J. C. (1952). *Bull. Amer. Math. Soc.*, **58**, 116–36.

Palis, J. (1969). *Topology*, **8**, 385–404.

Palis, J. jr. and de Melo, W. (1978). *Introducao aos sistemas dinamicos*, Instituto de Matematica Pura e Aplicada, IMPA, Brasil.

Palis, J. and Smale, S. (1970). *Global Analysis, Proc. Symp. Pure Math.*, **14**, Amer. Math. Soc., Providence, R. I., 223–32.

Peixoto, M. (1962). *Topology*, **1**, 101–20.

Pełczyński, A. (1968). *Diss. Mathem.* LVIII, PWN, Warsaw.

Poincaré, H. (1892–99). *Les méthodes nouvelles de la mécanique céleste*, Gauthier-Villars, Paris.

Pugh, Ch. (1967a). *Amer. J. Math.*, **89**, 956–1009.

Pugh, Ch. (1967b). *Amer. J. Math.*, **89**, 1010–21.

Pugh, Ch. (1969). *Amer. J. Math.*, **91**, 363.

Przytycki, F. (1976). *Stud. Math.*, **58**, 249–85.

Robbin, J. (1971). *Ann. of Math.*, **94**, 447–93.

Rohlin, V. A. (1962). *Amer. Math. Soc. Translations, ser. 1*, **10**, 1–54.

Sacksteder, R. (1973). *Lecture Notes in Mathematics*, 392, Springer-Verlag, Berlin–Heidelberg–New York, 179–93.

Sataev, E. A. (1975). *Izv. Akad.Nauk SSSR Ser. Mat.*, **39**, No. 4, 860–78 (Russian), *MR*, **52**, # 12005.

Schwartz, A. J. (1963). *Amer. J. Math.*, **85**, 453–58.

Shub, M. (1969). *Amer. J. Math.*, **91**, 175–99.

Siegel, C. L. (1945). *Ann. of. Math.*, **46**, 3, 423–8.

Siegel, C. L. and Moser, J. (1971). *Lectures on Celestial Mechanics*, Springer-Verlag, Berlin.

Sinai, Ya. G. (1959). *Dokl. Akad. Nauk SSSR*, **124**, 768–71 (Russian), *MR*, **21**, # 2036a.

Sinai, Ya. G. (1963). *Soviet Math. Dokl.*, **4**, 1818–22; *MR*, **35**, # 5576.

Sinai, Ya. G. (1970). *Dynamical Systems*, Lecture Notes, ser. No. 23 of Aarhus Univer., Aarhus.

Sinai, Ya. G. (1976). *Introduction to Ergodic Theory*, Princeton Univ. Press, Princeton.

Smale, S. (1961). *Ann. of Math.*, **74**, 199–206.

Smale, S. (1963). *Ann. Scuola Normale Superiore Pisa*, **17**, 97–116.

Smale, S. (1966), *Amer. J. Math.*, **88**, 491–6.

Smale, S. (1967). *Bull. Amer. Math. Soc.*, **73**, 747–817.

Smale, S. (1970). *Global Analysis, Proc. Symp. Pure Math.*, **14**, Amer. Math. Soc., Providence, R. I., 277–87.

Spanier, E. H. (1966). *Algebraic Topology*, McGraw–Hill, New York.

Springer, G. (1957). *Introduction to Riemann Surfaces*, Addison-Wesley Publ. Co., Reading, Mass.

Sternberg, Sh. (1955). *Amer. J. Math.*, **77**, 526–34.

Sternberg, Sh. (1964). *Lectures on Differential Geometry*, Prentice-Hall, Englewood Cliffs.

Sternberg, Sh. (1969). *Celestial Mechanics*, Benjamin, New York.

Strelcyn, J. M. (1970a). *C.R. Acad. Sci. Paris*, **271**, 903–05.

Strelcyn, J. M. (1970b). *Stud. Math.*, **35**, 261–72.

Šklover, M. D. (1967). *Izv. Vyssh. Uchebn. Zaved. Matematika*, **10** (65), 113–24 (Russian), *MR*, **32**, # 1737.

Štilman, M. S. (1971). *Mat. Zametki*, **9**, 291–302 (Russian), *MR*, **43**, # 7592.

Szlenk, W. (1974). *Dynamical Systems*, Lecture Notes, ser. No. 41, Aarhus University, Aarhus.

Totoki, H. (1970). *Ergodic Theory*, Lecture Notes, ser. No. 14, Aarhus University, Aarhus.

Wallace, A. H. (1968). *Differential Topology*, Benjamin, New York–Amsterdam.

List of Symbols

Subject Index

Index

Autho